Lecture Notes in Physics

Edited by H. Araki, Kyoto, J. Ehlers, München, K. Hepp, Zürich
R. L. Jaffe, Cambridge, MA, R. Kippenhahn, Göttingen, D. Ruelle, Bures-sur-Yvette
H. A. Weidenmüller, Heidelberg, J. Wess, Karlsruhe and J. Zittartz, Köln
Managing Editor: W. Beiglböck

389

J.C. Miller   R.F. Haglund, Jr. (Eds.)

Laser Ablation

Mechanisms and Applications

Proceedings of a Workshop
Held in Oak Ridge, Tennessee, USA
8–10 April 1991

Springer-Verlag
Berlin Heidelberg New York London Paris
Tokyo Hong Kong Barcelona Budapest

Editors

John C. Miller
Chemical Physics Section
Oak Ridge National Laboratory
Post Office Box 2008
Oak Ridge, TN 37831-6125, USA

Richard F. Haglund, Jr.
Department of Physics and Astronomy
Vanderbilt University
Nashville, TN 37235, USA

ISBN 0-387-97731-7 Springer-Verlag New York Berlin Heidelberg
ISBN 3-540-97731-7 Springer-Verlag Berlin Heidelberg New York

This work is subject to copyright. All rights are reserved, whether the whole or part of the material is concerned, specifically the rights of translation, reprinting, re-use of illustrations, recitation, broadcasting, reproduction on microfilms or in other ways, and storage in data banks. Duplication of this publication or parts thereof is only permitted under the provisions of the German Copyright Law of September 9, 1965, in its current version, and a copyright fee must always be paid. Violations fall under the prosecution act of the German Copyright Law.

© Springer-Verlag Berlin Heidelberg 1991
Printed in Germany

Printing: Druckhaus Beltz, Hemsbach/Bergstr.
Bookbinding: J. Schäffer GmbH & Co. KG., Grünstadt
2153/3140-543210 – Printed on acid-free paper

Preface

Everyone who has ever used a moderate- to high-power laser has observed laser ablation at one time or another. This may have taken the form of damaged lenses or windows or huge ion, electron, or photon signals when misaligning the laser onto electrodes or chamber walls. Or, we may simply give the "Hey, look at this!" response when tightly focusing a laser to make a loud pop and very visible plume from a piece of black paper or metal. When the laser beam is very bright, it is easier to find the focal point with a card by listening rather than looking!

The diverse applications of laser ablation, ranging from materials applications to laser medicine, have bred an equally diverse group of scientists and engineers who make use of the technique. Coming from different backgrounds, working in different applications, and equally spread out among universities, national laboratories, and industry, many practitioners of the art (black art?) are unknown to each other and unaware of much relevant work outside their field. Yet all of these people and topics are related by the need to understand the fundamental physics of laser-solid interactions.

It was this realization by several Department of Energy program managers and scientists during successive contractors' meetings on "Advanced Laser Technology for Chemical Measurements" that made us think such an interdisciplinary workshop centered on the physics of laser ablation would be useful. In East Tennessee alone, over a dozen groups at the Oak Ridge National Laboratory, Vanderbilt University, and commercial laboratories such as Atom Sciences were found to be using laser ablation techniques. From these groups, the organizing and program committees were drawn, and the meeting began to take shape.

Held in the Garden Plaza Hotel in Oak Ridge, Tennessee, April 8-10, 1991, the workshop was judged to be successful by all attendees. Fifty-three oral or poster presentations were given, highlighted by 18 invited lectures. Over 80 participants from the U.S., Japan, Germany, France, Soviet Union, and Sweden attended the three-day workshop. The reception, evening poster session, and banquet/excursion at the Museum of Appalachia in Norris, Tennessee, were dominated by "shop talk" as the participants got to know each other.

The present proceedings of the conference provide a written record of the workshop and a snapshot of a rapidly moving field. Contributions from each of the major applications areas of film deposition, biological molecules, elemental analysis, laser optics damage, lithography and micromachining, and medical applications are included. As in the workshop, they are grouped in the general subject areas of Thin Films and Superconductors, Polymers/Medicine, Biological Molecules, and Fundamental Physics. The last topic, reflecting the focus of the workshop, had the most contributions, but presentations in other sessions also emphasized fundamentals.

In organizing any meeting, many people and organizations contribute to its success. The financial support of the Office of Health and Environmental Research of the Department of Energy is gratefully acknowledged as is the encouragement and help from Gerald Goldstein, Paul Duhamel, and Richard Gordon of DOE. The organizing and program committees did a great job of compiling a mailing list, setting the focus, and persuading the renowned (and busy) invited speakers to attend. These committee members were E. T. Arakawa, C. H. Chen, R. N. Compton, D. H. Lowndes, R. H. Ritchie (all of Oak Ridge National Laboratory), R. F. Haglund, Jr., N. H. Tolk (both Vanderbilt University), and A. P. Duhamel (Department of Energy). Thanks are also due to Richard Haglund of Vanderbilt University for co-editing the present volume and to Thomas von Foerster of Springer Verlag for his editorial support. Finally, I wish to personally thank Nancy Currence, the conference secretary, and her colleagues, Darlene Holt and Denise Henderson, as well as Bonnie Reesor of the Oak Ridge National Laboratory Conference Office, for their meticulous attention to the "nuts and bolts" of the conference infrastructure which is usually invisible unless it begins to tumble down.

Of course, the ultimate success of the conference rested on the importance of the research performed, the uniformly high quality of the presentations, and the enthusiastic interplay between all of the participants.

The unanimous consensus of the organizers and participants is that a similar meeting should be held in the future (spring/summer 1993), building on the foundation detailed in the present volume.

Oak Ridge, ORNL, USA
April 1991

John C. Miller
Conference Chairman

Contents

Part I Thin Films and Superconductors

Diagnostic Studies of YBa$_2$Cu$_3$O$_{7-\delta}$ Laser Ablation
By N. S. Nogar, R. C. Dye, R. C. Estler, S. R. Foltyn, R. E. Muenchausen,
and X. D. Wu ... 3

Pulsed Laser Deposition of High Temperature Superconducting Thin Films
and Hetero-Structures
By T. Venkatesan, Q. Li, and X. X. Xi ... 12

In-Situ Monitoring of Laser Ablation of Superconductors
By C. H. Chen, R. C. Phillips, P. W. Morrison, Jr., D. G. Hamblen,
and P. R. Solomon ... 16

Spectroscopic and Ion Probe Characterization of Laser Produced Plasmas
Used for Thin Film Growth
By D. B. Geohegan ... 28

Synthesis of SiO$_2$ Thin Films by Reactive Excimer Laser Ablation
By E. Fogarassy, C. Fuchs, A. Slaoui, and J. P. Stoquert ... 38

Characteristics of Laser-Material Interactions Monitored by Inductively Coupled
Plasma-Atomic Emission Spectroscopy
By Wing Tat Chan and R. E. Russo ... 53

Part II Desorption

Trace Surface Analysis Using Ion and Photon Desorption with Resonance
Ionization Detection
By M. J. Pellin, C. E. Young, W. F. Calaway, K. R. Lykke, P. Wurz,
D. M. Gruen, D. R. Spiegel, A. M. Davis and R. N. Clayton ... 63

Pulse Rate Dependence of Laser Desorption and Ionization of Molecules on
Thin Metal Films: Mathematics of Laser Heating and Pulse Rate Dependence
By J. R. Millard and J. P. Reilly ... 68

Photodesorption of Metal Atoms by Collective Electron Excitation
By W. Hoheisel, M. Vollmer, and F. Träger ... 77

Desorption of Al, Au, and Ag Using Surface Plasmon Excitation
By E. T. Arakawa, I. Lee, and T. A. Callcott ... 82

Threshold Fluence UV Laser Excitation of W(100) and $O_2,H_2,F/W(100)$: Photoejected Ion KE Distributions
By HyunSook Kim and H. Helvajian ... 87

Excimer Laser Ablation of CdTe
By P. D. Brewer, J. J. Zinck, and G. L. Olson ... 96

Part III Polymers/Medicine

IR-Laser Ablation in Medicine: Mechanisms and Applications
By T. F. Deutsch ... 109

Pulsed Laser Ablation of Biological Tissue: Review of the Mechanisms
By A. A. Oraevsky, R. O. Esenaliev, and V. S. Letokhov ... 112

Etching Polymer Films With Continuous Wave Ultraviolet Lasers - The Photokinetic Effect
By R. Srinivasan ... 123

Mechanistic Insight in the Laser-Pulse Sputtering of Polymers by Combined Photography and Gas-Dynamic Analysis
By R. Kelly, B. Braren, and K. G. Casey ... 133

Part IV Biological Molecules

Laser Desorption and Multiphoton Ionization of Some Smaller Biomolecules: Recent Results and Prospects
By C. Köster, J. Lindner, G. R. Kinsel, and J. Grotemeyer ... 139

Matrix-Assisted Laser Desorption and Ionization of Biomolecules
By B. T. Chait and R. C. Beavis ... 149

Laser Ablation of Intact Massive Biomolecules
By P. Williams, D. Schieltz, Cong-Wen Luo, R. M. Thomas, and R. W. Nelson ... 154

Applications of Matrix-Assisted Laser Desorption Fourier Transform
Mass Spectrometry for Biomolecules
By R. L. Hettich and M. V. Buchanan 160

Comparison of Atomization Processes: Trace Element Analysis Using RIS of
Laser-Irradiated and Ion-Bombarded Biological and Metal Surfaces
By H. F. Arlinghaus and N. Thonnard 165

Laser Desorption of Peptide Molecules and Ions Using 193 nm Radiation
By J. P. Speir, G. S. Gorman, and I. J. Amster 174

Ablation of Material by Front Surface Spallation
By R. S. Dingus and R. J. Scammon 180

Part V Fundamental Physics

Laser Ablation and Optical Surface Damage
By L. L. Chase, A. V. Hamza, and H. W. H. Lee 193

Laser Induced Photodissociation, Desorption and Surface Reaction Dynamics
By T. J. Chuang 203

Mechanisms of Laser Ablation of Monolayers as Determined by
Laser-Induced Fluorescence Measurements
By R. W. Dreyfus 209

Laser-Induced Particle Emission from Surfaces of Non-Metallic Solids:
A Search for Primary Processes of Laser Ablation
By Noriaki Itoh, Ken Hattori, Yasuo Nakai, Jyun'ichi Kanasaki, Akiko Okano,
and R. F. Haglund, Jr. 213

Charged Particle Emission by Laser Irradiated Surfaces
By G. Petite, P. Martin, R. Trainham, P. Agostini, S. Guizard, F. Jollet,
J. P. Duraud, K. Böhmer, and J. Reif 224

Laser Ejection of Ag^+ Ions from a Roughened Silver Surface: Role of
the Surface Plasmon
By M. J. Shea and R. N. Compton 234

A Surface Plasmon Model for Laser Ablation of Ag^+ Ions from a
Roughened Ag Surface
By R. H. Ritchie, J. R. Manson, and P. M. Echenique 239

UV Laser Ablation from Ionic Solids
By R. F. Haglund, Jr., M. Affatigato, J. Arps, and Kai Tang 246

Physics of Pulsed Laser Ablation at 248 nm: Plasma Energetics and
Lorentz Interactions
By L. Lynds and B. R. Weinberger 250

Excimer Laser Ablation of Ferrite Ceramics
By A. C. Tam, W. P. Leung, and D. Krajnovich 260

Part VI Poster Presentations

Charge Emission from Silicon and Germanium Surfaces Irradiated with KrF
Excimer Laser Pulses
By M. M. Bialkowski, G. S. Hurst, J. E. Parks, D. H. Lowndes, and
G. E. Jellison, Jr. 265

Pulsed Laser Deposition of Tribological Materials
By M. S. Donley, J. S. Zabinski, V. J. Dyhouse, P. J. John, P. T. Murray,
and N. T. McDevitt 271

Ion-Molecule Reactions of Carbon Cluster Anions
By R. L. Hettich 280

Doubly Charged Negative Ions of Bucky Ball - C_{60}^{2-}
By R. L. Hettich, R. N. Compton, and R. H. Ritchie 285

Evaporation as a Diagnostic Test for Hydrodynamic Cooling of
Laser-Ablated Clusters
By C. E. Klots 294

Desorption of Large Organic Molecules by Laser-Induced Plasmon Excitation
By I. Lee, E. T. Arakawa, and T. A. Callcott 297

Simultaneous Bombardment of Wide Bandgap Materials with UV Excimer
Irradiation and keV Electrons
By J. T. Dickinson, S. C. Langford, and L. C. Jensen 301

Superconducting Transport Properties and Surface Microstructure for
$YBa_2Cu_3O_{7-\delta}$ -Based Superlattices Grown by Pulsed Laser Deposition
By D. P. Norton, D. H. Lowndes, X. Zheng, R. J. Warmack, S. J. Pennycook,
and J. D. Budai 311

CPSIA information can be obtained
at www.ICGtesting.com
Printed in the USA
LVHW111737190419
614855LV00001B/44/P

Monitoring Laser Heating of Materials with Photothermal Deflection Techniques
By M. A. Shannon, A. A. Rostami, and R. E. Russo 320

Studies of Laser Ablation of Graphite: $C_n^{+/-}$ Ion Kinetic Energy Distributions
By M. J. Shea and R. N. Compton 328

Infrared Laser Induced Ablation and Melting in Model Polymer Crystals
By B. G. Sumpter, D. W. Noid, and B. Wunderlich 334

Chemical Characterization of Microparticles by Laser Ablation in an Ion Trap Mass Spectrometer
By J. M. Dale, W. B. Whitten, and J. M. Ramsey 344

Photophysical Processes in UV Laser Photodecomposition of $Bi_2Sr_2Ca_1Cu_2O_8$ and $YBa_2Cu_3O_{x+6}$
By L. Wiedeman, HyunSook Kim, and H. Helvajian 350

Influence of Liquefaction on Laser Ablation: Drilling Depth and Target Recoil
By A. D. Zweig 360

Part I

Thin Films and Superconductors

Diagnostic Studies of YBa$_2$Cu$_3$O$_{7-\delta}$ Laser Ablation

N. S. Nogar,[a] R. C. Dye[a], R. C. Estler[a,c], S. R. Foltyn[a], R. E. Muenchausen[b] and X. D. Wu[b]
[a]Chemical and Laser Sciences Division, MS G738
[b]Exploratory Research and Development Center
Los Alamos National Laboratory
Los Alamos, New Mexico 87545
[c]Permanent Address: Department of Chemistry, Fort Lewis College
Durango, Colorado 81301

Introduction.

Pulsed laser ablation for the deposition of thin films is a technology that has been explored in some detail. This process has many advantages relative to conventional deposition technologies,[1-3] including rapid growth rates, the ability to evaporate congruently multicomponent targets, and the production of high-energy (eV) atoms, molecules, ions and clusters that enhance the production of epitaxial films. Pulsed laser deposition (PLD) has been used for the production of dielectric, ferroelectric, and piezoelectric films, semiconductor layers, superlattices, and most recently, for the production of high-temperature superconducting (HTSC) thin films[4-8].

The need for high quality HTSC thin films has sparked renewed interest in this technology. Laser ablation has shown great promise in producing thin films with good epitaxy, and high critical currents and transition temperatures. In this process, the bulk material is vaporized by a laser pulse in an oxygen atmosphere, and subsequently condensed on a heated substrate to produce a superconducting film. Although laser evaporation and ablation have been studied for over twenty years, the tertiary and quaternary metal oxide HTS systems are the most complex systems to which this deposition technique has been applied. The process of film production can be thought of as occuring in three separate steps: 1, absorption of photons and ablation at the target surface; 2, transport and gas-phase reaction of the ablated material; 3, condensation and crystal growth on the substrate.

Because of the difficulty of monitoring in real time either the ablation process itself, or thin film growth, most studies have relied on interrogation of the ablated plume to measure directly properties in step 2, and to infer information about the other steps. A variety of optical spectroscopies have been used with some success as diagnostics[9-13]. In addition, mass spectrometric techniques have also been used as diagnostic probes to understand better and model the ablation event[14-18].

In this manuscript, we describe two experiments related to the physics and chemistry of the ablated plume. In the first case[19], we use Resonance Ionization Mass Spectrometry (RIMS) to detect and map the velocity, internal energy and chemical speciation of ablated neutrals (Cu, Y, BaO) from $YBa_2Cu_3O_{7-x}$ under vacuum conditions. In the second experiment[20], optical emission is used to monitor the reaction of ablated yttria under film deposition conditions. In addition, we review and analyze the literature on diagnostics for laser ablation of $YBa_2Cu_3O_{7-x}$.

Experiment.

The RIMS measurements took place in the source region of a time-of-flight (TOF) mass spectrometer[19,21]. The Q-switch synch-out from a Nd:YAG laser was used to master the timing sequence. This laser was the ablation source and operated at either the fundamental frequency, 1.06 microns, or at the third harmonic, 355 nm. The laser output was focused to varying spot sizes on the target material; the angle of incidence was 60 degrees. Most of the ablation studies used a peak laser fluence of $\approx$ 1 J/cm^2.

Pulses from an excimer-pumped dye laser propagating parallel to the target surface, but displaced 3.2 cm (i.e., in the center of the ionization region), were used to interrogate the ablated neutral species. Ionization of the neutrals from ground electronic states was effected by a variety of multiphoton transition sequences. Dye-laser pulses (2mJ, 15 nsec) produced at an adjustable delay relative to the ablation laser were spatially filtered and loosely focused through the ionization region. By varying this time delay on successive shots, the velocity distribution could be mapped out. Detection electronics consisted of a channel electron multiplier, a preamplifier, a boxcar integrator and/or a transient recorder.

The experiments on laser ablation of yttria (Y_2O_3) were performed in the deposition apparatus. This deposition chamber consists of a modified six way 6" dia. stainless steel cross, with load locks provided for target and substrate introduction[22]. The target can be rotated in conjunction with horizontal rastering of the laser beam. Processing gases can be introduced into the deposition chamber using mass flow control to maintain a constant pressure. The excimer laser produced 250 mJ, 20 nsec pulses at 308 nm and a repetition rate of 20 Hz. Additionally, for the experiments described here, the plume optical emission was collected and analyzed. Light was collected by a 7.5 cm. diameter, 500 cm focal length quartz lens, and focused onto the end of a (200 µm diameter) multi-mode fiber optic cable (FOC), 10 m in length. The distal end of the fiber optic cable was coupled to an 0.5 m monochromator equipped with a gated, intensified CCD camera for spectrally resolved detection.

Results.

Laser Ablation/RIMS

Arrival-time distributions[19] were recorded for neutral BaO, Cu and Y ablated from $YBa_2Cu_3O_x$ pressed pellets at 355 nm. For Cu, this was also compared to ablation at a wavelength of 1.06 microns. In each case where 355 nm was used for ablation, the signals peak at approximately the same flight time (~20-22 µs), indicating similar velocities for each species, regardless of the mass. For Cu ablated at 1.06 microns, a thermal Maxwellian velocity distribution provides the best fit, indicating a most probable velocity of 9.8×10^4 cm/s and temperature of ≈ 3700 K. At 355 nm, the Cu signal peaks at the same time-of-flight, but possesses a substantially narrower width, although the same approximate ablation laser fluence was used, Figure 1. Here, a center-of-mass velocity of 13×10^4 cm/s and a thermal width velocity of 4.9×10^4 cm/s are obtained from a Knudsen-layer analysis[23,24]. The skewed distribution suggests a greater influence of collisions at 355 nm, indicating improved efficiency of material removal at 355 nm compared to 1.06 µm. This may be due to a higher effective energy density, resulting from a shallower penetration depth, at 355 nm. In addition, this provides evidence that the ablation process is in part photochemical as opposed to strictly evaporative. In addition, if these temperatures are representative of a local surface temperature, target morphology and stoichiometry changes should be anticipated since they exceed the 1288 K incongruent melting point for $YBa_2Cu_3O_x$, which forms a Y_2BaCuO_5, barium cuprate and copper oxide melt.

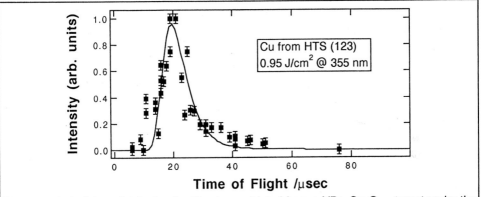

Figure 1. Arrival time distribution for Cu atoms ablated from a $YBa_2Cu_3O_{7-x}$ target under the conditions indicated. The solid line is a fit to a Knudsen layer treatment.

For BaO at 355 nm, similar TOF signals were observed. Here again, the best data fit was obtained with a shifted-Maxwellian velocity distribution. In this case, however, little statistical difference existed between the fits, partly as a result of the relatively low-intensity signals. A most probable velocity of 11×10^4 cm/s was measured from the data, while a surface temperature of 6800 K is calculated from the Knudsen layer model for BaO. If Cu and BaO shared a common surface temperature history, the ratio of their velocities ($v \propto \sqrt{T/m}$) should be 1.55:1.00 (Cu:BaO). This was definitely not the case in this instance.

Optical Emission/Yttrium reactions

It is interesting that in the experiment described above, and several others, few oxide species were detected, suggesting that the oxygen incorporated into the HTSC thin films during the laser deposition process originates in the processing atmosphere, rather than in the solid target material. To explore this further, we examined the process at a variety of oxygen pressures from 0.001 to 0.4 Torr.

Emission spectra were obtained from the plume in the spectral region near 600 nm, where both atomic yttrium emission (Y^*) and yttrium monoxide emission (YO^*) could be observed. Figure 2 displays a spectrum obtained at an oxygen pressure of 2×10^{-1} Torr, characteristic of HTSC film deposition conditions. The detector was gated to integrate signal for the duration of the emission pulse, ≈ 10 μsec. The two starred transitions in Fig. 2 correspond to Y^* while most of the remaining features are due to YO^*. The Y^* emissions are from the $^2D_{5/2}$ and $^2D_{3/2}$ states to the $^2D_{3/2}$ ground electronic state, while the YO^* emission is from the A $^2\Pi_{3/2,1/2}$ levels to the $^2\Sigma^+$ ground state.

The ratio of YO^* to Y^* emission was found to change dramatically with pressure of the ambient oxygen atmosphere. It was observed that the ratio YO^*/Y^* increased linearly with pressure at low oxygen pressures, and approached a limiting value at pressures ≥ 0.4 Torr. No YO^* emission was observed at oxygen pressures less than 10^{-3} Torr, indicating that YO^* is not formed directly from the solid material. To interpret these results, a kinetic model was developed, based on the production of yttrium atoms (both ground and excited states) in the ablation event, but no YO^* molecules. Subsequent reaction of Y with O_2 was assumed to produce all observed YO^*; quenching of YO^* and Y^* was assumed due to molecular oxygen. Applying the steady-state approximation to the model results in the following expression: $[YO^*]/[Y^*] = \{A * [O_2]\} / \{B + C * [O_2]\}$, where the constants A, B, and C are sums of products of the rates and rate constants, and $[O_2]$ represents the oxygen pressure. This result obeys the same limiting forms as the experimental data. That is, at zero pressure, $[YO^*]/[Y^*] \rightarrow 0$, at low pressure $[YO^*]/[Y^*] \propto (A/B)[O_2]$, and at high pressure, $[YO^*]/[Y^*] \propto (A/C) = $ constant.

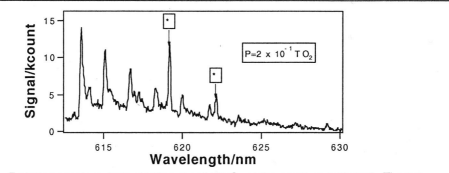

Figure 2. Emission spectrum from ablation of solid Y_2O_3 at the pressure indicated. The two starred features are atomic yttrium transitions, while the remainder of the spectrum is due to yttrium monoxide emission.

To characterize further the plume chemistry and dynamics, the YO* vibrational temperature was obtained from the vibrational progression seen in Fig 2. The intensity of each vibronic transition is assumed proportional to the peak height, and the peak heights corrected for Franck-Condon factors. The slope of this plot yields an effective temperature for the vibrational manifold of ≈2000 K. This temperature is lower than the effective translational temperatures, ≈4400 K reported above. It is likely that the principal reason for the discrepancy in measured temperatures is that the time-of-flight measurements where conducted in vacuum (p ≤ 10^{-7} Torr), while the present studies examined reactions with reactive oxygen gas. Thus, in the mass spectrometric experiments, no collisions with background gas slowed the expansion away from the surface: intra-plume collisions tended to convert random thermal energy into forward-directed translational energy, [24-25] yielding a higher effective temperature. In the reactive processes currently under study, collisions with the oxygen present in the chamber will penetrate the supersonic expansion from the Knudsen layer.

Discussion.

The results presented above on plume characteristics can be compared with other work on beam and deposition diagnostics. A compilation of recent results from laser ablated $YBa_2Cu_3O_{7-\delta}$ is shown in Table 1. This listing is not meant to be exhaustive; it is simply a sampling of representative results. Several conclusions can be drawn from this data. The first is that the result is highly dependent on the method of observation. As an example, very few of

Table 1. Diagnostic measurements on laser-ablated $YBa_2Cu_3O_{7-\delta}$.

Wavelength/nm Fluence	Experiment[1]	Distance Pressure	Species or measurement	Velocity[2] (cm/sec)	Comments (ref)
308 nm 0.1-1 J/cm^2	MS	1 x 10^{-6} Torr	Y^+, YO^+, Cu, Cu^+, Ba, Ba^+, BaO, BaO^+	6-20 % ions	ions ↑ as Fluence ↑ (17)
248 nm ≤ 4 J/cm^2	OES Ion probe	≈ 0.01, 3.2 cm ≈ 10^{-6} Torr	Y, Y^+, Cu, Ba, Ba^+	1 x 10^6 6-13 x 10^3 K	Velocity ∝ fluence (13)
308 nm 3 J/cm^2	OES	0.2 cm 10^{-5} Torr	Y,Ba,Cu,Y^+, Ba^+	Temperature = 8000 K	no molecules (10)
1.06 µm, 532 nm 0.1-10 J/cm^2	MS OES	1 x 10^{-6} Torr	Y,Ba,Cu,YO O^+, O_2^+, Y^+, YO^+, Ba^+, BaO^+		(11)
248 nm 0.25-19 J/cm^2	Ion ToF	12.7 cm ≈10^{-6} Torr	Ions etch depth	2 x 10^6 2-8 % ions	ions ↓ as Fluence ↑ (27)
530 nm 10 J/cm^2	OES	sample surface	Y,Y^+,Cu,Cu^+, Ba	25 ps[3] 1 ns	ps excitation (9)
266 nm 25 J/cm^2	LMS	0.06 cm	Y, Cu, Ba,Ba^+	1 x 10^5	No clusters (14)
193 nm 1-7 J/cm^2	OES	7.2 cm 1 x 10^{-5} Torr	Y,Ba,Cu,O, Ba^+	1 x 10^6	Free-jet distribution (28)
193 nm 1 J/cm^2	OES OAS	0.5 cm 2 x 10^{-6} Torr	Y, Ba, Cu, Ba^+	5 x 10^5	Neutrals slower than ions (12)
1.06 µm, 532 nm 1-4 J/cm^2	LMS		Cu^+, Y^+,Ba^+, Clusters		Clusters ↓ as Fluence ↑ (16)
351, 248, 193 nm ≤ 0.28 J/cm^2	MS	7 cm ≤10^{-7} Torr	Cu, Y, YO, Ba, BaO	5 x 10^5	Threshold; ions & neutrals (25)
532 nm 7.5 J/cm^2	Velocity filter SEM	10^{-6} Torr	1 µm particles	1 x 10^4	(26)
193, 248 nm 3.6, 1.2 J/cm^2	OES/MS	≤ 0.4 cm ≤ 5 Torr	Ba, Ba, Cu^+, Y, Y^+, YO	1 x 10^6	Electron/plasma excitation (29)
1.06 µm, 4.6 J/cm^2	framing photography	≤ 1 atm.	plasma emission	≈2 x 10^6	shock front formation (30)
1.06 µm, 355 nm 1.0 J/cm^2	LMS	3.2 cm 1 x 10^{-6} Torr	Cu, Cu_2, CuO Y, Ba, BaO	1.1 x 10^5	Knudsen layer T ≈ 4000K (19)

1. OES, optical emission spectroscopy; MS, mass spectrometry; LMS, laser mass spectrometry; OAS, atomic absorption spectroscopy; Ion ToF, ion time-of-flight; SEM, electron microscopy
2. The most probable velocity.
3. Temporal components of emission at surface.

these experiments report the production of clusters or particulates; at the same time, it is clear from film morphology that clusters are being formed and transported to the surface. The reason for the lack of reports on cluster formation is that optical emission and quadrupole mass spectrometry are ill-suited for the detection of clusters. At the same time, experiments capable of detecting heavy charged particles[16,26] do report the presence of such species, and are able to provide useful information concerning their disposition: in particular, cluster formation seems to be enhanced by long wavelength and low-fluence irradiation.

Second, the velocity measured for neutral atoms and molecules is almost universally found to fall in the range 10^5-10^6 cm/sec. In addition, many authors report non-thermal velocity distributions, suggesting that collisional processes play a significant role in plume evolution. Lastly, in some instances the velocity distribution evolves in time (or space) as the plume propagates away from the surface. Particle velocities are considerably slower, as might be expected.

Third, under typical conditions for film formation, the fraction of ions in the plume lies in the range 1-10%. Although there is some disagreement[17,27] whether this fraction increases or decreases with laser fluence, clearly ion-initiated processes will play a non-trivial part in plume evolution and film formation, and that the ion fraction is dependent on the wavelength of ablation[25]. In addition, it appears that the ion fraction is generally traveling faster than the neutral atoms and molecules. This suggests that a significant source for energy deposition in the film may be due to ion impact.

Lastly, most experiments report a predominance of free metal atoms and ions in the plume, as opposed to oxide and other molecular fragments. This indicates that the primary source of oxygen in the resultant films is not due to oxygen trapped in the target, but rather to oxygen incorporated by reaction with the processing gas or in post-deposition processing.

Gas phase collisions clearly play a major role in the laser deposition process. This includes both intra-plume collisions, which strongly effect the velocity and angular distributions of the plume, and plume-gas interactions, which will effect deposition rate and homogeneity. The latter can be seen by simple consideration of gas-kinetic effects. Under typical deposition conditions, the target-substrate distance will be ≈ several centimeters, and the pressure of processing gas will be a fraction of a Torr. Gas kinetic theory predicts that this will lead to 1-10 collisions for laser-ablated species between evolution from the target surface and deposition on the substrate. Since ≈10 collisions are typically sufficient to relax translationally excited atoms and small molecules[31], this means that at the upper end of the range for pressure-distance products that the deposition plume will be diffusing toward the substrate rather than being "sprayed" on as part of a well-directed plume. This places an upper limit on the pressure that can be used for optimum film growth.

On the other hand, numerous measurements of chemical speciation in the plume, as well as our recent measurement of chemical reactivity with the processing gas, mandate that a reactive source of oxygen be present in the deposition atmosphere. This, in turn, places a lower limit on the oxygen pressure that can be used for the production of in-situ HTSC films.

Conclusions.

The use of a variety of diagnostics can offer significant insight into the mechanisms of laser ablation and film growth. It seems clear that under conditions of optimum film growth that significant collisional processes occur both within the laser-desorbed plume, and between the plume and the processing gas. These collisional processes are responsible for the macroscopic characteristics of the plume, including velocity and spatial distributions, the ratio of ions and particles to neutral atoms and molecules, and the incorporation of oxygen into the film. It is also obvious that real-time diagnostics, applicable under conditions relevant to film production, could provide useful process monitors and feedback control for the production of HTSC thin films.

References.

1. J. T. Cheung, H. Sankur, *CRC Crit. Rev. Sol. St. Mat. Sci.* **15**, 63-109 (1988).
2. H. Sankur, J. T. Cheung, *Appl. Phys. A* **A47**, 271-284 (1988).
3. D. J. Ehrlich, J. Y. Tsao, *Laser Microfabrication (Thin Film Processes and Lithography)* (Academic Press, Boston, 1989).
4. T. Venkatesan, *et al.*, *IEEE J. Quantum Electron* **25**, 2388-93 (1989).
5. J. Narayan, N. Biunno, R. Singh, O. W. Holland, O. Auciello, *Appl. Phys. Lett.* **51**, 1845-7 (1987).
6. C. C. Chang, *et al.*, *Appl. Phys. Lett.* **53**, 517-19 (1988).
7. B. Roas, L. Schultz, G. Endres, *Appl. Phys. Lett.* **53**, 1557-1559 (1988).
8. G. Koren, A. Gupta, E. A. Giess, A. Segmuller, R. B. Laibowitz, *Appl. Phys. Lett.* **54**, 1054-1056 (1989).
9. K. M. Yoo, R. R. Alfano, X. Guo, M. P. Sarachik, L. L. Isaacs, *Appl. Phys. Lett.* **54**, 1278-9 (1989).
10. O. Auciello, *et al.*, *Appl. Phys. Lett.* **53**, 72-4 (1988).
11. C. H. Chen, M. P. McCann, R. C. Phillips, *Appl. Phys. Lett.* **53**, 2701-3 (1988).
12. D. B. Geohegan, D. N. Mashburn, *Appl. Phys. Lett.* **55**, 2345-2347 (1989).

13. P. E. Dyer, R. D. Greenough, A. Issa, P. H. Key, *Appl. Phys. Lett.* **53**, 534-6 (1988).
14. T. Venkatesan, *et al.*, *Appl. Phys. Lett.* **53**, 1431-3 (1988).
15. H.-J. Dietz, S. Becker, *Int. J. Mass Spec. Ion Proc.* **82**, R1-R5 (1988).
16. C. H. Becker, J. B. Pallix, *J. Appl. Phys.* **64**, 5152 (1988).
17. C. H. Chen, T. M. Murphy, R. C. Phillips, *Appl. Phys. Lett.* **57**, 937-9 (1990).
18. P. K. Schenck, D. W. Bonnell, J. W. Hastie, *J. Vac. Soc.* **A7**, 1745-1749 (1989).
19. R. C. Estler, N. S. Nogar, *J. Appl. Phys.* **69**, 1654-9 (1991).
20. R. C. Dye, R. E. Muenchausen, N. S. Nogar, *Chem. Phys. Lett.* **in press**, (1991).
21. N. S. Nogar, R. C. Estler, C. M. Miller, *Anal. Chem.* **57**, 2441-4 (1985).
22. R. E. Muenchausen, *et al.*, *Appl. Phys. Lett.* **56**, 578-80 (1990).
23. R. W. Kelly, R. W. Dreyfus, *Nucl. Instr. and Meth.* **B32**, 321-348 (1988).
24. R. Kelly, *J. Chem. Phys.* **92**, 5047-56 (1990).
25. L. Wiedeman, H. Helvajian, in *Materials Research Society* (MRS, 1990), pp. 217-222.
26. H. Dupendant, *et al.*, *Appl. Surf. Sci.* **43**, 369-376 (1989).
27. R. A. Neifeld, *et al.*, *J. Appl. Phys.* **69**, 1107-9 (1991).
28. J. P. Zheng, Z. Q. Huang, D. T. Shaw, H. S. Kwok, *Appl. Phys. Lett.* **54**, 280-282 (1989).
29. T. J. Geyer, W. A. Weimer, *Appl. Spec.* **44**, 1659-1664 (1990).
30. K. Scott, J. M. Huntley, W. A. Phillips, J. Clarke, J. E. Field, *Appl. Phys. Lett.* **57**, 922 (1990).
31. R. C. Tolman, *The Principles of Statistical Mechanics*, Dover, New York, 1979, Chapter VI.

PULSED LASER DEPOSITION OF HIGH TEMPERATURE SUPERCONDUCTING THIN FILMS AND HETERO-STRUCTURES

T. Venkatesan, Q. Li and X. X. Xi
Center for Superconductivity Research
University of Maryland
College Park, MD 20742

ABSTRACT

Pulsed laser deposition has become an ubiquitous technique for the stoichiometric deposition of thin films of complex, multi-elemental compounds. In addition, the system enables rather simple embodiments for the fabrication of sophisticated materials structures such as superlattices of different materials. In this talk the various issues relating to this deposition process would be illustrated for the case of the 90k superconductor, $YBa_2Cu_3O_{7-x}$ and its hetero-structures with cation substituted layers.

INTRODUCTION

The deposition of multi-elemental compounds with conventional techniques such as muti-hearth electron beam evaporation or sputtering become difficult when stringent stoichiometric control is needed. Pulsed laser evaporation was shown for the first time[1] to preserve the composition of high temperature superconducting material such as $YBa_2Cu_3O_{7-x}$ during the thin film depostion process.

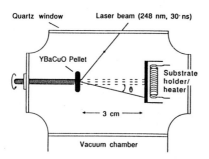

Figure 1. Schematic of a pulsed laser deposition system for in-situ epitaxial films.

The deposition was accomplished with a pulsed excimer laser (248 nm, 30 ns pulse width) irradiating a stoichiometric target of the superconductor placed in a chamber with an oxygen background pressure of about 100 mTorr and the substrate held at a temperature of about 750 C (Fig.1). Highly oriented films of the superconductor could be grown under such conditions. The film properties as measured by a variety of techniques have been some of the best to date[2]. The nature of the pulsed laser interaction with the target is not completely understood and a number of fascinating features have been observed with respect to the the material ejection process. For example, it was shown[3] that there was a strong angular dependence to the ejected materials and that there were two distinct components, one a highly forward directed non-equilibrium component (with a $\cos^x(\theta)$ dependence with x>11) and the other an equilibrium thermal evaporative component with a $\cos(\theta)$ dependence (Fig 2). The stoichiometry of the emitted species was well preserved in the forward directed component but the same was not true for the evaporative component. The two components indicate the presence of a non-equilibrium process in the laser surface interaction and the preservation of the composition in the evaporant requires this non-equilibrium process to dominate.

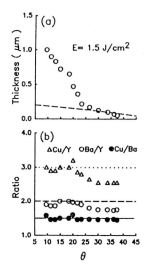

Figure 2. At 1.5 J/cm² laser energy density, the angular distribution of the deposited film (a) thickness (the dashed line is the $\cos(\theta)$ fit) and (b) composition.

One of the unique features of the laser deposition process is the easy adaptability of the process for the preparation of sophisticated materials structures. By using a segmented target consisting of two semi-circular segments of $YBa_2Cu_3O_{7-x}$ (superconductor) and $PrBa_2Cu_3O_{7-x}$ (a semi-conductor) high quality superlattice structures were prepared by simply spinning the target in the laser beam and controlling the duration of the irradiation on each of the segments to control the relative layer thicknesses. These superlattices have been useful in exploring the transport properties of these superconductors at thicknesses approaching unit cell dimensions. For example, it was concluded from such a study[4] that the interlayer coupling between the various unit cells of the superconductors was of at most importance in determining the transition temperature of the superconducting layers as evidenced by the drop in the transition temperature with the reduction in the superconductor layer thickness.

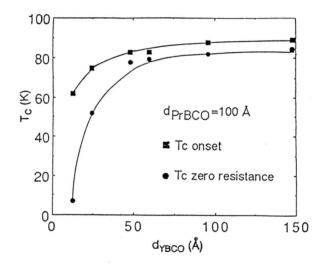

Figure 3. Superconducting transition temperature as a function of the superconductor layer thickness in the superlattice structure.

In this talk we will try to demonstrate the versatility of the pulsed laser deposition process and particularly its value to researchers working in the area of multi-component film deposition. The basic process of the laser-solid interaction resulting in the unusual material evolution from the surface is a subject of intense study as it richly deserves.

REFERENCES

1. D. Dijkkamp, T. Venkatesan, X. D. Wu, S. A. Shaheen, N. Jisrawi, Y. H. Lee, W. L. Mclean, and M. Croft, Appl. Phys. Lett.51, 619 (1987)

2. D.Kalokitis, A. Fathy, V. Pendrick, E. Belohoubek, A. Findikoglu, A. Inam, X. X. Xi, T. Venkatesan, Appl. Phys. Lett., 58, (1991), M. J. Ferrari, M. Johnson, F. C. Wellstood, J. Clark, A. Inam, X. D. Wu, L. Nazar, and, T. Venkatesan, Nature, (1989)

3. T. Venkatesan, X. D. Wu, A. Inam and J. B. Wachtman, Appl. Phys. Lett. **52**, 1193 (1988).

4. Qi Li, X. X. Xi, X. D.Wu, A. Inam, S. Vadlamannati, W. L. Mclean, T. Venkatesan, R. Ramash, D. Hwang, L. Nazar, Phys. Rev. Lett. 64, 3086 (1990).

In-Situ Monitoring of Laser Ablation of Superconductors

C. H. Chen and R. C. Phillips

Chemical Physics Section, Oak Ridge National Laboratory
P.O. Box 2008, Oak Ridge, Tennessee 37831-6378

and

P. W. Morrison, Jr., D. G. Hamblen, and P. R. Solomon

Advanced Fuel Research, Inc.
87 Church Street, East Hartford, Connecticut 06118

1. INTRODUCTION

In 1986, Bendorz and Muller[1] discovered that metallic oxide with a perovskite structure in the La-Ba-Cu-O system exhibited superconductivity at 20°K. In early 1987, Chu and his co-workers[2] made a $YBa_2Cu_3O_{7-x}$ compound which had a T_c higher than the temperature of liquid nitrogen. Since these discoveries, the research and development of high-T_c superconductors has aroused unpecedented attention, not only from the scientific community, but also from various industries and governments. As these new superconducting materials are in general very "brittle" and difficult to machine, it is extremely difficult to fabricate them into thin wires. In addition, practical use of new high-T_c superconductors will require materials with very high critical current density. Thus thin film preparation is particularly important for many applications of superconductors, such as electronics for the computer industry. Films with high superconducting transition temperatures have been prepared by electron beam evaporation, organometallic chemical vapor deposition, DC sputtering, molecular beam epitaxy, and laser evaporation and deposition processes. For example, Rice et al.[3] used electron beam and thermal evaporation to produce Ca-Sr-Bi-Cu-O superconducting films from CaF_2, SrF_2, Bi, and Cu targets. Berry et al.[4] produced a superconductive film by organometallic chemical vapor deposition. Hellman et al.[5] used molecular beam epitaxy to produce a $DyBa_2Cu_3O_{7-x}$ film on a $SrTiO_3$ substrate. Lynds et al.[6] used a Nd-YAG laser beam to do laser evaporation and deposition to prepare a $YBa_2Cu_3O_{7-x}$ thin film from targets of Y_2O_3, Ba_2CO_3, and CuO. Kwok et al.[7] applied a homogeneous excimer laser to obtain a film with the right stoichiometry of $YBa_2Cu_3O_{7-x}$. However, most of the films in the works cited above need to be annealed in an oxygen environment in order to produce superconductivity. Recently, several research groups obtained superconductive films of $YBa_2Cu_3O_{7-x}$ by raising the temperature of the substrate and introducing oxygen into the chamber during the thin film preparation process.[8] However, no superconductive films of the Bi-Sr-Ca-Cu-O or Tl-Sr-Ba-Cu-O systems were obtained without oxygen annealing. Nevertheless, the right ratio of various metal elements is one of the key factors in achieving superconductivity. Superconductive films are usually obtained based on repeated trials and adjustments. Venkatesan et al.[9] observed two distinct components during pulsed laser deposition of high-T_c superconductive film. It was pointed out that the quality of the superconductive film depends on controlling the film stoichiometry. The role of introducing oxygen into the thin film preparation chamber has never been fully understood. Thus, the need for a real-time monitor for high-T_c thin film preparation is quite critical.

In general, the preparation of a superconductive thin film involves the vaporization of various chemicals. The species deposited on substrates include neutral atoms or molecules in ground states, atoms and molecules in excited states, ions in ground states and ions in excited states. An instrument which has the capability of the quantitative determination of atoms, molecules, and ions in various electronic states should provide very detailed information on the mechanism of formation of superconductive films. A position sensitive monitor of various oxide compounds relative to the flow and pressure of oxygen should reveal the role of oxygen in superconductive film preparation. An ideal method of monitoring the deposition process is to continuously track the deposition *in situ* and automatically maintain optimum film growth. With this goal in mind, we have investigated several diagnostic methods. In this work, fluorescence spectra are presented which indicate suitability for monitoring excited atoms, molecules, and ions; mass analysis of laser desorbed ions is described; and the application of resonance ionization spectroscopy (RIS) to monitor desorbed neutral atoms and molecules is discussed. FTIR spectroscopy is used to measure substrate temperatures and the components on the surfaces.

One of the "peculiar" features of these high-T_c materials is the noninteger number of oxygen atoms. The oxygen content is extremely critical to the superconductive properties. All of the bulk metal oxide superconductors need to be annealed in an oxygen environment for an extensive period of time in order to achieve superconductivity. The oxygen content is equally critical to bulk and thin film superconductors. Take $YBa_2Cu_3O_{7-x}$ as an example. Its superconductive properties disappear whenever x is larger than 0.5. The samples not only need to be sintered in the oxygen environment, but also need to be cooled down slowly to take up oxygen.[10]

Batlogg et al.[11] and Leary et al.[12] observed a very small isotopic effect by replacing ^{16}O with ^{18}O for the $YBa_2Cu_3O_{7-x}$ system and concluded that weak electron-phonon coupling cannot be used to explain this observation. The crystal structures of $YBa_2Cu_3O_{7-x}$ have been studied very extensively by X-ray diffraction and neutron diffraction.[13] It was confirmed as orthorhombic for x = 0 in $YBa_2Cu_3O_{7-x}$. However, the crystal structure becomes tetragonal when $x \geq 0.5$. It was widely speculated that the change in crystal structure was responsible for the loss of superconductivity. However, experimental results by Xiao et al.[14] indicate that the crystal structure is not the key for superconductivity. It was pointed out that the T_c of $YBa_2Cu_3O_{7-x}$ is determined mostly by the actual oxygen content rather than its configuration and that the oxygen-vacancy order was found to be insignificant for high-T_c superconductors. From the viewpoint of valence numbers for the $YBa_2Cu_3O_{7-x}$ compound, the number of oxygen atoms should be 6.5 (i.e., $YBa_2Cu_3O_{6.5}$). Unfortunately, it has been proven experimentally that a single crystal of $YBa_2Cu_3O_{6.5}$ does not demonstrate superconductivity. The number of O atoms needs to be higher than 6.5 for $YBa_2Cu_3O_{7-x}$ to achieve superconductivity. A similar situation is true for all other high-T_c metal oxide superconductors. Thus the existence of Cu^{+3} in these high-T_c superconductors was suggested in order to account for the chemical structure. However, Sarma and Rao[15], as well as Horn et al.[16] used X-ray photoelectron spectroscopy (XPS) and Auger electron spectroscopy (AES) to investigate the chemical state of the copper ion. They found that Cu^{+3} was not present in these superconductors. Instead, Cu^+ was observed. It is well known that high-T_c superconductors in an oxygen-free atmosphere will lose oxygen. It has been assumed that the oxide anion (O^{-2}) in the CuO plane is in equilibrium with oxygen molecules in the gas phase at the same temperature and a fixed partial pressure. It is known that these metal oxide superconductors are, in general, very porous and that there are many deficit sites in the crystal. However, the possible existence of trapped O_2 molecules in the bulk or thin film materials has been ignored. During the development of the present monitor for superconductive films preparation, we observed the existence of trapped O_2 in superconductors.[17] The trapped O_2 gas can possibly have significant implications for the nature of the superconductivity of these high-T_c materials.

2. EXPERIMENTAL RESULTS AND DISCUSSION

A schematic of the experimental apparatus is shown in Fig. 1. A Nd:YAG laser beam was used to do laser ablation of samples. A quadrupole mass spectrometer was used to monitor the desorbed ions directly. The desorbed neutral atoms or molecules were ionized by electron impact or by a laser resonance ionization method. Fluorescence spectra were obtained from excited atoms and molecules.[18] Different samples of superconductors were used in this work. One was produced from an aqueous solution containing Y, Ba, and Cu salts and oxalic acid via a homogeneous coprecipitation technique using urea. Others were made by a standard solid-state chemistry approach. Transition temperatures of samples for Y − Ba − Cu − O and Bi − Sr − Ca − Cu − O were measured as $\sim 92°K$ and $\sim 125°K$, respectively. Most fluorescence data were taken using laser ablation of the superconductor sample by the fundamental of a Nd:YAG laser beam. Most secondary ion spectra were obtained using the second harmonic of the Nd:YAG laser beam. However, the data obtained with different harmonics show very little dependence on the frequency of the laser beam. The laser beam power density on the superconductor target was varied from 5×10^5 W/cm^2 to 1×10^9 W/cm^2. In general, breakdown occurs when the laser power density is higher than 1×10^7 W/cm^2. Insufficient photons could be detected to monitor fluorescence in the absence of laser-induced breakdown. However, the fluorescence spectra remain constant as long as laser-induced breakdown occurs. The laser energy per pulse was measured by a Scientech calorimeter. The laser beam pulse duration was 5–7 ns which was measured by a fast response photodiode. A Pellin-Broca prism was used to separate different harmonics of the laser beams. Since this laser utilized an unstable resonator, the typical laser beam distribution was a doughnut shape. Fluorescence spectra were obtained by using a GCA McPherson spectrometer which can be scanned automatically. The resolution of the monochromator was fixed at 4 cm^{-1}. A photomultiplier (RCA 31034) which has an almost flat response for visible and near-infrared (IR) photons was used to detect the wavelength-resolved photons. Since the laser beam was pulsed, the fluorescent light detected by the photomultiplier was gated for 20 μs so that the background from room light or dark current of the phototube could be neglected. The output of the phototube went through a charge-sensitive preamplifier followed by a pulse-shaping amplifier. It then went to a sample and hold circuit which drove a recorder.

The entire fluorescence spectra in the visible and near-IR region for $YBa_2Cu_3O_{7-x}$ and Tl-Ba-Ca-Cu-O are shown in Fig. 2 and Fig. 3. The fluorescence spectra of CuO, Y_2O_3, CaO, and $BaCO_3$ were also obtained as an aid in the identification of the peaks in the fluorescence spectra of these two superconductors. The emission due to different metal elements was identified from atomic spectra. Many different transitions for Y, Ba, and Ca were observed. However, no definite indication of transitions due to copper atoms, which had been suggested as the key element for superconductive properties, was observed. No clear spectra corresponding to CuO transitions were identified. The complete fluorescence spectra of high-T_c superconductors are quite complicated. However, the sharp transition lines, especially in the red and near-IR spectral regions, should be convenient for use in monitoring the concentration of most metal elements, since the broad continuum emission due to the plasma can be neglected in this region. The fluorescence band for YO was also observed in the Y-Ba-Cu-O system. Absolute fluorescence quantum yields can be obtained by using the fluorescence of a dye solution excited by the frequency doubled Nd:YAG laser beam.[19]

Elements which provide low emission intensity such as Cu and Tl can be monitored by desorbed ion spectra. Ion spectra of laser-ablated Y-Ba-Cu-O and Tl-Ca-Ba-Cu-O are shown in Fig. 4(b) and Fig. 4(c) with a laser power density of 2×10^7 W/cm^2. YO$^+$ signals were observed to be much stronger than the signal of BaO$^+$ for the Y-Ba-Cu-O system for a modest ablation laser energy. No CuO$^+$ signals were observed for the Y-Ba-Cu-O or Tl-Ca-Ba-Cu-O systems. The crystal structure of Y-Ba-Cu-O has been extensively studied by X-ray diffraction and neutron scattering methods.[20] There were no oxide atoms on the Y plane. Thus the formation of YO$^+$ was thought to be either from peroxide or from trapped oxygen molecules. For the Tl-Ca-Ba-Cu-O system, CaO$^+$ was observed, although X-ray diffraction data indicate that there are no oxides on the Ca plane. The formation of YO$^+$ and CaO$^+$ can possibly come from the chemical reaction between Ca or Y atoms with oxygen in the plume just above the surface. Secondary ion spectra obtained in this work are much simpler than the results from atom probe experiments.

The major species detected following laser ablation of superconductors are neutral atoms and molecules in the ground state. However, it is difficult to detect most of these desorbed species by a conventional mass spectrometer with an electron impact ionizer due to the low density of the desorbed neutral molecules. However, RIS has been demonstrated to have the capability of detecting a single neutral metal atom.[21] Recently, the detection of Bi atoms from the Bi-Sr-Ca-Cu-O system was demonstrated. Different laser excitation schemes for various metal elements are given in Table 1. It shows that commercially available lasers can be used to detect all desorbed neutral atoms and molecules. With an arrangement of multiple laser beams to do resonant excitation and ionization, several metal elements can be probed at the same time. Then, a real-time monitor for a superconductive thin film preparation can be achieved.

Table 1
Resonance ionization schemes for various metal elements in metal oxide superconductors

Scheme	λ
Ba $(6s^2)$ $\xrightarrow{h\nu}$ Ba $(6s\,7p)$ $\xrightarrow{h\nu}$ Ba$^+$	324.5 nm
Tl $(6s^2\,6p)$ $\xrightarrow{h\nu}$ Tl $(6s^2\,7s)$ $\xrightarrow{h\nu}$ Tl$^+$	377.6 nm
Pb $(6s^2\,6p^2)$ $\xrightarrow{h\nu}$ Pb $(6s^2\,6p\,7s)$ $\xrightarrow{h\nu}$ Pb$^+$	283.3 nm
Bi $(6p^3)$ $\xrightarrow{h\nu}$ Bi $(6p^2\,7s)$ $\xrightarrow{h\nu}$ Bi$^+$	306.8 nm
Ca $(4s^2)$ $\xrightarrow{h\nu}$ Ca $(4s\,5p)$ $\xrightarrow{h\nu}$ Ca$^+$	273.6 nm
Y $(4d\,5s^2)$ $\xrightarrow{h\nu}$ Y $(4d\,5s\,5p)$ $\xrightarrow{h\nu}$ Y$^+$	359.4 nm
Cu $(3d^{10}\,4s)$ $\xrightarrow{h\nu}$ Cu $(3d^{10}\,4p)$ $\xrightarrow{h\nu}$ Cu$^+$	324.8 nm
Sr $(5s^2)$ $\xrightarrow{h\nu}$ Sr $(5s\,6p)$ $\xrightarrow{h\nu}$ Sr$^+$	295.2 nm

To provide in-situ monitoring of substrate condition on thin film production, a Fourier Transform Infrared (FTIR) spectroscopy can be used. Reflection, transmission, and emission spectroscopy are all valuable in providing the kinetic information on thin films production. Emission FTIR yields substrate temperature while reflection or transmission FTIR provides information on film thickness, morphology, and crystalline phase. For example, very different reflection spectra were observed between superconductive $YBaCuO_{7-x}$ films and nonsuperconductive YBCO film; (Fig. 5a-b); the reflectance from the $SrTiO_3$ substrate is included for reference (Fig. 5c). The as-deposited films do not display metallic behavior reflectance (increasing reflectance at low wavenumbers). After annealing, however, the ablated film becomes a metal and has a spectrum similar to the annealed bulk. Figure 6 contains in-situ measurements of the reflectance of a growing film at 0, 1, 2, and 3 minutes of ablation (corresponding to thicknesses of 0, 0.5, 1.0, and 1.5 μm). The background for these spectra is a gold mirror mounted over the $SrTiO_3$ substrate. The reference mirror must be slightly misaligned, since the reflectance exceeds 100% at some wavenumbers. Note that the reflectance at wavenumbers above 5000 cm^{-1} slowly decreases with thickness. There is also a broad peak that moves as the YBCO thickness increases. The source of this feature is unknown, but it could be scattering from grains in the film. Annealing these films show a similar change as in Fig. 5. For transmission spectra, the superconductive film is almost completely opaque while the non-superconducting sample transmits. Emission spectroscopy yields surface temperature of substrate materials like $SrTiO_3$ by using the Christiansen effect (Fig. 7). From Fig. 7a, one can see that a 1 mm thick substrate of $SrTiO_3$ has a transmittance of zero below 1400 cm^{-1}. In addition, the reflectance of $SrTiO_3$ drops to zero at 870 and 482 cm^{-1} (Fig. 7b). The dispersion relation for the index of refraction (n) of $SrTiO_3$ shows that n is equal to unity at these two points; this is called the Christiansen effect. As a result of these two observations, radiation at 870 cm^{-1} and 482 cm^{-1} is neither reflected nor transmitted and must be completely absorbed; i.e., the emissivity is unity. By measuring the radiance (emission) at those two wavenumbers and using $\epsilon(870$ cm$^{-1}) = \epsilon(482$ cm$^{-1}) = 1$, one can solve for the substrate temperature. Once the temperature is known, one can also determine ϵ at all the other wavenumbers. Figure 7c shows quantitative measurements of radiance and temperature for a $SrTiO_3$ substrate. The prediction in Fig. 7c is the blackbody function at 542 K ($H(\nu,542K) - H(\nu,300K)$). The thermocouple measurement on the surface of the substrate is about 12K lower than the temperature determined by the fit; the fit is accurate to ±5 K.

The above feasibility study demonstrates that FTIR has great promise as an in-situ monitor for the deposition of superconducting films. Using transmission, emission, and reflection configurations, FTIR can monitor: substrate temperature (±5 K), phase changes (ablated/annealed/de-annealed), film thickness (0–2 μm), water contamination, infer oxygen content, and infer grain size. In-situ FTIR has also been demonstrated on a laser ablation reactor. Coupling the spectrometer to a preexisting deposition chamber is relatively straightforward. In-situ reflection and emission measurements of the film yield valuable processing information in real time. Further development of the FTIR monitor could have a large scientific impact on understanding the relationship between film properties and processing conditions. By combining FTIR, fluoresance spectra, and mass spectra an on-line, in-situ monitor can thus be developed to completely monitor the superconductive film preparation.

It was observed that a large amount of O_2 was desorbed during the laser ablation process. the amount of desorbed O_2 was estimated as $\sim 10^{10}$ molecules per pulse from laser ablation of a $YBa_2Cu_3O_{7-x}$ target at 22°K with a very modest laser power density of 1×10^6 W/cm^2. With this laser power level, there should be no significant destruction of any chemical bonds. No significant numbers of atoms, molecules, or ions other than O_2 were observed, as shown in Fig. 4(a). This process of desorption should be due to a heating effect. Since the desorption of O_2 was observed for laser ablation by the fundamental Nd:YAG laser beam at 1064 nm, it is very unlikely the desorption was due to bond breaking of a metal oxide by the one-photon absorption process. The bond strength of Cu-O and Cu-O$^-$ were estimated by Sawyer[22] to be 2.8 and 3.1 eV, respectively. Thus, the present results indicated that O_2 molecules were loosely trapped in these superconducting materials. Even a mild heating process is enough to drive out significant quantities of oxygen molecules trapped in these materials. The average velocity of desorbed O_2 was estimated as $\sim 7 \times 10^4$ cm/sec which indicated that the surface temperature was less than 300°C. There was a concern that the observed neutral O_2 molecules could be due to the adsorption of O_2 on the surface of the superconductor sample. To minimize this concern, a power density of 1×10^9 W/cm^2 was used to vaporize the top layers of material for several minutes before making measurements of desorbed O_2. We also carefully checked to see that there was no H_2O, N_2, or CO desorbed from the target to assure that the laser was probing below the original surface layer. These procedures rule out the possiblity of O_2 being adsorbed just on the surface. Therefore, the fact that oxygen molecules are trapped in $YBa_2Cu_3O_{7-x}$ and Tl-Ca-Ba-Cu-O superconductors is confirmed.

It has been known for two decades that O_2 can be trapped in an alkali-halide crystal.[23] Consequently, it should be much more likely that O_2 can be trapped in loosely packed and very porous superconductor materials, which were produced and sintered either in air or in a pure oxygen environment. From our results, the existence of trapped O_2 in the bulk of these superconductive materials is confirmed. The implication of trapped O_2 inside of the bulk of superconductors can be quite significant. It has long been considered that the existence of Cu^{+3} accounts for the excess of oxygen in $YBa_2Cu_3O_{7-x}$. The nominal composition suggests that 33% of the copper is in the +3 state. However, no Cu^{+3} ions were observed by X-ray photoelectron spectroscopy. With the trapped molecular oxygen, there is no need to have Cu^{+3} account for the number of oxygen atoms to be higher than 6.5. This should have an important effect on the calculation of the electronic structure of these crystals. Since a very small isotope effect was observed for $YBa_2Cu_3O_{7-x}$, conventional BCS theory regarding electron-phonon coupling can not be used to account for this observation. However, if the trapped O_2 molecule plays an important role in superconductivity, the isotopic effect can be expected to be much smaller if trapped O_2 has only weak coupling to the crystal lattice. If the trapped O_2 is critical for the superconductivity, the conservation of O_2 in superconductive film becomes very critical for any practical application for these newly invented superconductors.

3. ACKNOWLEDGEMENTS

This research was sponsored by the Superconductor Pilot Center at Oak Ridge National Laboratory and by the Office of Health and Environmental Research, U.S. Department of Energy, under contract number DE-AC05-84OR21400 with Martin Marietta Energy Systems, Inc. The Small Business Innovation Research Program of the Strategic Defense Initiative Organization has sponsored the work at Advanced Fuel Research, under contract number DASG60-88-C-0083.

"The submitted manuscript has been authored by a contractor of the U.S. Government under contract No. DE-AC05-84OR21400. Accordingly, the U.S. Government retains a nonexclusive, royalty-free license to publish or reproduce the published form of this contribution, or allow others to do so, for U.S. Government purposes."

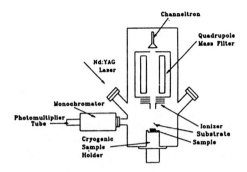

Fig. 1. Experimental schematic for fluorescence spectra, desorbed ion spectra, and resonance ionization spectra for laser ablation of the superconductors.

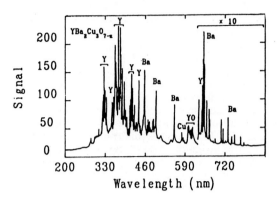

Fig. 2. Emission spectrum of laser ablation of $YBa_2Cu_3O_{7-x}$.

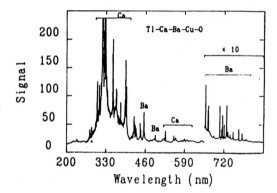

Fig. 3. Emission spectrum of laser ablation of Tl-Ba-Ca-Cu-O superconductors.

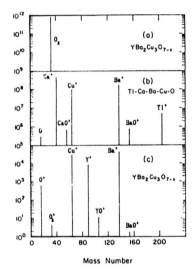

Fig. 4. Desorbed mass spectra from a laser ablation process of superconductor targets. (a) represents mass spectrum from $YBa_2Cu_3O_{7-x}$ target with laser density at 2×10^6 W/cm^2, the desorbed neutral O_2 molecules were ionized by electron impact ionization. No other desorbed ions or molecules were observed. (b) and (c) represent desorbed ion spectra from $YBa_2Cu_3O_{7-x}$ and Tl-Ca-Ba-Cu-O targets, respectively. The ablation laser power density was 2×10^7 W/cm^2. Filament was off.

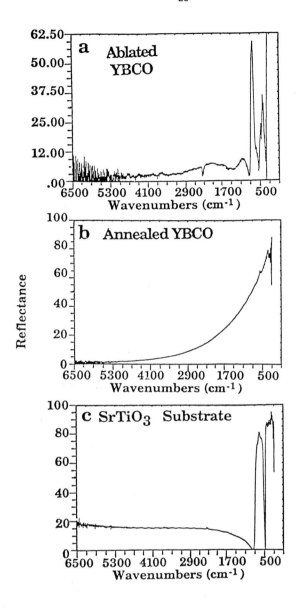

Figure 5. Reflectance From a) Ablated YBCO Film, b) Annealed YBCO Film, and c) SrTiO$_3$ Substrate.

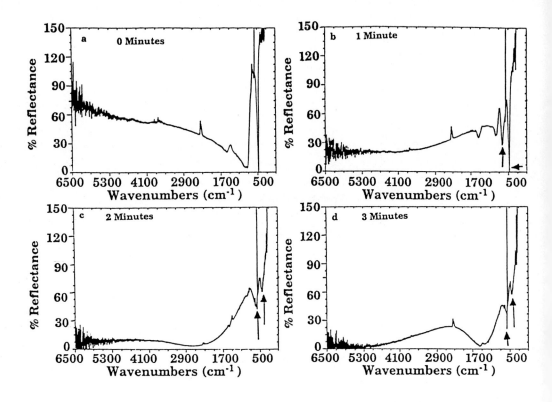

Figure 6. In-Situ Reflectance Measurements During Laser Ablation: a) 0 Minutes, b) 1 Minute, c) 2 Minutes, and d) 3 Minutes.

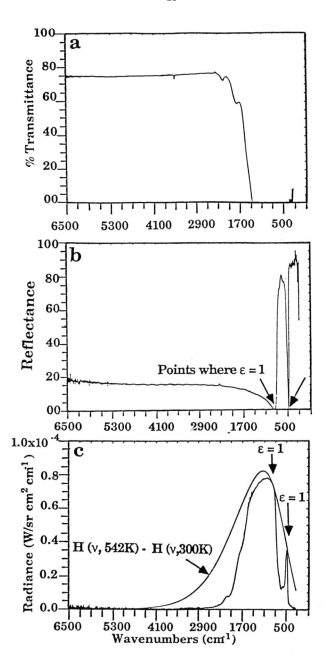

Figure 7. a) Transmittance, b) Reflectance, and c) Radiance of a 1 mm Thick Slab of $SrTiO_3$ Crystal.

4. REFERENCES

1. J. G. Bednorz and K. A. Müller, *Z. Phys. B*, 64 (2), pp. 189-193, September 1986.
2. M. K. Wu, J. R. Asburn, C. J. Torng, P. H. Hor, R. L. Meng, L. Gao, Z. J. Huang, Y. Q. Wang, and C. W. Chu, *Phys. Rev. Lett.*, 58 (9), pp. 908-910, March 1987.
3. C. E. Rice, A.F.J. Levi, R. M. Fleming, P. Marsh, K. W. Baldwin, M. Anzlowar, A. E. White, K. T. Short, S. Nakahara, and H. L. Stormer, *Appl. Phys. Lett.*, 52 (21), pp. 1828-30, May 1988.
4. A. D. Berry, D. K. Gaskill, R. T. Holm, E. J. Cukauskas, R. Kaplan, and R. L. Henry, *Appl. Phys. Lett.*, 52 (20), pp. 1743-45, May 1988.
5. E. S. Hellman, D. G. Schlom, N. Missert, K. Char, J. S. Harris, Jr., M. R. Beasley, A. Kapitulnik, T. H. Geballe, J. N. Eckstein, S.-L. Weng, and C. Webb, *J. Vac. Sci. Technol. B*, 6 (2), pp. 799-803, March/April 1988.
6. L. Lynds, B. R. Weinberger, G. G. Peterson, and H. A. Krasinski, *Appl. Phys. Lett.*, 52 (4), pp. 320-322, January 1988.
7. H. S. Kwok, P. Mattocks, L. Shi, X. W. Wang, S. Witanachchi, Q. Y. Ying, J. P. Zheng, and D. T. Shaw, *Appl. Phys. Lett.*, 52 (21), pp. 1825-1827, May 1988.
8. H. S. Kwok, J. P. Zheng, S. Witanachchi, P. Mattocks, L. Shi, Q. Y. Ying, X. W. Wang, and D. T. Shaw, *Appl. Phys. Lett.*, 52 (13), pp. 1095-1097, March 1988.
9. T. Venkatesan, X. D. Wu, A. Inam, and J. B. Wachtman, *Appl. Phys. Lett.*, 52 (14), pp. 1193-1195, April 1988.
10. X. D. Chen, S. Y. Lee, J. P. Golben, S. Lee, R. D. McMichael, Y. Song, T. W. Noh, and J. R. Gaines, *Rev. Sci. Instrum.*, 58 (9), pp. 1565-1571, September 1987.
11. B. Batlogg, R. J. Cava, A. Jayaraman, R. B. van Dover, G. A. Kourouklis, S. Sunshine, D. W. Murphy, L. W. Rupp, H. S. Chen, A. White, K. T. Short, A. M. Mujsce, and E. A. Rietman, *Phys. Rev. Lett.*, 58 (22), pp. 2333-2336, June 1987.
12. K. J. Leary, H. Loye, S. W. Keller, T. A. Faltens, W. K. Ham, J. N. Michaels, and A. M. Stacy, *Phys. Rev. Lett.*, 59 (11), pp. 1236-1239, September 1987.
13. J. D. Jorgensen, H. B. Schüttler, D. G. Hinks, D. W. Capone, K. Zhang, M. B. Brodsky, and D. J. Scalapino, *Phys. Rev. Lett.*, 58 (10), pp. 1024-1027, March 1987.
14. G. Xiao, M. Z. Cieplak, A. Gavrin, F. H. Streitz, A. Bakhshai, and C. L. Chien, *Phys. Rev. Lett.*, 60 (14), pp. 1446-1449, April 1988.
15. D. D. Sarma and C.N.R. Rao, *Solid State Commun.*, 65 (1), pp. 47-49, January 1988.
16. S. Horn, J. Cai, S. A. Shaheen, Y. Jeon, M. Croft. C. L. Chang, and M. L. denBoer, *Phys. Rev. B*, 36 (7), pp. 3895-3898, September 1987.
17. C. H. Chen, R. C. Phillips, and M. P. McCann, *Phys. Rev. B*, 39 (4), pp. 2744-2747, February 1989.

18. C. H. Chen, M. P. McCann, and R. C. Phillips, *Appl. Phys. Lett.*, 53 (26), pp. 2701-2703, December 1988.
19. C. H. Chen and M. P. McCann, *Chem. Phys. Lett.*, 153 (4), pp. 338-340, December 1988.
20. J. D. Jorensen, M. A. Beno, D. G. Hinks, L. Soderholm, K. J. Volin, R. L. Hitterman, J. D. Grace, Ivan K. Schuller, C. U. Segre, K. Zhang, and M. S. Kleefisch, *Phys. Rev. B*, 36 (7), pp. 3608-3616, September 1987.
21. G. S. Hurst, M. G. Payne, S. D. Kramer, and C. H. Chen, *Phys. Today*, 33 (9), pp. 24-29, September 1980.
22. Donald T. Sawyer, *J. Phys. Chem.*, 92 (1), pp. 8-9, January 1988.
23. J. Rolfe, F. R. Lipsett, and W. J. King, *Phys. Rev.*, 123 (2), pp. 447-454, July 1961.

SPECTROSCOPIC AND ION PROBE CHARACTERIZATION OF LASER PRODUCED PLASMAS USED FOR THIN FILM GROWTH

David B. Geohegan
Solid State Division, Oak Ridge National Laboratory
Oak Ridge, Tennessee 37831-6056

ABSTRACT

Laser ablation for in situ film growth of 1:2:3 superconductors requires high laser energy densities and results in a dense, multicomponent laser plasma which expands into high background pressures (~200 mTorr) of oxygen. The physics of the transport process in the film growth regime involves a variety of high-density collisional phenomena which one attempts to avoid in near-threshold ablation studies: laser-plasma heating and acceleration, thermalization, attenuation and reactive scattering of the plume by background gas collisions, and shock formation. Three in situ techniques are employed and reviewed to characterize these effects: optical emission spectroscopy for viewing the excited states of atoms and molecules resulting from plasma kinetics, optical absorption spectroscopy for viewing the non-emitting ground states of atoms, ions, molecules, and fast ion probes for viewing the flux of positive ions in the plasma. In this paper, spatially- and temporally-resolved optical absorption and emission spectroscopy and fast ion probes are applied to study the formation and acceleration of the laser plasma formed by KrF laser irradiation of $Y_1Ba_2Cu_3O_x$ in vacuum and the propagation of the plume through background gases of oxygen and argon at pressures and laser energies typically used for thin film growth.

INTRODUCTION

Pulsed-laser deposition has become a very attractive method for the growth of thin films due to the relative ease with which the congruent transfer of material from target to substrate can be achieved. This is thought to result from the high surface temperatures which are transiently attained using high laser energy densities. In many cases, as for the 1:2:3 superconductors, these energy densities also result in the formation of a visible laser plasma which extends to the substrate. In addition, in situ deposition of 1:2:3 superconducting thin films must be carried out in background pressures (10–300 mTorr) of reactive gases, such as oxygen, in order to form the film in a crystalline phase.

In situ monitors of the composition, kinetic energy and gas phase reactions within the plume during actual film growth conditions are essential for an understanding of film growth mechanisms. Near-threshold studies of ablation products and kinetic energy distributions are not extendable to the film growth regime. Velocity distributions of the various components change from nearly collisionless supersonic expansion distributions to a composite distribution of material moving with a common center-of-mass velocity. The extent of laser interaction with the ablated products during the laser pulse is not yet known, but is invoked to explain the rapid increase in ionization with laser energy density.

EXPERIMENTAL

The experimental apparatus has been described before [1] and consists of a stainless steel, turbopumped bell jar which was evacuated to 2×10^{-6} Torr prior to backfilling with argon or oxygen. A Lambda Physik EMG 210 excimer laser provided 50ns FWHM pulses of 248-nm radiation which were apertured and focused by a single cylindrical lens (focal length 50 cm) at a 60° incident angle onto a rotating, high density $Y_1Ba_2Cu_3O_{7-x}$ target. The beam was focused to a line of 1.0 cm width on the target, with a burn area of 0.04 cm^2. Typically, 100 mJ was transmitted onto the pellet, producing energy densities of 2.5 J cm^{-2} and peak intensities of 50 MW cm^{-2}. The beam energy was monitored on a shot to shot basis with a beamsplitter and pyroelectric detector (Gentec ED-200). The line focus geometry produced a plume which expanded little along the long dimension of the line, providing a nearly constant 1 cm absorption length for the optical absorption measurements.

Optical absorption was detected by the same spectrometer/ photomultiplier system in conjunction with a pulsed Xe flashlamp, which provided a structured continuum spectrum for wide spectral coverage and scans. The collimated lamp beam was 0.5 mm wide across the absorption region. A digital delay generator controlled the time delay between the laser and boxcar averager gate (typically 50 ns width) which was set on the peak of the lamp pulse. Alternatively, a CW copper hollow cathode lamp was used to detect ground-state Cu in the plume in conjunction with another photomultiplier with interference filter. Using this arrangement, the temporal absorption of Cu on a single shot basis was recorded on a transient digitizer.

The visible emission from the plume was spatially and spectrally resolved by imaging the plume onto the entrance slits of a 1-meter spectrometer with photomultiplier. The time response of the emission, CW lamp absorption or ion probe current was captured by a transient digitizer (Tektronix 7912AD). Ten waveforms were averaged for a given collection condition. The time responses of the photomultiplier and the ion probe circuit were 0.3 and 3 ns, respectively.

Two types of ion probes were employed for measurements of the positive ion flux along the target normal. A copper planar probe housed 2 mm behind a grounded Cu plane with 0.02 cm^2 aperture was employed with different meshes attached. Alternatively, a cylindrical copper Langmuir probe of the same area provided identical results. Under most of the conditions described here, the plasma density was sufficiently high to prevent separating the plasma with even the finest meshes. The ion probe was floated with respect to the chamber and pellet by terminating the coaxial signal line at the digitizer and by providing the bias with a completely isolated and shielded 1 mfd-capacitor-bypassed 300 V battery and potentiometer in series with the 50 ohm signal line.

RESULTS AND DISCUSSION

Near ablation threshold in vacuum, optical absorption spectroscopy detects ground state neutrals and ion probes detect positive ions emitted from a YBCO target before a plume becomes visible extending from the surface. For resonance lines with strong oscillator strengths, optical absorption spectroscopy is very sensitive, allowing for densities of ~10^9 cm^{-3} to be observed over path lengths of 1 cm.[2] At higher energy densities, collisions in the plume continually populate excited states of species with short lifetimes (~10 ns) several microseconds following the laser pulse arrival at the target, producing visible fluorescence.

The majority of the monatomic atoms and ions are in the ground states, however, during transport. Optical absorption allows one to observe the ground states directly, via strong resonance lines of the atoms and ions. A high-resolution (0.04 Å) spectrometer reveals that the spectral lines are broadened sufficiently[1] to permit absorption measurements with a pulsed, broadband Xe-lamp which also allows rapid wavelength switching. For $Y_1Ba_2Cu_3O_x$, absorption by atomic Y, Ba, Cu, Y^+, and Ba^+ was measured.

Optical absorption spectroscopy reveals that the plume transport lasts much longer than that inferred by the observed fluorescence. Figure 1 compares the temporal histories of the measured absorption of ground state Ba and Ba^+ at $d = 1$ cm with fluorescence from excited Ba*. The fluorescence is observed to follow the ground-state ion population, and is nearly gone by the peak of the ground-state neutral population. Measurable absorption lasts approximately an order of magnitude longer than the emission, revealing a slow component to the transport of the ground state species.

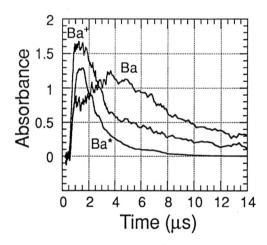

Figure 1: Comparison between the temporal profiles of Ba and Ba^+ ground-state absorption (at 553.5 nm and 455.4 nm, respectively) and Ba* fluorescence (553.5 nm, not to scale) monitored at $d = 1.0$ cm following 2.7 J cm^{-2} irradiation of $Y_1Ba_2Cu_3O_x$ at 10^{-5} Torr.

When time-of-flight velocity profiles are plotted from the measured absorption of ground state Ba and Ba$^+$ (as shown in Fig. 2 for two energy densities) several key features are displayed. Up to 1.0 J cm^{-2}, the profiles for Ba$^+$ and Ba are very similar in magnitude, shape and leading edge velocity. Spectral widths measured at the peak of the 1.0 J cm^{-2} absorbances, however, imply that Ba$^+$ ions outnumber Ba neutrals by 5 to 1.

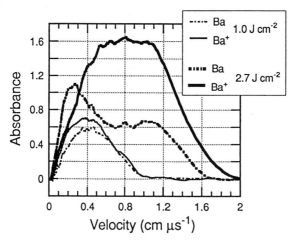

Figure 2: Measured absorbances vs. time-of-flight velocity for ground state Ba (553.5 nm) and Ba$^+$ (455.4 nm) at $d = 1$ cm from a $Y_1Ba_2Cu_3O_x$ pellet at 10^{-5} Torr following KrF irradiation at 1.0 J cm^{-2} and 2.7 J cm^{-2}.

At higher energy densities, the Ba$^+$ ions grow more rapidly in number, and appear on the leading edge of the expanding plume, with the Ba neutrals lagging behind. In addition, the neutral distribution takes on a two-component appearance with a fast component trailing the ion fast component, and a slow component. From the relative Ba and Ba$^+$ absorbances and spectral widths, the Ba in the plume is highly ionized in vacuum. The excited state fluorescence appears to feed the fast component of the ground state neutral population as one would expect from recombination of ions on the leading edge of the pulse or from resonant charge exchange in collisions between like neutrals and ions. Similar results were obtained for Y and Y$^+$.

Fast ion probes confirmed that the plume is a nearly neutral, weakly ionized plasma with ions and electrons traveling at common velocity. Attempts to electrostatically separate the ions from the electrons in the plume with finely meshed screens failed even at low fluences implying Debye lengths <10 μm at film growth distances. It was found that a bias of a few negative volts repelled most of the electrons from reaching the probe, and that complete saturation of the ion current resulted from biases more negative than -70 V. The ion probes were operated in this ion saturation regime with biases of typically -100 V.

The increase in ionization and acceleration of the plume with increasing laser energy density were examined with the ion probes along the normal to the

KrF-irradiated YBCO region. Series of digitized ion probe waveforms (e.g., Fig. 5, 0.01 mTorr waveform) were recorded in vacuum at a fixed distance of 2.9 cm. The total positive charge collected by the probe was obtained by integrating the ion current signal and is plotted versus KrF energy density in Fig. 3. Two distinct regions are observed. Near threshold, from 0.11 to ~0.37 J cm^{-2}, the positive charge in the plume increases with a $\Phi^{7.4}$ multiphoton dependence. In this region, a visible laser plasma is just beginning to appear. Above ~0.37 J cm^{-2}, the visible emission becomes much brighter and the ion current changes slope to a $\Phi^{1.5}$ dependence.

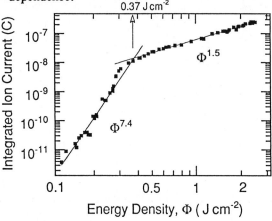

Figure 3: Total positive charge collected 2.9 cm along the normal from a $Y_1Ba_2Cu_3O_x$ pellet through a 0.125 cm^2 aperture at 10^{-5} Torr with a Cu planar ion probe.

The velocity increase of the expanding plasma was simultaneously measured by recording the arrival times of the peak and the leading edge (10% of the peak) of the positive ion probe pulse. The time of flight velocities are given in Fig. 4 as a function of 248 nm energy density. The velocity increase is well-described by a power law dependence given by $v = a (\Phi - \Phi_{th})^n$ where $n = 0.35$, $\Phi_{th} = 0.21$ J cm^{-2} (leading edge) and $n = 0.37$, $\Phi_{th} = 0.11$ J cm^{-2} (peak).

The increases in ionization and velocity of the plume with laser energy density presented in Figs. 3 and 4 seems to support a laser interaction with the dense plasma formed within the laser pulse at the target. The coincidence between the appearance of a visible plasma and the change in energy density dependence in Fig. 3 may signal the onset of plasma absorption. Absorption in laser plasmas can occur from inverse Bremsstrahlung, free-free transitions of electrons in the presence of ions. The increase in velocity of the laser plasma edge can be modeled to result from the increased heating of the laser plasma followed by adiabatic expansion with varying values of the exponent n depending upon assumptions of the extent of ionization and other factors.[3,4]

The Y, Ba, and Cu are observed to travel with nearly common velocity, rather than common energy, in the plume. This is likely a result of the many equilibrating collisions experienced in a dense Knudsen layer near the target.[5] For most-probable velocities of 1 cm μs^{-1}, this corresponds to kinetic energies of ~40–70 eV. These hyperthermal energies are an order of magnitude larger than the measured plasma temperatures at the pellet. This supports the notion of adiabatic conversion of thermal to kinetic energy.

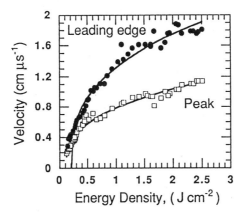

Figure 4: Time-of-flight velocity of the expanding plasma front, 2.9 cm along the normal from a $Y_1Ba_2Cu_3O_x$ pellet in vacuum as measured with an ion probe. Two time intervals were obtained from digitized waveforms, corresponding to the peak and the leading edge (defined as 10% of the peak) ion flux. Curve fits to the data are of the form $v = a (\Phi - \Phi_{th})^n$ where $n = 0.35$ (leading edge) and $n = 0.37$ (peak).

Kinetic energies corresponding to the leading edge of plasma expansion are often several hundred eV (e.g. Ba^+ traveling 2 cm μs^{-1} in Fig. 2 has kinetic energy of 286 eV). Such high kinetic energies are capable of damaging surface layers of a deposited film. On the other hand, impinging kinetic energies in the 1–10 eV range are thought to aid epitaxy by increasing adatom mobility.

Pulsed laser deposition of 1:2:3 superconductors is typically carried out in 50–300 mTorr of oxygen. Under these conditions, the plume fluorescence increases, oxides are preferentially formed and the deposition rate falls due to scattering collisions between the plume and the background gas [6-8]. Atoms and ions with most-probable energies of ~40–70 eV are reduced to energies of <1 eV at film growth distances [9]. Accurate modeling is extremely complicated. Individual cross sections for the various collision processes are highly velocity dependent and generally unknown.

The transmission of positive ions through background pressures of 50–300 mTorr of both argon and oxygen was measured using a Langmuir probe along the normal to the target for distances from 2.0 to 10.0 cm. The maximum range of the ions was limited by the exponential drop in transmitted intensity with distance and the observed slowing. Ground state Cu neutrals, observed by optical absorption, were also slowed similarly. In general, the transmission of the plume as a whole was well represented by the ion probe transmission.

Figure 5a shows the attenuation and slowing of the ion pulse at $d = 2.0$ cm for various pressures of argon and oxygen up to 300 mTorr. At each distance, the total positive charge transmitted to the probe dropped exponentially with pressure as suggested by the areas under the curves in Fig. 5a. While the magnitude of the ion probe signal transmitted through argon was typically 2 to 4 times greater than when oxygen was used, Fig. 5b shows that the shape and delay of the ion pulse is nearly identical for the two gases.

It was found that the overall ion attenuation rates are nearly equal in oxygen and argon for distances >2 cm, however the magnitude of the number of ions reaching a given distance is significantly reduced in the reactive oxygen environment. While the emission from neutrals, ions and oxides increased due to increased collisions in either background gas, gated emission spectra showed greater increase in oxide fluorescence vs. atomic lines and approximately twice as much integrated oxide fluorescence from 50–400 mTorr with oxygen instead of argon.

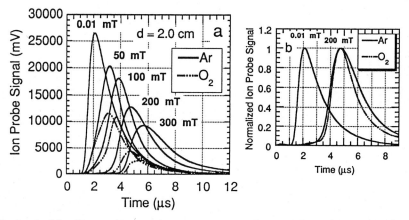

Figure 5: (a) Comparison of ion current waveforms 2.0 cm from 2.5 J cm^{-2} KrF-irradiated $Y_1Ba_2Cu_3O_7$ in 0.01 and 50, 100, 200, and 300 mT background pressures of argon and oxygen. (b) Normalized 0.01 and 200 mT waveforms showing similarity in collisional slowing effects for oxygen and argon.

The integrated ion charge delivered to the ion probe for various distances are shown in Fig. 6a. In vacuum the ion flux drops as $d^{-1.8}$, in near agreement with the expected d^{-2} drop for a detector of constant size at distance d from the origin of the nearly two-dimensional expansion. Using the vacuum density for I_0, the attenuation of the integrated charge were fit to $I = a\,I_0\,\exp(-bd)$, where $b=N\sigma$ represents the attenuation coefficient. The attenuation coefficient was found to depend linearly on background pressure where the slope of the curve yields a general scattering cross section for ion-argon $\sigma_{i\text{-}Ar} = 2.1 \times 10^{-16}$ cm^2 and ion-oxygen $\sigma_{i\text{-}O_2} = 2.3 \times 10^{-16}$ cm^2 interactions for background pressures up to 300 mTorr. Although not shown explicitly here, this attenuation appears to be mimicked by the amount of collected film deposited at a given distance, yielding a good measurement of the relative deposition rate under different background pressure conditions.

The ion probe signal monotonically decreases with increasing pressure and distance, while the inelastic processes result in conversion of the ions into excited states and oxides, peaking their densities at a given distance and pressure. The leading edge fluorescence of all observed species were found to follow the leading edge of the ion pulse at pressures >200 mTorr. Above 200 mTorr and distances of 2 cm, the onsets of fluorescence from the ions, neutrals and oxides appear to converge into a weak shock structure with common velocity, collisionally equilibrated and coincident with the ion probe signal.

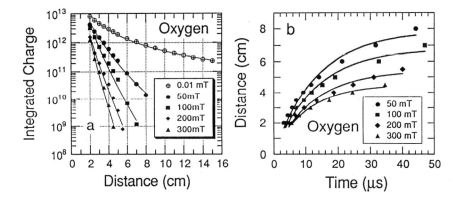

Figure 6: (a) Transmitted total positive-ion charge through oxygen backpressures of 0.01, 50, 100, 200, and 300 mTorr. Fits to the data follow the form $d^{-1.8}e^{-bd}$, where b is the attenuation coefficient and d is the distance from the pellet. (b) Position-time plots of the plasma propagation through 50, 100, 200, and 300 mT oxygen and argon following KrF laser ablation of $Y_1Ba_2Cu_3O_7$ at 2.5 J cm^{-2}. Fits to the curves are of the form $x = x_f[1 - \exp(-\beta t)]$.

The slowing of the plasma front in oxygen and argon was measured by recording the arrival time of the ion probe signal peak at various distances [Fig. 6b]. The slope of a smooth curve drawn through the points gives the instantaneous velocity of the plasma peak at a given time and distance.

The ions show a continuous velocity decrease with distance and exponential drop with time, with the maximum range of the ions saturating at smaller distances with increasing pressure. At high background pressures, a blast wave model has been applied to the luminous shock front caused by the expansion of laser ablation plasmas [10]. This model strictly applies when the mass of the ejected products is small compared with the mass of the background gas set in motion by the shock, a situation which does not apply here. Nevertheless, the propagation of a shock front through the background gas is described by the relation $d = \xi_0 (E/\rho_0)^{1/5} t^{2/5}$, where ξ_0 is a constant, E is the energy deposited in the explosion, ρ_0 is the density of the background gas and d and t give the position and time [10]. The d, t plots of Fig. 6b can be fit by power law dependences, but do not saturate sufficiently at low pressures. The best-fit values of the exponent also vary from 0.54 at 50 mTorr to 0.45 at 300 mTorr suggesting a better agreement at higher pressures.

A classical drag force model shows better agreement at low pressures and curve fits to the data are shown in Fig. 6b. The ejected pulse of ablation products is regarded as an ensemble which experiences a viscous force proportional to its velocity through the background gas. The equations of motion are given by

$$a = -\beta v$$
$$v = v_0 \exp(-\beta t)$$
$$x = x_f (1 - \exp(-\beta t))$$

giving
$$v = \beta (x_f - x)$$
$$a = -\beta^2 (x_f - x),$$

where β is the slowing coefficient, and $x_f = v_0/\beta$ is the stopping distance of the plume. From these equations of motion and the coefficients given in Table 1 one can obtain the velocity and deceleration of the plasma boundary as it propagates through the background gas.

Table 1

p (mTorr)	$1/\beta$ (μs) O_2	$1/\beta$ (μs) Ar	x_f (cm) O_2	x_f (cm) Ar
50	13.0	13.2	7.9	8.0
100	12.9	13.4	6.8	6.8
200	12.1	12.1	5.4	5.4
300	10.2	11.2	4.5	4.5

CONCLUSIONS

The combination of optical absorption and emission spectroscopy with fast ion probes provides a versatile set of in situ diagnostics which operate in the dense plasmas and high background gas pressures found in the film growth regime. The plume is characterized as a partly ionized plasma with some of the monatomic species predominantly singly ionized at the higher laser energy densities. Increasing laser energy density at the target results in acceleration of the plasma boundary in which the ions precede the neutrals with evidence of recombinational feeding of the ground state neutral population. Laser interaction with the plasma plume is indicated by the change in ion production rate and the onset of visible plasma fluorescence with increasing laser energy density. The variation of the velocity of the plasma front has also been characterized in vacuum.

The penetration of the KrF/YBCO laser plasma through background gases of argon and oxygen has been characterized using spatially resolved ion probe and optical emission measurements. As one might expect from scattering theory, the number of ions in the plume are attenuated exponentially with distance or pressure. The gross features of the attenuation are described by scattering cross sections which do not take into account the velocity-dependence of the scattering. Reactive scattering, converting ions to oxides is indicated by preferential loss of ions and increased oxide fluorescence in oxygen compared to argon.

A weak shock structure appears to be formed at the pressures used for thin film growth. The ion probe waveform represents the plasma front as it moves through the background gas. At pressures ~150 mTorr and d > 2cm, the leading edge of the plasma front becomes coincident with the onset of emission from the ions, neutrals and oxides in the plume and the visible emission is enhanced as the shock front interacts with the gas ahead of it. In this pressure range, the mass of the ablated products exceeds the mass of the background gas in its path and the collisional slowing appears better represented by a viscous drag force rather than a rigid blast wave. The drag force model determines the range and velocity of the plume from distance-time plots of the plasma front.

ACKNOWLEDGEMENTS

The technical assistance and physics discussions with Doug Mashburn are gratefully acknowledged as well as research discussions with D.P. Norton and D.H. Lowndes. This research was sponsored by the Division of Materials Sciences, U.S. Department of Energy under contract DE-AC05-84OR21400 with Martin Marietta Energy Systems, Inc.

REFERENCES

1. D. B. Geohegan and D. N. Mashburn, *Appl. Phys. Lett.* **55**, 2345 (1989).
2. A. D. Akhsakhalyan, Yu. A. Bityurin, S. V. Gaponov, A. A. Gudkov, and V. I. Luchin, *Sov. Phys. Tech. Phys.* **27**, 969 (1982).
3. H. Puell, *Z. Naturforsch.* **25 a**, 1807 (1970).
4. J. F. Ready, *Effects of High Power Laser Radiation*, Academic Press, London (1971).
5. Roger Kelly and R. W. Dreyfus, *Surf. Sci.* **198**, 263 (1988).
6. T. Venkatesan et al., *Appl. Phys. Lett.* **53**, 1431 (1988).
7. C. Girault, D. Damiani, J. Aubreton, and A. Catherinot, *Appl. Phys. Lett.* **55**, 182 (1989).
8. D. B. Geohegan and D. N. Mashburn, in *Laser Ablation for Materials Synthesis*, Materials Research Society, San Francisco, California, 1989, p. 211.
9. D. N. Mashburn and D. B. Geohegan, in *Processing of Films for High-T_c Microelectronics*, *SPIE*, Vol. 1187, p. 172.
10. Ya. B. Zel'dovich and Yu. P. Raizer, in *Physics of Shock Waves and High-Temperature Hydrodynamic Phenomena*, (Academic Press, New York 1966), vol.1, p. 94.

"The submitted manuscript has been authored by a contractor of the U.S. Government under contract No. DE-AC05-84OR21400. Accordingly, the U.S. Government retains a nonexclusive, royalty-free license to publish or reproduce the published form of this contribution, or allow others to do so, for U.S. Government purposes."

SYNTHESIS OF SiO_2 THIN FILMS BY REACTIVE EXCIMER LASER ABLATION

Eric Fogarassy, C. Fuchs, A. Slaoui and J.P. Stoquert

Centre de Recherches Nucléaires (IN2P3), Laboratoire PHASE

(UPR du CNRS n°292)

23, rue du Loess, F-67037 Strasbourg Cedex

ABSTRACT

Silicon oxide films are deposited by laser ablation from silicon, silicon monoxide and silicon dioxide (quartz) targets, performed under vacuum and in oxygen atmosphere, with a high power pulsed ArF (λ = 193 nm) excimer laser. The specific influence of oxygen in the chamber during the laser processing on the synthesis of stoichiometric SiO_2 films is demonstrated.

INTRODUCTION

Laser ablation has been successfully used to deposit a wide range of materials including dielectrics, semiconductors, metals, and superconductors [1-3]. This technique, which results from the interaction of a high intensity laser with a solid target is found to present several unique advantages, especially when using pulsed lasers, over other conventional or photo-assisted deposition techniques. For most compounds, laser ablation deposition under vacuum gives film stoichiometry close to the target. In addition, modification in the composition of the deposit and synthesis of new compounds can be achieved by working in reactive atmosphere [4-6] (reactive laser ablation).

The final properties of the deposits strongly depend on the basic phenomena which take place during the interaction of the laser pulse with the bulk target. The overall process can be divided into three main steps : surface vaporization, plasma generation and subsequent material ejection under vacuum or in a reactive gas, leading to thin film growth.

In this work, a series of target materials with different linear optical properties such as silicon, silicon monoxide, and silicon dioxide (quartz) were irradiated under vacuum and in oxygen atmosphere with a high power pulsed ArF excimer laser. The deposition by laser ablation from these targets of silicon oxide layers was investigated in various experimental conditions. The synthesis of stoichiometric SiO_2 films appears strongly related to the presence of oxygen in the processing chamber able to strongly react with ablated species in the gas phase and at the surface of the growing oxide.

EXPERIMENTAL

High purity silicon (Si), silicon monoxide (SiO) and fused silica (a-SiO_2, SUPRASIL quartz) bulk targets were irradiated under vacuum ($\approx 10^{-5}$ torr) or in pure argon or oxygen atmosphere ($\leq 10^{-1}$ torr), with a pulsed ArF excimer laser (Lambda Physik EMG 201 MSC), which provides, at the 193 nm wavelength, 300 mJ pulses of 20 nsec duration at a repetition rate of 50 Hz. The laser beam was focused onto the targets under normal incidence to give energy densities ranging between 0.5 and 10 J/cm^2. In addition, the targets were rotated during the laser irradiation in order to avoid the formation of deep craters able to modify material ejection. Emitted species, in a collection angle of about 15°, were deposited onto crystalline (<100> orientation) Si substrates set parallel to the target and located at a distance of 3 cm (scheme in Fig. 2). The substrate temperature was varied between 20 and 450°C, during the deposition.

Silicon oxide (SiO_x, 1 < x < 2) films were characterized by infrared (IR) absorption spectroscopy using a Perkin Elmer 938 G double-beam grating spectrometer at room temperature. The half-width and position of the asymetric SiOSi stretching (ASM) band are strongly influenced by the bonding character, stoichiometry, density, porosity, and strain of the film [7]. The ASM infrared peaks for thermally grown SiO and SiO_2 layers, used as references, are located at around 980 and 1080 cm^{-1}, respectively. In addition, the small absorption band observed at 940 cm^{-1} is generally attributed to a Si-OH bending vibration with a non bridging oxygen atom [7-9].

The oxygen and silicon depth profiles in the oxide films and their atomic composition were deduced from Rutherford Backscattering (RBS) experiments ($^4He^+$, 1-2 MeV) by simulating the RBS spectra with the "Système d'Analyse Monospectre" [10] program (same results as RUMP [11] simulation). In addition, a subtraction of the background corresponding to multiple scattering effects gives us the possibility to decompose the experimental spectra in separate contributions due respectively to Si and O in the oxide layers. Taking into account the Rutherford cross sections for scattering and the stopping powers of $^4He^+$ tabulated by Ziegler [12], we are able to transform the energy peaks into depth profiles.

The elastic recoil detection technique (ERDA) has been used to determine the hydrogen profiles in the oxide films. A monoenergetic 2.9 MeV ^{4}He ion beam, collimated to a spot size of 1 mm^2, is incident on a solid target under a grazing angle α (10°). The hydrogen nuclei recoil at an angle $\alpha + \beta$, where β (10°) is the glancing angle between the surface and the detector direction. In these geometrical conditions, the elastically scattered ^{4}He beam can be stopped by an absorber (Mylar 13 μm thickness) while only the hydrogen atoms can be detected, allowing the spectrum which corresponds to the hydrogen profile in the target to be recorded [13]. Film thicknesses were also measured with a mechanical stylus (TALYSTEP) and their surface morphology was examined by scanning electron microscopy (SEM).

RESULTS and DISCUSSION

We have summarized in Table I the laser ablation thresholds (E_T expressed in J/cm^2), at 193 nm (6.4 eV), for Si, SiO and SiO$_2$ targets, as deduced from various experiments [6, 14, 15].
It is interesting to notice that a threshold close to 1 J/cm^2 (laser intensity of 5 x 10^7 W/cm^2) is found for these three materials which present very different linear optical absorption in the UV range (Fig. 1).

TARGET	LASER (ArF, 20 nsec) ABLATION THRESHOLD
SILICON	1.3 J/cm^2 [14]
SILICON MONOXIDE	0.7 - 1 J/cm^2 [6]
QUARTZ	1.5 J/cm^2 [15]

Table I

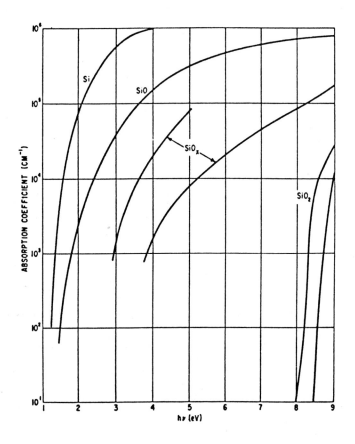

Figure 1 : The spectral dependence of the absorption coefficient of amorphous Si, SiO, SiO$_x$ (x ≈ 1.5) and SiO$_2$. The curve on the far right is for crystalline SiO$_2$ (from Philipp [16]).

By opposition to silicon and silicon monoxide, which are strongly absorbing at 6.4 eV, SUPRASIL quartz is transparent at this photon energy. However various authors [17-19] demonstrated that the 193 nm irradiation of fused silica, at high fluence, is able to produce color centers (oxygen vacancy defects) and macroscopic volume change (compaction), via a two photon absorption process [20] leading to the formation of superficial sub-stoichiometric SiO_x (x < 2) layers. Subsequently, this oxide is able to be strongly coupled with the 6.4 eV photons from the ArF (Fig. 1) through a one photon absorption mechanism, as for silicon and silicon monoxide.

For $E_T > 1$ J/cm^2, and whatever the angle of incidence of the laser beam relative to the target normal, each pulse produces a bright plasma plume (white under vacuum, blue in O_2 atmosphere) in a direction normal to the irradiated surface, responsible on film deposition. The deposition rate (V_R), measured at a collection angle of ≈ 15°, increases with the laser fluence to reach at $E_T ≈ 10$ J/cm^2 a maximum value of 0.5 Å/pulse with the SiO target, under vacuum.

TARGET	SILICON	SILICON MONOXIDE	QUARTZ
under vacuum (≈ 10^{-5} torr)	0.04	0.43 - 0.57	0.07 - 0.1
in oxygen (≈ 10^{-1} torr)	0.013-0.017	0.20 - 0.27	0.053

Deposition rate (Å/pulse)

Table II

As shown in Table II, for identical conditions of irradiation, lower values are measured for Si and SiO targets. It is also interesting to notice that V_R is significantly reduced when laser ablation takes place in presence of oxygen atmosphere (Table II).

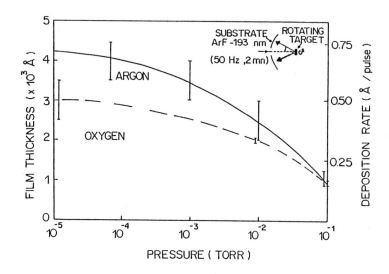

Figure 2 : Evolution of film thickness and deposition rate of the oxide film deposited by laser ablation from a SiO target under Ar and O_2 atmosphere at various pressures.

The influence of Ar and O_2 on the ablation process of SiO target has been investigated on a wide range of gas pressures (10^{-5} to 1 torr). As shown in Fig. 2, the continuous decrease of film thickness and deposition rate between 10^{-5} and 10^{-1} torr, which does not seem to significantly depend on the gas nature, can be mainly correlated to collisional effects between the background gases (Ar, O_2) and ejected species, of the laser generated plasma arriving at the substrate.

Finally, at pressures in the order of 1 torr and above, we observe the deposition of powdery films, which present a poor adhesion.

Figs. 3, 4 and 5 show the IR spectra of oxide films deposited at ambient temperature, under vacuum and in O_2 atmosphere, by laser ablation from Si, SiO and SiO_2 bulk targets, respectively. Whatever the target used, the films deposited under vacuum, appear non-stoichiometric (SiO_x, with x < 2). It is also important to notice that, when using Si target, the deposited layer appears partly oxidized as revealed by the 980 cm^{-1} absorption band (Fig. 3) which is characteristic of a Si/O ≈ 1 composition. By contrast, the films deposited in O_2 are fully converted into stoichiometric SiO_2, as confirmed by the peak position at 1080 cm^{-1}. However, we observe a pronounced shoulder on the high frequency side of the ASM band, which extends up to 1300 cm^{-1}, already observe in laser grown and irradiated film oxides [21]. Here, this feature could be attributed to disorder and/or defects induced, for one part, by the very fast deposition rate (up to 1000 Å/min at 50 Hz) and, for another part, the energetic ions of the laser generated plasma which bombard the substrate and the growing film during the deposition. In addition, the half-width of the ASM infrared bands (80 to 150 cm^{-1}), larger than for high temperature thermally grown SiO_2 (≈ 70 cm^{-1}) is indicative of the porosity of these oxides deposited at ambient temperature. We observe a significant reduction of this parameter by depositing the films above 400°C substrate temperatures [22]. Film porosity is also evidenced by the presence, at 940 cm^{-1}, of hydrogen bonded silanol groups (Si-OH) in the IR spectra of Figs. 3, 4 and 5. Si-OH groups could be incorporated into low temperature deposited SiO_2 films by two

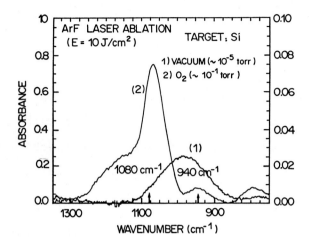

Figure 3 : IR absorption spectra of oxide films deposited onto silicon by laser ablation from a silicon target under vacuum and in O_2.

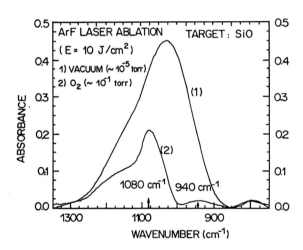

Figure 4 : IR absorption spectra of oxide films deposited onto silicon by laser ablation from a silicon monoxide target under vacuum and in O_2.

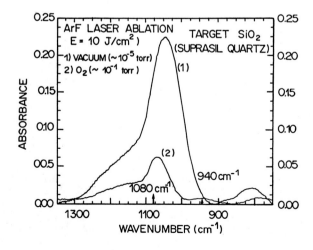

Figure 5 : IR absorption spectra of oxide films deposited onto silicon by laser ablation from a quartz target under vacuum and in O_2.

different ways : (i) via reactions with OH containing species that occur during the laser ablation deposition process, (ii) by ex-situ exposure to atmospheric H_2O that diffuses into porous films and reacts to form Si-OH groups. In our case, hydrogen, contained into the targets, especially silicon monoxide and quartz, and present as a residual contaminant in the deposition chamber, under the form of OH radicals or H_2O molecules, is believed to be the main responsible for the Si-OH group incorporation.

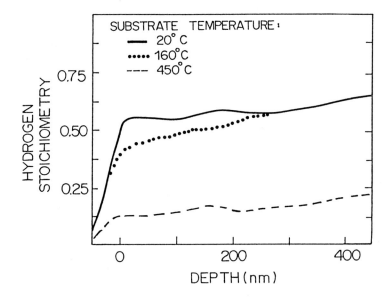

Figure 6 : Concentration and in-depth profile of hydrogen in oxide films deposited by laser ablation from SiO target at various substrate temperature.

The concentration and in-depth distribution of hydrogen in the oxide films were deduced from ERDA experiments. Fig. 6 shows the influence of the substrate temperature on the final amount of H incorporated into the SiO_2 layer deposited by laser ablation from a SiO target in O_2 atmosphere. By heating the substrate, we observe a strong reduction of hydrogen incorporated into the deposit (from 15 % at 20°C to less than 3 % at 450°C) in good agreement with the infrared measurements.

The oxide film composition has been precisely deduced from RBS experiments. Fig. 7 is a typical example, showing the in-depth profile of O and Si into thermally and laser grown (from SiO target) oxide films. The 2500 Å thick oxide layer, grown under O_2 atmosphere, presents a nearly uniform in-depth distribution with a well defined atomic composition (SiO_x with $x \approx 2$), by comparison to the oxide deposited under vacuum which is non-stoichiometric ($x \approx 1.5$).

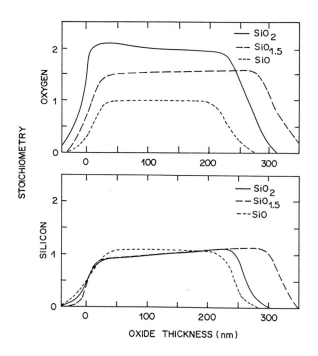

Figure 7 : In-depth profiles of O and Si into oxide films deposited onto silicon by thermal evaporation and laser ablation from SiO target under vacuum and in O_2.

The basic mechanisms leading to the growing of stoichiometric SiO_2 films by laser ablation, in oxygen atmosphere, must be strongly correlated to the nature of emitted species in the laser generated plasma, where a high degree of ionization is expected at the highest fluences. Spectroscopic investigations demonstrated recently the generation in the laser produced plasma from silicon [14] and quartz [23] targets of a high quantity of excited neutral and ionized (singly, doubly) Si atoms able to strongly react with oxygen molecules in the gas phase and at surface of the growing oxide. By contrast, the direct photodissociation of O_2 molecules into monoatomic oxygen species at the 193 nm wavelength of the excimer laser does not seem to be able to play an imporant role in the process because at the pressures used in our experiments ($\leq 10^{-1}$ torr), the one photon absorption coefficient of O_2 at 6.4 eV is low ($K \approx 10^{-4}$ $cm^{-1}.atm^{-1}$).

The last point to mention regarding the laser ablation process is the generation with atomic species of macroscopic particles (0.1 to 1 μm in size) which seems difficult to avoid whatever the ablated material and irradiation conditions. Density and size of these particles increase with increasing laser fluences and also depend on the target properties and especially their density and compacity.

The density and capacity of the target seems to play an important role. For silicon and silicon monoxide target, these particles mainly consist of small size ($\approx$ 0.1 μm in diameter) spherical grains, as shown on the SEM micrograph of Fig. 8 (a). Their formation is believed to be related to the ejection of liquid material in the form of droplets from the molten zone of the irradiated target which solidify by arriving onto the substrate [24]. For the quartz suprasil target, the situation appears a little different since additional irregular shaped solid grains (≥ 1 μm in size) are also observed [Fig. 8 (b)]. The latter may have been ejected from the target as solid particles.

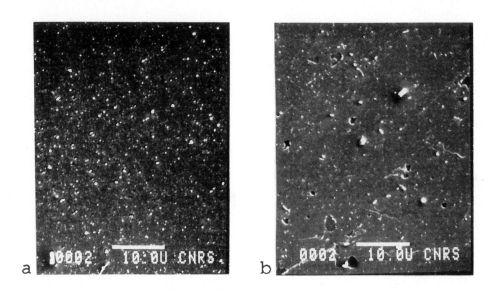

Figure 8 : Scanning Electron Micrograph of a SiO_2 film deposited by laser ablation (E ≈ 10 J/cm2) from silicon monoxide (a) and quartz (b) targets in O_2.

CONCLUSION

In this work, we have demonstrated the possibility to deposit stoichiometric SiO_2 thin films by excimer laser ablation from silicon, silicon monoxide and quartz targets, in oxygen atmosphere. The basic mechanism leading to the synthesis of SiO_2

must be correlated to the strong reaction of oxygen molecules in the gas phase and at surface of the growing film with the neutral and ionized emitted species from the laser generated plasma above the irradiated target area.

REFERENCES

1. See *Laser Ablation for Materials Synthesis*, edited by D.C. Paine and J.C. Bravman, Mat. Res. Soc. Symp. Proc. Vol. 191 (Materials Research Society, Pittsburgh, 1990).
2. H. Sankur and J.T. Cheung, Appl. Phys. A 47, 271 (1988).
3. D. Bäuerle, Appl. Phys. A 48, 527 (1989).
4. H. Oesterreicher, H. Bittner and B. Kothari, J. Solid State Chem. 26, 97 (1978).
5. M. Hanabusa and M. Suzuki, Appl. Phys. Lett. 39, 431 (1981).
6. E. Fogarassy, C. Fuchs, A. Slaoui and J.P. Stoquert, Appl. Phys. Lett. 57 (7), 664 (1990) and Appl. Surf. Sc. 46, 195 (1990).
7. W.A. Pliskin, J. Vac. Sci. Technol. 8, 1064 (1977).
8. W.A. Pliskin and H.S. Lehman, J. Electrochem. Soc. 112, 1013 (1965).
9. J.A. Theil, D.V. Tsu, M.W. Watkins, S.S. Kim and G. Lucovsky, J. Vac. Sci. Technol. $A8$, 1374 (1990).
10. J.P. Stoquert, unpublished.
11. L.R. Doolittle, Nucl. Instr. Meth. $B9$, 344 (1985).
12. J.F. Ziegler, Handbook of Helium Stopping Powers and Ranges in All Elements, Vol. 4 (Pergamon Press, New York, 1977).
13. J.P. Stoquert, M. Hage-Ali, J.L. Regolini, C. Ganter and P. Siffert in *Advanced Materials for Telecommunication*, edited by P.A. Glasow, J.-P. Noblanc and J. Speight, Europ. Mat. Res. Soc. Symp. Proc. Vol. 13 (Editions de Physique, Les Ulis, 1986) p. 159.
14. G.B. Shinn, F. Steigerwald, H. Stiegler, R. Sauerbrey, K.F. Tittel and W.L. Wilson Jr., J. Vac. Sci. Technol. $B4$ (6), 1273 (1986).
15. B. Braren and R. Srinivasan, J. Vac. Sci. Technol. $B6$ (2), 537 (1988).
16. H.R. Philipp, J. Phys. Chem. Sol., $V32$, 1935 (1971).
17. R.A.D. Devine and C. Fiori in *Defect Properties and Processing of High Technology Non Metallic Materials*, edited by Y. Chen, W.D. Kingery, R.J. Stokes, Mat. Res. Soc. Symp. Proc.

Vol. 60 (Materials Research Society, Pittsburgh, 1986) p. 303.
18. K. Arai, H. Imai, H. Hosono, Y. Abe, H. Imagawa, Appl. Phys. Lett. 53 (20, 1891 (1988).
19. M. Rothschild, D.J. Ehrlich and D.C. Shaver, Appl. Phys. Lett. 55 (13), 1276 (1989).
20. R.K. Brimacombe, R.S. Taylord and K.E. Leopold, J. Appl. Phys. 66 (9), 40035 (1989).
21. A. Slaoui, E. Fogarassy, C.W. White and P. Siffert, Appl. Phys. Lett. 53, 1832 (1988).
22. A. Slaoui, E. Fogarassy, C. Fuchs and P. Siffert, Submitted to J. Appl. Phys. (1991).
23. H. Sankur, J.G. Nelson, A.T. Pritt Jr. and W.J. Gunning, J. Vac. Sci. Technol. A5 (1), 15 (1987).
24. C. Fuchs and E. Fogarassy in *High Tempeature Superconductors : Fundamental Properties and Novel Materials Processing*, edited by D. Christen, J. Narayan, L. Schneemeyer, Mat. Res. Soc. Symp. Proc. Vol. 169 (Materials Research Society, Pittsburgh, 1990) p. 517.

CHARACTERISTICS OF LASER–MATERIAL INTERACTIONS MONITORED BY INDUCTIVELY COUPLED PLASMA–ATOMIC EMISSION SPECTROSCOPY[†]

Wing Tat Chan and Richard E. Russo
Applied Science Division
Lawrence Berkeley Laboratory
Berkeley, California 94720

Inductively coupled plasma–atomic emission spectroscopy was used for the study of the explosive removal of solid material by high power pulsed laser radiation. The effect of sample transport characteristics, laser power density, and the pressure pulse generated by the expansion of laser plasma are discussed.

Introduction

Understanding laser–material interactions (LMI) is an important prerequisite to the application of laser sampling for chemical analysis. In this work, inductively coupled plasma–atomic emission spectroscopy (ICP–AES) was employed for studying the behavior of the LMI. The ICP is a high temperature source widely used in analytical atomic spectroscopy because of its ability to dissociate refractory chemical species, and excite the resultant atomized vapor to optical emission [1]. Depending on the spectroscopic detection arrangement, simultaneous multi–element quantitative analysis can be performed.

In this study, the ICP operating parameters are held constant and the effect of laser operating parameters on the LMI is monitored by observing the ICP emission signal temporally and spatially. The effect of sample transport characteristics due to different particle size, laser power density on the amount of sampled material, the extent of differential vaporization, and the pressure pulse generated by the expansion of laser induced plasma will be described.

Experimental

A schematic diagram of the experimental system is shown in Figure 1. A KrF excimer laser (Questek model 2680, λ = 248 nm) was used. The laser beam was focused onto the samples with a plano–convex singlet lens (focal length = 18.3 cm). The laser has a pulse width of 20 ns FWHM and pulse energy of 200 to 550 mJ. Typical power density at the target surface

---

[†] This research was supported by the U.S. Department of Energy, Office of Basic Energy Sciences, Division of Chemical Sciences, under contract #DE-AC03-76SF00098.

was in the range of 10^7 to 10^9 W/cm^2. A Q-switched Nd:YAG laser has also been used in preliminary studies.

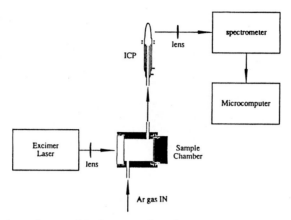

Figure 1. Experimental setup of the laser sampling ICP–AES system. A PMT-based monochromator and a PDA spectrometer were used separately as the spectrometer.

The ICP was a Plasma Therm 2500D, operating at a forward power of 1.25 kW. A teardrop shaped plasma is maintained at the load coil by a tangential flow of argon gas. A carrier gas, flowing through the central tube of the ICP torch, punches through and maintains a central channel in the plasma. Samples are introduced by the carrier gas into the central channel of the plasma through the central tube, where it is vaporized/atomized and excited in the plasma. The emission signal is then measured from the side of the plasma. Two types of spectrometers have been used in this study. A 0.3 m Jarrell-Ash monochromator coupled with a 1P28A photomultiplier tube (PMT) was employed to study fast temporal behavior of the emission signal. A photodiode array (PDA) spectrometer (Instruments SA HR-320 coupled with an EG&G model 1412 silicon PDA detector) was used to monitor multiple wavelengths simultaneously or to study the vertical profile of an emission line throughout the ICP discharge.

A hollow chamber with a water cooled mount was constructed to accommodate the samples. Disc-shaped samples with diameter of 2 cm and thickness of 1 to 2 mm are machined from pure copper sheet (Aldrich Chemicals) and commercial brass and bronze rods. Copper and brass samples were chosen for their simple matrices. The sample was held in a fixed position during repetitive pulsed laser sampling so that the temporal evolution of the process as well as a *steady state* signal could be monitored.

Result and Discussion

Typical temporal behavior of ICP emission signal for repetitive pulsed laser sampling is shown in Figure 2. The result was obtained by focusing a Q-switched Nd:YAG laser ($\lambda = 1.06$ µm, average power = 0.1 W at 10 Hz) onto NIST #661 steel. Each temporal profile in the figure represents repetitive laser sampling at a different spot on the sample. A large peak always precedes the constant or *steady state* signal response. The *steady state* signal provides better reproducibility (RSD= 5.4 %) compared to the maximum peak response (RSD= 23.6 %) for the five separate locations on the surface. In this work, we report intensity obtained when the temporal behavior approximates *steady state* during repetitive laser sampling at each position.

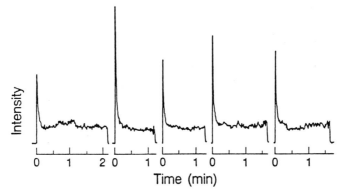

Figure 2. Temporal profiles of Ni emission at 352.45 nm for the laser sampling of NIST #661 steel. A Q-switched Nd:YAG laser was used at pulse rate of 10 Hz.

Laser sampling of metals generates a large number of particles ranging in size from 10 μm to less than 1 μm [2, 3]. Particles of size 2 μm or smaller account for about 90 % of the particle population but only a few percent of the total mass, for a wide selection of metals with diverse physical characteristics [3]. As the sampled material is entrained into the carrier gas, most of the particles larger than 2 μm will be left behind because of low transport efficiency [2, 3]. This effect is demonstrated in Figure 3. The constant signal above the baseline mainly originates from the ensemble of smaller particles. As larger particles are entrained, random spikes in the emission signal are observed because the particles are small in number but large in mass. The observed width of the spikes, however, is due to undersampling of the emission signal; a low sampling rate (50 Hz) was used and the particle residence time in the ICP is in the order of milliseconds. Particle size distribution is an important parameter in understanding the processes underlying the LMI. Under correct sampling conditions (fast ADC rates) it may be possible to correlate emission intensity of the spikes to the particle diameters, and work is currently underway to use ICP-AES for this purpose.

At constant laser energy, the quantity of material removed varies with the laser power density (Figure 4). The power density was increased by increasing the lens to sample distance, while keeping this distance shorter than the focal length of the lens in order to avoid gas breakdown at the focal point. The rate of copper and zinc removal from a brass sample increases with power density until ~1 GW/cm^2 and then levels off at higher power density. Laser sampling of copper has similar behavior. Power density of ~1 GW/cm^2 appears to be a transition point for two distinct removal mechanisms, consistent with the thermal and non-thermal mechanisms as described by Ready [4, 5]. The plateau region is not due to sample loading or optical self–absorption in the ICP. The plateau region is likely due to plasma screening of the laser energy from the surface.

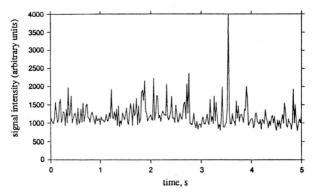

Figure 3. Temporal profile of Cu I emission at 324.8 nm for the laser sampling of copper. The laser pulse rate was 5 Hz. The monochromator–PMT was used (ADC rate= 50 Hz). The laser power density at the sample surface was approximately 10^8 W/cm^2.

Brass is an excellent material to demonstrate a change in mechanisms at this transition point. The typical composition of yellow brass is 60 % copper (bp 2567 K) and 40 % zinc (bp 907 K [6]). Because of the large difference in their boiling points, preferential vaporization of zinc is expected and has been demonstrated [7]. In this study, the Zn/Cu emission ratio reduces as power density increases, until it approaches the characteristic of the actual brass composition at power density ~ 1 GW/cm^2. Once again, power density of ~1 GW/cm^2 appears to be the transition point for the thermal versus non–thermal removal mechanisms, which correlates with the power density at which the sample removal rate levels off (Figure 3).

Another characteristic of brass is that the zinc is inhomogeneously distributed as inclusions in the copper matrix. During the sampling process, the melted zinc inclusions may be ejected from the surface explosively, resulting in a momentary increase of the zinc content in the vapor phase. In Figure 5, the spikes on the zinc signal indicate the ejection of the zinc inclusions, while there are no corresponding spikes on the copper signal measured simultaneously using the PDA spectrometer. These observations demonstrate that there are several processes operating concurrently during the high peak power LMI.

During a laser pulse, the laser induced plasma expands rapidly and exerts pressure on its environment. The effect of this pressure pulse on the flow of the sampled material can be observed using ICP–AES. The pressure pulse accelerates the carrier gas momentarily, resulting in the sampled material passing through the ICP discharge at a higher velocity for a brief period of time. The observed emission signal will be reduced and the emission intensity maximum of the ICP discharge will be shifted to a higher position. This effect is shown in Figure 6 for the laser sampling of copper. The PDA was oriented vertically to monitor the ICP emission of the Cu I line. The emission signal was in *steady state* except during the laser induced pressure pulse, at which time the intensity dropped. The vertical emission profiles of the ICP between two laser pulses, extracted from the same set of data, are shown in Figure 7. The profiles changed periodically after each laser pulse, with intensity reduced by ~50 % and the position of emission maxima shifted up by ~2 mm during the laser induced pressure pulse. The extent of signal intensity reduction is proportional to laser pulse energy.

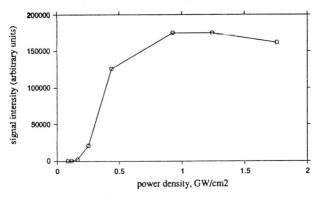

Figure 4. *Steady state* signal intensity of Cu I emission at 327.4 nm vs laser power density for brass sample. Laser repetition rate was 5 Hz. The diameter of the unfocused laser beam to 0.6 cm and the laser energy was fixed at 19 mJ. The PDA spectrometer was used, integration time = 0.3 to 1 s. The intensity was normalized to 1 s integration time in the figure.

Conclusion

Laser–material interaction involves complex processes, in which the amount of material generated and the particle size distribution change with laser fluence and power density. Changes in the particle size distribution are expected to be significant when low thermal conductivity, brittle materials are sampled, and minimized for the case of high thermal conductivity, ductile materials as employed here. In spite of the convolution of the laser sampling process with sample transport/excitation processes, several characteristics of laser sampling were described and interpreted using ICP-AES. Spikes were observed on the emission signal due to the entrainment of large particles and it may be possible to correlate the signal spikes with particle size. At constant laser energy, the amount of sampled material increased with the laser power density until ~1 GW/cm^2, at which point it leveled off. The roll–off is probably due to a change of the laser–material interaction mechanism and plasma shielding. Differential vaporization of zinc and copper from brass sample was observed at lower power density, while stoichiometric sample was obtained at power densities larger than ~ 1 GW/cm^2. The correlation of the transition point for both sampled volume and differential vaporization indicates that a change in the mechanism of the LMI may occur. We have also observed the effect of a laser induced pressure pulse on the emission signal. The strength of the pressure pulse appears to correlate to the power density of the laser beam. From these observations, it is apparent that ICP–AES is a versatile tool for studying laser sampling itself.

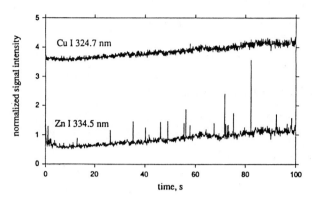

Figure 5. Normalized signal of Cu I and Zn I. (The Cu I signal was offset by 3 for clarity.) Laser energy was 350 mJ and repetition rate was 5 Hz. Laser power density was 3.7 GW/cm^2. The PDA spectrometer was used, with integration time of 0.09 s.

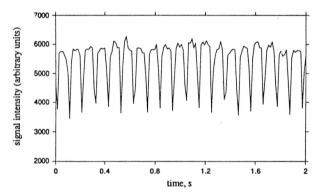

Figure 6. Temporal profile of Cu I emission signal at 327.4 nm for the laser sampling of copper. Observation height was 8 mm above load coil. Laser energy was 350 mJ and repetition rate was 10 Hz. Laser power density was 1.4 GW/cm^2. The PDA spectrometer was used with integration time of 0.014 s. The origin of the time axis is arbitrary.

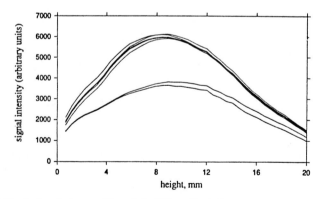

Figure 7. Vertical emission profiles of the ICP at Cu I 327.4 nm for the laser sampling of copper. Laser energy was 350 mJ and repetition rate was 10 Hz. Laser power density was 1.4 GW/cm^2. The PDA spectrometer was used with integration time of 0.014 s.

References

1. P.W.J.M. Boumans, ed., *Inductively Coupled Plasma Emission Spectroscopy*, Part 1 and 2, John Wiley & Sons, New York (1987)
2. P. Arrowsmith and S.K. Hughes, *Appl. Spectrosc.* **42**, 1231 (1988)
3. M. Thompson, S. Chenery, and L. Brett, *J. Anal. At. Spectrom.* **5**, 49 (1990)
4. J.F. Ready, *J. Appl. Phys.* **36**, 462 (1965)
5. J.F. Ready, *Effect of High-Power Laser Radiation*, Academic Press, New York (1971)
6. R.C. Weast, M.J. Astle, and W.H. Beyer, Editors, *CRC Handbook of Chemistry and Physics*, 64th ed., CRC Press, Florida (1983)
7. J.M. Baldwin, *Appl. Spectrosc.* **24**, 429 (1970)

Part II

Desorption

Trace Surface Analysis Using Ion and Photon Desorption with Resonance Ionization Detection*

M. J. Pellin, C.E. Young, W.F. Calaway, K. R. Lykke, P. Wurz, and D.M. Gruen
Argonne National Laboratory, Argonne, IL USA

D. R. Spiegel, A. M. Davis, and R. N. Clayton
Enrico Fermi Institute, University of Chicago, Chicago, IL 60637, USA

ABSTRACT

Surface Analysis by Resonant Ionization of Sputtered Atoms (SARISA) has demonstrated the ability to detect trace elements at concentrations below 100 ppt. Use of energetic primary ions to desorb surface species is uniquely suited for elemental surface analysis because the ratio between the number of incident energetic ions and the number of ejected surface atoms is easily quantifiable (Sputter Yield). Molecular surface analysis by ion desorption does not possess this advantage, however. In this case, laser desorption followed by resonant or near resonant ionization is often a better analysis tool. Here, the power of resonant ionization detection of desorbed species is demonstrated on hibonite samples (for elemental analysis) and on fullerene samples (for molecular analysis).

INTRODUCTION

Ultrasensitive, *surface* analysis is difficult both because of the few analyte atoms or molecules available and because of interference from the vast excess of bulk material. Surface Analysis by Resonant Ionization of Sputtered Atoms (SARISA) has demonstrated the ability to detect trace elements at concentrations below 100 ppt.[1-5] While the detectable concentration levels are impressive, recent results show that the strength of the SARISA technique may be found in analysis of naturally occurring samples with their wide complement of minor and trace elements. Analysis of such samples is often complicated by several factors including sample charging due to poor sample conductivity, isobaric interferences, and lateral inhomogeneities. Here we will focus on the measurement of Ti in a naturally occurring oxide material (e.g., hibonite). These results demonstrate the ability of SARISA to measure insulating samples without reduction in the Ti useful yield when compared to that of Ti metal.

The development of a straightforward method for the synthesis of macroscopic quantities[6,7] of the truncated icosahedron called Buckminsterfullerene[8], C_{60}, along with other stable closed shell clusters (e.g., C_{70} and C_{84}) has triggered substantial interest in carbon cluster chemistry. Analysis of solid samples of fullerenes is, therefore, interesting. Unfortunately, the relatively large mass of these species makes study by ion induced desorption or sputtering difficult. This report details time-of-flight mass spectrometry using laser desorption followed by laser ionization with a second laser. Of particular interest, is a velocity distribution of C_{60} obtained by suitably delaying the desorbing laser with respect to the ionizing laser.

---

*Work supported by the U.S. Department of Energy, BES-Materials Sciences, under Contract W-31-109-ENG-38.

EXPERIMENTAL

Experiments involving Ti atoms were carried out in the SARISA time-of-flight instrument which has been described in detail elsewhere[5]. Three tunable wavelengths were generated using XeCl excimer-pumped dye lasers. For Ti, several schemes have been investigated. The most suitable used a triply resonant scheme with ω_1 =19323 cm^{-1} ($a^3F_2 \rightarrow z^3F_2^o$) and ω_2 =18216 cm^{-1} ($z^3F_2^o \rightarrow e^3F_2$). The third laser was then used to pump the excited populations into autoionizing resonances above the ionization potential of the Ti atom (55010 cm^{-1}). These resonances have not been assigned and were found by scanning the third laser color and monitoring the photoion-induced intensity in the presence of the first two lasers. Many resonances were observed; the most intense at ω_3 =17668 cm^{-1} was utilized in these studies. Two resonant schemes investigated are shown in Figure 1.

Hibonite is a naturally occurring mineral with the composition of $CaAl_{12-2x}Ti_xMg_xO_{19}$, where x is the concentration of Ti which varies depending on the origins of the mineral. The Ti content of the hibonite sample used in these experiments was 1.2 at.% as verified by electron-induced x-ray fluorescence spectroscopy.[9]

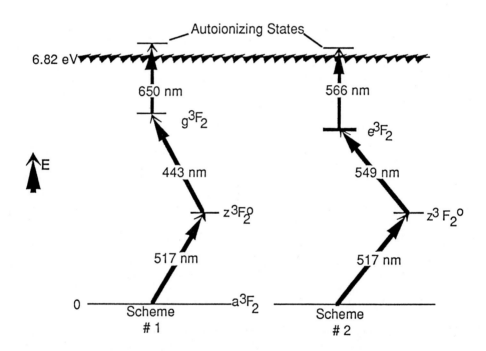

Fig. 1. Two different 3-color resonance ionization schemes which have been used to photoionize Ti.

A Marz-grade Ti metal sample was also mounted in the chamber. The primary ion beam used for these measurements had a diameter of about 200 μm. Samples were prepared by pressing chips of hibonite mineral into a gold foil that was subsequently pressed onto an Al electron microscope stub. The hibonite samples had dimensions between 90 μm and 300 μm.

Fullerene samples were prepared, extracted, and separated using a procedure that has been described elsewhere.[10] Hexane solutions of pure C_{60} were allowed to evaporate on a stainless steel sample probe. The films produced in this manner were optically transparent and nonuniform.

For the fullerene measurements, a XeCl excimer laser was used for desorption. A frequency-doubled Coumarin 540A dye laser pumped by a second excimer laser was used for ionization. Both lasers are operated at 1 to 50 Hz repetition rate with laser pulse widths of aproximately 20 ns. The wavelengths are 308 nm and 270 nm for desorption and ionization, respectively. The fluence of the desorption laser is held constant at approximately 10-100 mJ/cm^2 for all experiments. The time delay between excimer laser firing times was controlled using a Stanford Instruments DG535 delay generator.

RESULTS

Figure 2 shows the RIMS spectrum of a small (80 μm x 100 μm) hibonite sample obtained using SARISA IV. The spectrum was obtained from an average of 2000 laser pulses and was obtained with very little sample consumption. Assuming that the sputtering yield (atoms emitted/incident ion) of hibonite is near unity, about 10^7 atoms are sputtered per pulse, i.e.,

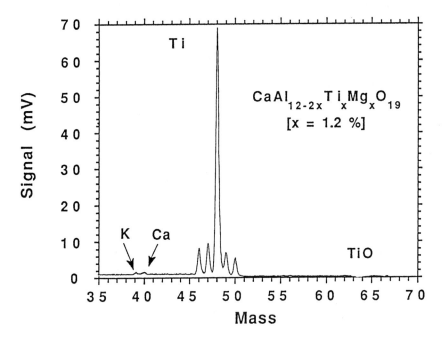

Fig. 2. RIMS spectrum of Ti from a 80 μm x 100 μm hibonite sample.

approximately 2 x 10^{10} atoms were removed for data acquisition. By analyzing multiple spectra acquired sequentially, we have demonstrated that the Ti isotope ratios can be measured with sub-1% reproducibility on ~100 μm hibonite grains.

A comparison of the Ti signal from a larger (250 μm x 350 μm) sample of hibonite and from pure Ti was made by alternately centering the Ti metal and then the hibonite sample under the primary ion beam in the SARISA apparatus. Comparison in signal could then be made without change in the transmission of the instrument. Table 1 lists the relevant information. One key to this measurement is calibrating the gain of the channel plate detection system.[2,3] The sputtering yield of hibonite was taken to be equivalent to that of rutile.[11] Presently, we are measuring precisely the sputtering yield of hibonite. The measurement presented here may be in error by as much as a factor of 2 because of the uncertainty in the sputtering yield.

Table 1. The relative useful yield of Ti from hibonite and from Ti metal.

Material	Signal Level	Detector Gain	Sputtering Yield	[Ti]	Relative Useful Yield
Ti	54.3	1	1.0	1.0	Y_{Hib}/Y_{Ti} =1.0
Hibonite	45.1	75	0.9	0.012	

In another set of measurements, we determined the velocity distributions for C_{60} clusters laser desorbed from a stainless steel substrate by allowing the desorbed neutral molecules to travel a short distance before being ionized and accelerated into the mass spectrometer. Several velocity distribution measurements were taken to insure reproducibility. These velocity distributions are fitted to a Maxwell-Boltzmann distribution giving reasonably good agreement with the measured data. Figure 3 shows a velocity distribution and a best fit with a Maxwell-Boltzmann distribution for each sample. Theoretical calculations are done to gauge the effect of angular and lateral distributions of the desorbed species on the measured velocity distribution for our particular setup. We found that these have only a small effect compared to other uncertainties encountered in the measurement, although there is a systematic underestimation of the temperature of about 50 K. Taking this into account, we obtain average values for the temperatures of 2300 ± 200 K for the C_{60} sample.

There are two distinct models that have been proposed to describe laser desorption by UV photons. These are (i) desorption due to excitation of an antibonding state of the parent or a fragment (cracking) and (ii) thermal desorption due to localized heating by the laser beam.[12-14] We conclude that in our experiment laser desorption is a thermal process for three reasons. First, the measured velocity distribution is well fit by a Maxwell-Boltzmann distribution; second, the temperature derived from this fit is comparable to desorption temperatures previously reported for polymers,[13] and finally, we do not observe fragmentation during the desorption process, which might be expected if a bond-breaking mechanism were active. A more complete description of this work may be found elsewhere.[13-15]

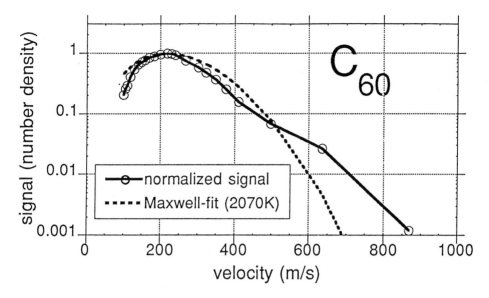

Fig. 3. Velocity distribution for laser desorbed neutral C_{60}. Dashed curve gives the best fit for a Maxwell-Boltzmann temperature distribution.

REFERENCES

(1) Pappas, D. L.; Hrubowchak, D. M.; Ervin, M. H.; Winograd, N. *Science* **1989**, *243*, 64.
(2) Blum, J. D.; Pellin, M. J.; Calaway, W. F.; Young, C.E..; Gruen, D. M.; Hutcheon, I. D.; Wasserburg, G. J. *Geochimica* **1990**, *54*, 875 - 881.
(3) Blum, J. D.; Pellin, M. J.; Calaway, W. F.; Young, C. E.; Gruen, D. M.; Hutcheon, I. D.; Wasserburg, G. J. In *Secondary Ion Mass Spectrometry SIMS VII*; A. Benninghoven, H. W. Huber, and H. W. Werner, Eds.; Wiley: New York, 1990; pp 312-322.
(4) Blum, J. D.; Pellin, M. J.; Calaway, W. F.; Young, C.E.; Gruen, D. M.; Hutcheon, I. D.; Wasserburg, G. J. *Analytical Chemistry* **1990**, *62*, 209-214.
(5) Pellin, M. J.; Young, C. E.; Calaway, W. F.; Whitten, J. E.; Gruen, D. M.; Blum, J. D.; Hutcheon, I. D.; Wasserburg, G. J. *Phil. Trans. R. Soc. Lond. A.* **1990**, *333*, 133-146.
(6) Krätschmer, W.; Fostiropoulos, K.; Huffman, D. R. *Chem. Phys. Lett.* **1990**, *170*, 167-170.
(7) Krätschmer, W.; Lamb, L. D.; Fostiropoulos, K.; Huffman, D. R. *Nature* **1990**, *347*, 354-358.
(8) Kroto, H. W.; Heath, J. R.; O'Brian, S. C.; Curl, R. F.; Smalley, R. E. *Nature* **1985**, *318*, 162-163.
(9) Davis, A. *private communication*
(10) Parker, D. H.; Wurz, P.; Chatterjee, K.; Lykke, K. R.; Hunt, J. E.; Pellin, M. J.; Hemminger, J. C.; Gruen, D. M.; Stock, L. M. *submitted to JACS*.
(11) Bach, H. *Nucl. Instruments and.Methods* **1970**, *84*, 4-12.
(12) Lazare, S.; Granier, V. *Laser Chem.* **1989**, *89*, 25-40.
(13) Danielzik, B.; Fabricus, N.; Rowekamp, M.; von der Linde, D. *Appl. Phys. Lett.* **1986**, *48*, 212-214.
(14) Srinivasan, R.; Braren, B. *Chem. Rev.* **1989**, *89*, 1303-1316.
(15) Wurz, P.; Lykke, K. R.; Pellin, M. J.; Gruen, D. M. *submitted to Appl. Phys.*

PULSE RATE DEPENDENCE OF LASER DESORPTION AND IONIZATION
OF MOLECULES ON THIN METAL FILMS:
MATHEMATICS OF LASER HEATING AND PULSE RATE DEPENDENCE

J. R. Millard* and J. P. Reilly
Chemistry Department
Indiana University
Bloomington IN 47405

I. Introduction

The use of pulsed laser beams to stimulate desorption of molecules has recently been developed as a powerful method of probing kinetic phenomena occurring at surfaces. Recent reviews on the use of laser desorption to probe diffusion and reaction kinetics provide an excellent introduction to this field[1-2].

In a typical laser desorption experiment to probe surface kinetics, a pulse of light rapidly heats the surface and reduces the concentration of molecules in the focal area. After the pulse, molecules adsorb from the gas phase (or in a diffusion experiment, diffuse in from the perimeter) and replenish the surface. A second light pulse then irradiates the surface and probes the new surface concentration. The surface coverage interrogated by the second pulse depends on the time delay between the two; by varying this delay, adsorption kinetics can be investigated. In addition to the study of adsorption phenomena[3-7], considerable laser desorption work has focussed on diffusion kinetics[8-17] and rates of chemical reactions[1,18-28] on surfaces.

We have utilized repetitively pulsed UV lasers to desorb and ionize molecules adsorbed to thin metal films supported on a fused silica prism[29-31]. In one study[31], we found that the ionization signal was independent of the laser pulse rate for both conventional gas phase ionization and for external reflection ionization. However, internal reflection ionization signals exhibited a nearly linear dependence on the time between laser pulses.[31-32] This demonstrated that internal reflection ionization probes adsorbed molecules, while external reflection ionization probes only gas phase molecules. A companion publication[32] describes other experiments involving laser pulse repetition rate to investigate surface concentrations. A method is also demonstrated therein for determining the residual, undesorbed population of molecules that remain on a surface immediately after a laser pulse. This technique involves irradiating the surface with one pulse to desorb some of the molecules, and with a second pulse nanoseconds to microseconds later to probe the fraction of molecules remaining. In the present work we develop the mathematics for interpreting the kinetics of adsorption and desorption.

In the section that follows, expressions that relate surface coverage to the time between laser pulses are derived. Because a single laser pulse does not necessarily desorb all of the molecules in the irradiated region, two types of experimental conditions are considered: (a) the laser completely desorbs the molecules, and (b) the laser pulse desorbs only a fraction of the adsorbed molecules. Previous analyses[3,10,15] of laser desorption experiments have generally neglected incomplete desorption. This was valid because the molecules were weakly bound, the laser pulse length was comparatively long (microseconds), or the surface was probed only after several laser pulses had completely cleaned it. However, the desorption of chemisorbed molecules by a single nanosecond pulse is more likely to be incomplete as we have recently demonstrated experimentally.[32]

II. Laser Repetition Rate Dependence Equations

A. Basic Kinetics of Adsorption and Desorption

The quantity of interest for these laser desorption studies is the number of molecules desorbed by the laser pulse. For first order desorption kinetics, this is proportional to the number of molecules on the surface at the time of the pulse. The fractional surface coverage (θ) as a function of the time between laser pulses (τ) depends upon: (1) the number of adsorbed molecules that are desorbed by each laser pulse, (2) the rate that molecules in the gas phase strike the surface and adsorb, and (3) the rate that adsorbed molecules spontaneously desorb. Note that our simulations do not include repopulation by diffusion of molecules in from the perimeter of the irradiated zone. Neglect of diffusion is valid because, at the pressures used in our experiments, repopulation of the surface by adsorption should be considerably faster than by diffusion. In laser desorption studies of diffusion kinetics, refilling of the depleted zone by diffusion typically requires minutes[8-17]. In comparison, a depleted surface in contact with a gas at a pressure of 10^{-5} torr will be repopulated in seconds. Thus a significant contribution to refilling by diffusion is not expected. We have demonstrated experimentally that, for the conditions of these simulations, surface diffusion of the molecules can indeed be neglected[33].

Consider the adsorption of a gas at a pressure P in contact with the surface. The absorption rate depends on the frequency that molecules strike the surface and the probability that they adsorb. The number of molecules that strike a unit area of the surface per unit time (the impingement rate Γ, in units of molecules/(cm$^2\cdot$sec)) is given by

$$\Gamma = (P/(2\pi M k_B T_g)^{1/2}) \tag{1}$$

where M is the molecular mass, k_B is Boltzmann's constant, and T_g is the temperature of the gas. It is useful to define the average frequency that molecules strike any one particular adsorption site, Γ'. This is accomplished by multiplying Γ, the number of molecules that strike a unit area per unit time, by the area (a) of an individual adsorption site, which is effectively the area occupied by a molecule adsorbed on the surface. Note that due to packing efficiencies, and repulsive or attractive interactions between adsorbed molecules, this area is not necessarily the same as the cross-sectional area of a gas phase molecule. Γ' is thus

$$\Gamma' = (Pa/(2\pi Mk_B T_g)^{1/2}) \qquad (2)$$

and has units of sec^{-1}. $1/\Gamma'$ is the time required for enough molecules to strike the surface to form a monolayer, if every one adsorbed.

The probability that a molecule striking the surface adsorbs (the sticking probability S) may be written as the product of the sticking probability of a molecule striking a bare surface, S_0, and a coverage dependence term $f(\theta)$, so that $S = S_0 \cdot f(\theta)$. The fractional number of unoccupied sites on the surface is $(1-\theta)$. If adsorption depends linearly on the number of vacant surface sites, then $f(\theta) = (1-\theta)$. Thus molecules striking unoccupied surface sites adsorb with a probability S_0, whereas molecules impinging on occupied sites do not adsorb. The first order rate of adsorption in this case is

$$\Gamma' \cdot S_0 \cdot (1-\theta) = k_a \cdot (1-\theta) \qquad (3)$$

where $k_a = \Gamma' \cdot S_0$ is the rate constant for adsorption.

For the moment, laser desorption will be considered to be instantaneous, since it occurs on a time scale much shorter than the time between laser pulses. Only the rate of spontaneous thermal desorption occurring between laser pulses will initially be considered.

Assuming that desorption is a first order process, its rate depends linearly on the fraction of occupied sites, yielding the following rate equation

$$-d\theta/dt = k_d \cdot \theta = n \cdot \theta \cdot \exp(-E_d/k_B T_s) \qquad (4)$$

where k_d is the rate constant for desorption, E_d is the desorption energy, n is a frequency factor, and T_s is the surface temperature (300 K).

B. Dependence on Laser Repetition Rate and Fraction Desorbed

From the rates of adsorption and spontaneous thermal desorption, an expression for the coverage as a function of the time between laser pulses can be derived. Consider the situation when an integral number n of laser pulses irradiate the surface. θ_n is the surface coverage prior to the nth laser pulse. Time t is referenced to the previous (n-1)th pulse; that is, t = 0 is the instant after pulse n-1 hits the surface. For first order adsorption and desorption kinetics, the rate that the fractional coverage θ changes with time is

$$d\theta_n/dt = k_a \cdot (1-\theta_n) - k_d \cdot \theta_n \qquad (5)$$

Equation 5 describes adsorption and desorption that occurs following the n-1th laser pulse and preceding the nth pulse. The temperature is taken to be constant in this interval; that is, at t = 0, the laser pulse instantaneously heats the surface, some or all of the molecules desorb, and the surface instantly cools to its original temperature. We emphasize that the desorption term $k_d\theta_n$ describes the rate of spontaneous, not laser stimulated, thermal desorption. Equations for the much faster kinetics of laser induced desorption are developed elsewhere.[33] Surface cooling is in fact very rapid relative to the repetition rate of the laser; at times longer than about 100 microseconds after a laser pulse the surface temperature is roughly constant.

Integration of equation 5 from t = 0 to τ, the time between laser pulses, gives

$$\theta_n(\tau) = \theta_n^0 \exp[-(k_a + k_d)\cdot\tau] + \left[\frac{k_a}{k_a + k_d}\right]\left[1 - \exp[-(k_a + k_d)\cdot\tau]\right] \qquad (6)$$

θ_n^0 is the coverage on the surface immediately after the n-1th laser pulse (i.e., at t = 0).

As τ approaches infinity, the surface coverage approaches its limiting (or "saturation") value, $\theta(\infty)$. As can be seen from equation 6, $\theta(\infty) = k_a/(k_a+k_d)$. Therefore as τ becomes large, the surface coverage (and hence the ionization signal) asymptotically approaches a constant value.

If desorption by each laser pulse is complete ($\theta_n^0 = 0$), the expression for the fractional coverage versus time between laser pulses simplifies to:

$$\theta(\tau) = \theta(\infty)\left[1 - \exp[-(k_a + k_d)\cdot\tau]\right] \qquad (7)$$

The subscript (n) has been omitted because for complete desorption $\theta(\tau)$ is independent of the number of pulses that have irradiated the surface previously.

If the laser does not completely desorb the molecules, θ_n^0 is nonzero with a value that depends on the number of molecules on the surface preceding the (n-1)th pulse. If laser desorption is also first order in coverage, after each laser pulse some fraction χ of the original surface coverage remains. Thus the coverage immediately after the (n-1)th laser pulse θ_n^0 is simply the coverage immediately before the (n-1)th laser pulse $\theta_{n-1}(\tau)$ multiplied by the fraction that remains:

$$\theta_n^0 = \chi\cdot\theta_{n-1}(\tau) \qquad (8)$$

Inserting (8) into (6) we obtain the solution for incomplete desorption,

$$\theta_n(\tau) = \chi\cdot\theta_{n-1}(\tau)\cdot\exp(-(k_a+k_d)\cdot\tau) + \theta(\infty)[1 - \exp(-(k_a+k_d)\cdot\tau)] \qquad (9)$$

Equation 9 is a recursion relation that expresses the fractional coverage immediately before the nth laser pulse in terms of the coverage before the (n-1)th pulse.

Let θ_0 be defined as the coverage before any laser pulses strike the surface. If we know θ_0 (and k_a, k_d, τ, and χ) then θ_n can be calculated. For example:

$$\theta_1(\tau) = \theta(\infty)[1 - \exp(-(k_a+k_d)\cdot\tau)] + \chi\cdot\theta_0\cdot\exp(-(k_a+k_d)\cdot\tau) \qquad (10A)$$

$$\begin{aligned}\theta_2(\tau) &= \theta(\infty)[1 - \exp(-(k_a+k_d)\cdot\tau)] + \chi\cdot\theta_1(\tau)\cdot\exp(-(k_a+k_d)\cdot\tau) \\ &= \theta(\infty)[1 - \exp(-(k_a+k_d)\cdot\tau)](1 + \chi\exp(-(k_a+k_d)\cdot\tau)) + \theta_0\cdot\chi^2\cdot\exp(-2\cdot(k_a+k_d)\cdot\tau)\end{aligned} \qquad (10B)$$

$$\begin{aligned}\theta_3(\tau) &= \theta(\infty)[1 - \exp(-(k_a+k_d)\cdot\tau)] + \chi\cdot\theta_2(\tau)\cdot\exp(-(k_a+k_d)\cdot\tau) \\ &= \theta(\infty)[1 - \exp(-(k_a+k_d)\cdot\tau)][1 + \chi\cdot\exp(-(k_a+k_d)\cdot\tau) + \chi^2\cdot\exp(-(k_a+k_d)\cdot\tau)] \\ &\quad + \chi^3\cdot\theta_0\cdot\exp(-3\cdot(k_a+k_d)\cdot\tau)\end{aligned} \qquad (10C)$$

and so on.

By inspection, the above expressions can be generalized to:

$$\theta_n(\tau) = \theta_0\, \chi^n \exp\left[-n\cdot\tau\cdot[k_a+k_d]\right] + \theta(\infty)\left[1 - \exp\left[-\tau\cdot[k_a+k_d]\right]\right] \sum_{i=0}^{n-1} \chi^i \exp\left[-i\cdot\tau\cdot[k_a+k_d]\right] \qquad (11)$$

In order to use equation 11, a value θ_0 is needed. Two representative cases will be considered.

Case 1. When the gas is in prolonged contact with the surface before the laser pulses begin irradiating the surface, the surface is saturated with molecules so $\theta_0 = \theta(\infty)$. This is the typical experimental situation.

Case 2. No molecules are on the surface, so $\theta_0 = 0$. This situation can be arranged in the laboratory using one or several intense laser pulses to completely clean off the surface. An example of this type of experiment is described elsewhere[32].

In the two cases above, the surface concentration starts at some initial coverage (0 or $\theta(\infty)$). Molecules from the gas phase adsorb on the surface while a repetitively pulsed laser perturbs the system. Equation 11 describes the approach of the perturbed system to a new equilibrium surface coverage. After a sufficiently large number of pulses, the same equilibrium is reached independent of the initial coverage. At equilibrium, the amount desorbed by a laser pulse will equal the amount adsorbed during the time interval τ. In this "steady state" situation, $\theta_n(\tau) = \theta_{n-1}(\tau)$, and equation 9 reduces to

$$\theta(\tau) = \theta(\infty) \left[\frac{1 - \exp[-(k_a + k_d) \cdot \tau]}{1 - \chi \exp[-(k_a + k_d) \cdot \tau]} \right] \tag{12}$$

For complete desorption ($\chi = 0$), the expression above simplifies to equation 7, as expected. In addition to its mathematical simplicity, this final case is the most easily implemented experimentally, because there is usually a delay of at least several seconds between the time that the laser starts irradiating the surface and the time that data are recorded, enabling the steady state to become established.

Experimentally, the surface coverage $\theta(\tau)$ is not measured directly. Instead, the laser-induced ion current is measured, and the signals from many laser pulses are accumulated to improve reproducibility. Assuming that laser ionization is directly proportional to surface coverage, the overall ion signal q at a laser repetition period τ is

$$q(\tau) = \sigma \cdot \sum_{n=i}^{j} \theta_n(\tau) \tag{13}$$

where $\theta_n(\tau)$ is given by equations 7, 9, 11, or 12 and σ is a proportionality constant. Data acquisition begins with the ith laser pulse ($i \geq 0$) and continues to the jth laser pulse ($j > i$). For complete desorption, or for the steady state case of incomplete desorption, computation of $q(\tau)$ is particularly simple, because $\theta(\tau)$ does not depend on n. In these cases, $q(\tau)$ is simply the number of laser pulses multiplied by $\sigma\theta(\tau)$. This simple form of $q(\tau)$ is more convenient to use for curve fitting purposes than equation 13.

From equation 12 it is evident that kinetic parameters such as S_0 or k_d can be derived by a mathematical fit of the experimental data $q(\tau)$ versus τ, provided that χ is known. Alternatively, if S_0 and k_d are known, χ can be calculated by the same method. Because σ is in fact an arbitrary constant it does not affect the fit of the data (it only determines the vertical scale), and thus it is not necessary that its value be known. Therefore, measurement of the ion signal as a function of the time between laser pulses enables the kinetics of adsorption and desorption to be probed.

To check the validity of our desorption/ionization model, its predictions were compared with measured UV laser multiphoton ionization signals for aromatic molecules ionized from thin metal films. These films were coated onto the hypotenuse of a fused silica prism mounted inside of an ultrahigh vacuum time-of-flight mass spectrometer and irradiated in an "internal reflection" geometry.[29-33] Mass spectra were recorded at various laser repetition rates and the yield of parent ions measured as a function of the time between laser pulses. A typical result is displayed in Figure 1. The points represent experimental data and the solid line is a graph of Equation (12) assuming a sticking probability of 1 and $\chi = 0.98$

(i.e., 2% of the irradiated molecules are desorbed by each laser pulse). The excellent agreement between the predicted and experimentally observed laser repetition rate dependence suggests that our simple picture describing this phenomenon is fundamentally sound.

In summary, equations have been derived that describe the laser repetition rate dependence of the desorption/ionization yield for volatile molecules adsorbed to thin metal films. These equations are appropriate for either complete or incomplete desorption by the laser. They are found to faithfully replicate the results of simple internal reflection laser repetition rate dependence experiments.

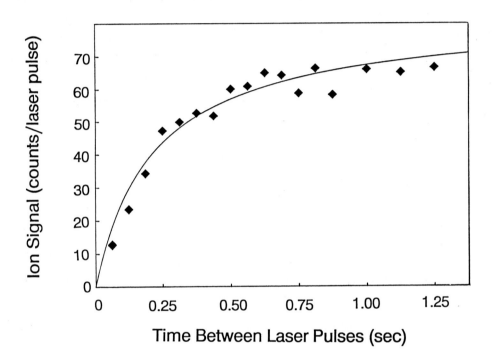

Figure 1. Laser repetition rate dependence of benzene desorption/ionization from a thin silver film. Experimental points are plotted as filled diamonds. The solid curve is obtained by a Simplex optimization least squares fit of the data assuming a sticking probability of $S_0=1$. With this sticking coefficient the optimized value of $\chi = 0.98$, which corresponds to 2% of the molecules desorbed by each laser pulse.

References

1. R.B. Hall and S.J. Bares, in **Chemistry and Structure at Interfaces: New Laser and Optical Techniques**, R.B. Hall and A.B. Ellis, eds. VCH Publishers: Deerfield Beach, FL., 1986. 85.

2. P.C. Stair and E. Weitz, J. Opt. Soc. Am. B $\underline{4}$ 255 (1987).

3. K. Christman, D. Schober, G. Ertl and M. Neumann, J. Chem. Phys. $\underline{60}$ 4528 (1974).

4. A. Pospieszyczck and J.A. Tagle, J. Nucl. Mater. $\underline{105}$ 14 (1982).

5. E.G. Seebauer, A.C.F. Kong and L.D. Schmidt, Surf. Sci. $\underline{176}$ 134 (1986).

6. P. Gupta, P.A. Coon, B.G. Koehler and S.M. George, J. Chem. Phys. $\underline{93}$ 2827 (1990).

7. J.L. Brand, M.V. Arena, A.A. Deckert and S.M. George, J. Chem. Phys. $\underline{92}$ 4483 (1990).

8. R. Viswanathan, D.R. Burgess, P.C. Stair and E. Weitz, J. Vac. Sci. Technol. $\underline{20}$ 605 (1982).

9. S.M. George, A.M. DeSantolo and R.B. Hall, Surf. Sci. $\underline{159}$ L425 (1985).

10. E.G. Seebauer and L.D. Schmidt, Chem. Phys. Lett. $\underline{123}$ 129 (1986).

11. C.H. Mak and S.M. George, Surf. Sci. $\underline{172}$ 509 (1986).

12. C.H. Mak, J.L. Brand, A.A. Deckert and S.M. George, J. Chem. Phys. $\underline{85}$ 1676 (1986).

13. R.B. Hall, T.H. Upton and E. Herbolzheimer, J. Vac. Sci. Technol. B $\underline{5}$ 1470(1987).

14. D.A. Mullins, B. Roop and J.M. White, Chem. Phys. Lett. $\underline{129}$ 511 (1986).

15. B. Roop, S.A. Costello, D.R. Mullins and J.M. White, J. Chem. Phys. $\underline{86}$ 3003 (1987).

16. E.D. Westre, M.V. Arena, A.D. Deckert, J.L. Brand and S.M. George, Surf. Sci. $\underline{233}$ 293 (1990).

17. M.V. Arena, A.A. Deckert, J.L. Brand and S.M. George, J. Phys. Chem. $\underline{94}$ 6792 (1990).

18. R.B. Hall and A.M. Desantolo, Surf. Sci. $\underline{137}$ 4211 (1984).

19. R.B. Hall, A.M. DeSantolo and S.J. Bares, Surf. Sci. $\underline{161}$ L533 (1985).

20. R.B. Hall, S.J. Bares, A.M. DeSantolo, F. Zaera, J. Vac. Sci. Technol. A $\underline{4}$ 1493 (1986).

21. R.B. Hall, J. Phys. Chem. $\underline{91}$ 1007 (1987).

22. A.A. Deckert and S.M. George, Surf. Sci. $\underline{182}$ L215 (1987).

23. A.A. Deckert, J.L. Brand, C.H. Mak, B.G. Koehler and S.M. George, J. Chem. Phys. $\underline{87}$ 1936 (1987).

24. B.G. Koehler, C.H. Mak and S.M. George, Surf. Sci. $\underline{221}$ 565 (1989).

25. C.L. Pettiette-Hall, D.P. Land, R.I. McIver and J.C. Hemminger, J. Phys. Chem. $\underline{94}$ 1948 (1990).

26. D.P. Land, D.T.-S Wang, T.-L. Tai, M.G. Sherman, J.C. Hemminger and R.T. McIver, in **Lasers and Mass Spectrometry**; Lubman, D.M., Ed. Oxford University Press: New York, 1990.

27. A.A. Deckert, M.V. Arena, J.L. Brand and S.M. George, Surf. Sci. $\underline{226}$ 42 (1990).

28. B.G. Koehler, C.H. Mak, D.A. Arthur, P.A. Coon and S.M. George, J. Chem. Phys. $\underline{89}$ 1709 (1988).

29. M. Yang, J.R. Millard and J.P. Reilly, Optics Commun. $\underline{55}$ 41 (1985).

30. J.R. Millard, M. Yang and J.P. Reilly, J. Phys. Chem. $\underline{91}$ 4323 (1987).

31. M. Yang and J.P. Reilly, Optics Commun. $\underline{71}$ 193 (1989).

32. J.R. Millard, M. Yang and J.P. Reilly, J. Phys. Chem. $\underline{95}$ xxx (1991).

33. J.R. Millard, Ph.D Thesis, Indiana University, 1989.

* Present address: JAYCOR, 11011 Torreyana Rd., San Diego CA 92121

Photodesorption of Metal Atoms by Collective Electron Excitation

W. Hoheisel, M. Vollmer and F. Träger
Fachbereich Physik, Universität Kassel
Heinrich-Plett-Str. 40, D-3500 Kassel, Germany

Desorption stimulated with laser light can take place by a variety of mechanisms. Most frequently it is simply promoted by thermal heating. Light is absorbed at the substrate surface, the temperature rises and the particles desorb. Bond breaking at the surface can also be stimulated by vibrational excitation of the adsorbate molecules with infrared laser light. Coupling of the vibrational excitation to the phonon bath of the substrate is the dominant channel for energy relaxation. As a result, the surface temperature rises causing desorption. This means, that the energy is absorbed by a resonant process initially, but the desorption is thermal, a phenomenon denoted as "resonant heating". Photodesorption can also be accomplished by electronic excitation with visible or UV laser light, a field that has recently attracted growing interest. The majority of such experiments have been performed with high intensity pulsed lasers to accomplish electronic excitation in the adsorbate, in the substrate or in the adsorbate-substrate complex. Major difficulties are to identify the electronic excitation and to understand the bond breaking mechanism in detail.

The main reason for a rapidly increasing number of experimental and theoretical studies in the field of laser desorption by electronic excitation is twofold: first, the underlying desorption mechanisms are of fundamental interest. Secondly, laser desorption or ablation processes offer a large variety of applications in such areas as microelectronics, thin film production, vaporization of large organic molecules or removal of tissue in biomedicine. Among the different experimental schemes and bond breaking mechanisms laser-stimulated desorption of metal atoms as a result of collective electron excitation [1-3] seems to be particularly interesting. Such processes have been observed by excitation of surface plasmons for example in small sodium particles [1]. Desorption of other metal atoms such as Al and Au has been reported also from thin films in which surface plasmons were excited with an attenuated total reflection scheme [4]. Also, laser detachment of Na and Cs atoms has been found [5].

The present article summarizes recent experiments on laser-induced desorption stimulated by electronic excitation in small metal particles [1-3]. These studies are different from earlier work for several reasons. First, the electronic excitation preceding the rupture of a surface chemical bond is a collective rather than a single electron excitation. Second, desorption can be stimulated with low intensity continuous-wave laser light in the visible spectral range rather than with high power pulsed lasers. Third, desorption of *metal* atoms is observed. The experiments illustrate that even on a metal surface the rupture of the bond by electronic excitation can compete with quenching and generation of heat in the substrate.

The principle of the experiment and the set-up have been described in detail elsewhere [1,2]. Briefly it consists of an ultrahigh vacuum system equipped with a quadrupole mass spectrometer for detection of the desorbed metal atoms. Small metal particles with sizes of the order of several hundred Ångstroms are generated on a LiF(100) single crystal surface by deposition and surface nucleation of metal atoms. For this purpose a thermal atomic beam of either Na, K or Ag atoms is directed onto the substrate. The average size of the generated clusters is characterized by measurement of their optical transmission spectra with a Xe arc lamp and a monochromator [2,3]. In addition, it can be determined from the known defect density of the surface and the number of deposited metal atoms which is measured with a quartz crystal microbalance. Desorption of atoms from the surface of the particles is accomplished by irradiation with an unfocussed Ar^+ of Kr^+ laser beam. Light powers of up to 10 W are used and a variety of laser wavelengths is applied. Instead of continuous wave irradiation, the particles can also be irradiated with light pulses of 3 µs duration that are made with a mechanical chopper. With these pulses time-of-flight measurements are carried out in order to determine the kinetic energy of the desorbed atoms.

The experimental results can be summarized as follows:
- Desorption of metal atoms can be readily detected when the light is incident on the sample. Even with a light intensity as low as 2 mW Na desorption can still be observed.
- Desorption starts immediately if the laser is turned on and stops promptly if the beam is blocked.
- The desorption rate depends linearly on the light intensity over a range of almost four orders of magnitude. No threshold for the desorption signal is found.

- The photodesorption yield depends resonantly on the laser frequency. This frequency dependence of the signal was examined by using Ar$^+$ ion and Kr$^+$ ion laser lines between $\lambda = 410$ and 752 nm.
- The signal depends on the particle size (see Fig. 1). It first increases with particle size, reaches a maximum and drops off for larger sizes.
- The time-of-flight measurements reveal a kinetic energy of the desorbed atoms of 0.4 eV for sodium and 0.13 eV for potassium. These translational energies do not depend on the laser intensity.
- A fraction of up to 80 % of the total coverage can be desorbed with the laser light at a given light intensity. The exact percentage depends on the particle size and on the excitation wavelength.
- A comparison of the optical extinction with the number of photodesorbed atoms gives a quantum efficieny on the order of 10^{-5}.

The high kinetic energy of the desorbed atoms, its independence of the laser power, the absence of a threshold, the linear dependence of the rate on the light power and only a very moderate temperature rise of the surface during laser irradiation indicate that the desorption process is *nonthermal*. The dependence of the desorption rate on the laser wavelength as well as on the particle size indicate that the process is stimulated by excitation of *surface plasmons*. This is further supported by comparison of the extinction spectra of the clusters [2,3] with spectra reported in the literature (see e.g. [6]).

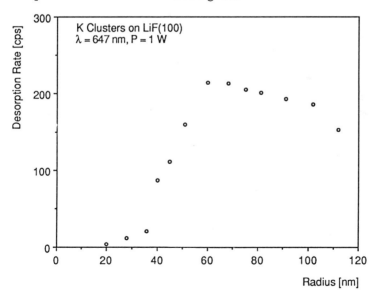

Fig. 1 Desorption rate as a function of the mean potassium particle size. Surface plasmon excitation was accomplished with laser light of $\lambda = 647$ nm.

The mechanism of the desorption process can be visualized with the schematic diagram depicted in Figure 2. The interaction of metal clusters with electromagnetic radiation is usually described by the classical Mie theory (see right hand side of Fig. 2). The clusters are treated as spheres with sharp edges and bulk optical constants are used. With the appropriate boundary conditions the near-field and scattered far-field can be calculated. The associated absorption causes generation of heat in the sphere. A more refined picture is based on a realistic surface potential which allows the electron density distribution to extend into the vacuum beyond the sharp edge of the ion cores. Furthermore, effects of non-local optics come into play (see e.g. [7]). Radiation is absorbed in a

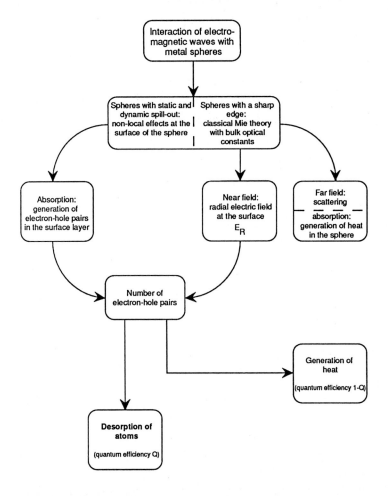

Fig. 2 Schematic diagram of the interaction of small particles with electromagnetic radiation, absorption in the bulk and in the surface layer and subsequent desorption of metal atoms.

thin surface layer thereby creating electron-hole pairs. The number of these excited electronic states is governed by the radial electric field at the particle surface which is strongly enhanced at the surface plasmon resonance [8]. The collective electron oscillation therefore couples to localized electronic states, i.e. electron-hole pairs, at the cluster surface. If these states are anti-bonding in nature, atoms with energies larger than thermal can be ejected. Again, quenching and generation of heat comes into play as a competing process. Since the atoms desorb preferably from sites with low coordination number the particles do not only shrink in size but may also undergo changes of their shape. Of course, details of the mechanism have to be investigated in future experimental and theoretical work. Particularly the repulsive single electron state from which the atoms finally desorb remains to be identified. A more detailed description of the model and the desorption mechanism will be published elsewhere [9].

The observations made in experiments on Na, K, Ag and other metals indicate that the underlying bond breaking mechanism is of very general nature. This will certainly stimulate future theoretical and experimental efforts with the main goal to clarify details of the mechanism and to further explore the possibilities for technological applications [10].

This work is supported by the Deutsche Forschungsgemeinschaft.

References
[1] W. Hoheisel, K. Jungmann, M. Vollmer, R. Weidenauer, and F. Träger, Phys. Rev. Lett. 60, 1649 (1988)
[2] W. Hoheisel, U. Schulte, M. Vollmer, F. Träger, Appl. Phys. A 51, 271 (1990)
[3] W. Hoheisel, M. Vollmer, F. Träger, Appl. Phys. A, July (1991)
[4] I. Lee, J.E. Parks, T.A. Callcott, E.T. Arakawa, Phys. Rev. B 39, 8012 (1989)
[5] A.M. Bonch-Bruevich, Yu.N. Maksimov, S.G. Przhibelsky, V.V. Khromov, Sov. Phys. JETP 65, 161 (1987)
[6] J.-C. Payan, D. Roux, Opt. Commun. 7, 26 (1973)
[7] S.P. Apell, J. Giraldo, S. Lundqvist, Phase Transitions 24-26, 577 (1990)
[8] R. Monreal, S.P. Apell, Phys. Rev. B 41, 7852 (1990)
[9] W. Hoheisel, M. Vollmer and F. Träger, to be published
[10] M. Vollmer, R. Weidenauer, W. Hoheisel, U. Schulte, F. Träger Phys. Rev. B 40, 12509 (1989)

DESORPTION OF Al, Au, AND Ag USING SURFACE PLASMON EXCITATION[*]

E. T. Arakawa, I. Lee[†], and T. A. Callcott[†]
Biological and Radiation Physics Section
Oak Ridge National Laboratory, Oak Ridge, TN 37831-6123

INTRODUCTION

This paper describes a promising method of photodesorption which greatly amplifies electric field strength in the surface region through the excitation of surface plasmons. We present data on the yield of desorbed Al, Au, and Ag atoms as a function of incident angle in the attenuated total reflection (ATR) geometry. We also present data on the kinetic energy distribution of the desorbed atoms. Finally, the mechanism responsible for the desorption will be discussed briefly.

EXPERIMENTAL METHODS

The metal thin films were deposited on the base of a 45° glass prism in a vacuum evaporator with a base pressure of 10^{-6} Torr. The thicknesses of the Al films in this experiment were 27 nm, the gold films were 38 nm, and the Ag films were 38 nm.

The desorption experiments (Fig. 1) were performed in a cryopumped vacuum system with a base pressure of 10^{-7} Torr. The metal-coated prism was placed between the acceleration plates of the time-of-flight mass spectrometer (TOF-MS). P-polarized light from the second harmonic of a Nd:YAG laser with wavelength 532 nm and pulse length 7 ns was passed through the prism onto the sample film. The YAG laser beam was attenuated with a series of glass slides to ~ 0.1 J/cm^2 or 10^7 W/cm^2. The angle of incidence was varied by rotating mirror M_r. This rotation also exposed a fresh area of the film. Desorbed neutral atoms were ionized by an XeCl excimer laser with a wavelength of 308 nm and a pulse length of 7 ns. The saturation power level for non-resonant multiphoton ionization of > 10^8 W/cm^2 was available from the excimer laser. The ions were detected by an electron multiplier in the TOF-MS. The triggering device of the Nd:YAG laser was connected to a delay generator which, after a chosen time delay, triggered the excimer laser. By varying the delay time between the two lasers, we were able to determine the kinetic energy of the desorbed neutrals. The Nd:YAG laser was set at a repetition rate of 10 Hz. The ion signal from the TOF-MS was amplified by a fast amplifier and recorded by a transient digitizer.

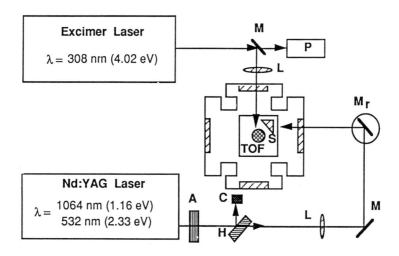

Fig. 1. Arrangement for desorption experiment, where A is the attenuation slides, H is the harmonic separator, C is the copper cavity, L is the focus lens for the laser beam, M is the mirror, M_r is the scanning mirror that directs the Nd:YAG laser light through the prism to the back of the metal film at different incident angles, S is the metal film deposited on a glass prism, TOF is the acceleration plate of a time-of-flight mass spectrometer (TOF-MS), and P is the photo-diode. The shaded area on the TOF plate is the aperture of the acceleration plate of the TOF-MS.

RESULTS AND DISCUSSION

The kinetic energy distribution for desorbed Ag, Au, and Al atoms shown in Figs. 2, 3, and 4 exhibit two peaks. The low energy peak at 0.081 eV is due to a thermal effect and corresponds to a translational temperature of approximately 900 °K. The low energy peak is found to increase in energy with laser fluence. The high energy peak does not vary with laser fluence and is observed at 0.51 eV, 0.33 eV, and 0.90 eV for Al, Au, and Ag, respectively. These kinetic energies are four to eleven times higher than the kinetic energy of the thermal peak. The desorption yields of the high-energy components, as functions of incidence angles, show sharp peaks near the resonance angle associated with surface plasmon excitation as illustrated in Fig. 5 for Ag . The angular distribution of desorbed neutral Ag atoms (Fig.5), shows that the yield is linear with the photon energy density at the metal-vacuum interface. The photon energy density profile of Ag film is shown in Fig. 6. Similar results were obtained for Au and Al. We believe that the high-energy peak is produced by electronic-excited desorption.

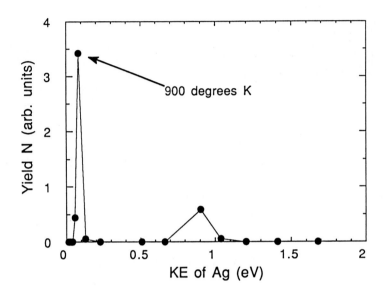

Fig. 2. Kinetic energy distribution of desorbed Ag atoms.

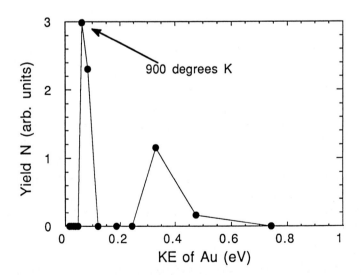

Fig. 3 Kinetic energy distribution of desorbed Au atoms.

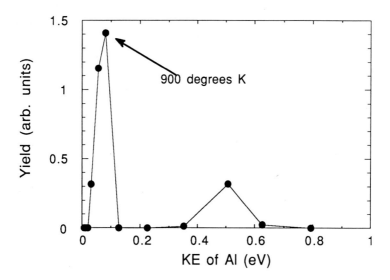

Fig. 4 Kinetic energy distribution of desorbed Al atoms.

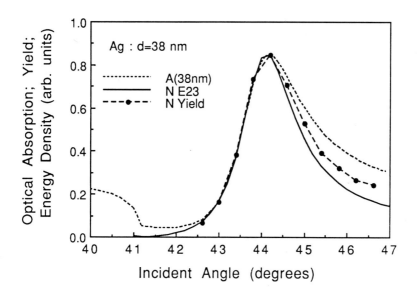

Fig. 5. Desorption yield N, total optical absorption A(38 nm), and energy density at metal-vacuum interface N E23 for the Ag film as a function of incidence angle. The laser fluence was 180 mJ/cm^2.

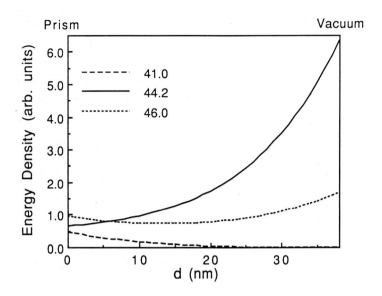

Fig. 6 The photon energy density profile of Ag film. The plasmon resonance angle is at 44.2 degree.

* Work supported by the U.S. Department of Energy under Contract DE-AC05-84OR21400 with Martin Marietta Energy Systems, Inc.

† Also at Department of Physics & Astronomy, University of Tennessee, Knoxville, TN 37996-1200

Threshold Fluence UV Laser Excitation of W(100) and $O_2,H_2,F/W(100)$: Photoejected Ion KE Distributions[a]

HyunSook Kim[b] and Henry Helvajian[c]
Laser Chemistry and Spectroscopy Department
Aerophysics Laboratory,
The Aerospace Corporation,
P. O. Box 92957, Los Angeles, California 90009

INTRODUCTION

The interaction of a pulsed laser field with an adsorbate covered surface generally promotes desorption of thermally excited products. However, if the incident laser fluence is maintained near the desorption threshold, the ejected product kinetic energy (KE) and population distributions may also include features characteristic of a nonthermal process. This phenomenon is especially prevalent when the laser wavelength is in the UV, and care is taken to ensure that the measured distributions are not perturbed by above surface collisional processes. Recent experiments have shown that low fluence UV laser/surface excitation may induce the ejection of high KE ion species [1, 2]. Furthermore, the measured ion KE distributions cannot be modeled by a thermally activated desorption process, i.e., non-Boltzmann. In addition, the measured high kinetic energies are not the result of a plasma induced acceleration process either. The recent observation of a wavelength dependence in the photoejected species population distribution provides additional support for the possibility of nonthermally activated ion desorption[3]. The photoejected ion yield, when measured over the range of UV wavelengths (3–6 eV) has a nonlinear dependence on the incident laser fluence [4]. For the specific case of Ag^+ ion ejection from crystalline silver, the fluence dependence suggests that the energy necessary for Ag^+ removal is comparable to a silver core electronic excitation (50–60 eV) [1].

To date, no comprehensive theory exists for explaining the pulsed laser induced ion ejection phenomenon. However, two theoretical concepts have successfully explained most results from the x-ray photon stimulated desorption (PSD) and electron stimulated desorption (ESD) studies. In the Menzel-Gomer-Redhead (MGR) model, species are desorbed via Franck-Condon type excitations of the adsorbate/substrate system [5]. Kinetic energy in the ejected species is gained at a cost to the potential energy of the excited repulsive electronic state. The Knotek-Feibelman model (KF) relies on an initial core electron-

---
[a] This project sponsored by the Aerospace Sponsored Research (ASR) Program.
[b] National Research Council Post Doctoral Fellow.
[c] Person to whom correspondence should be addressed.

ic excitation process, which is followed by an interatomic Auger decay and dissociation via coulombic repulsion forces [6]. The KF model has been adequate in explaining the results from maximal valency compound systems. Both theories predict formation of high KE ion desorption, although the KF model is more amenable in explaining ejected species with multi-eV kinetic energy.

Our experiment tries to compare the results from the pulsed UV laser excitation (UV-PLE) with those from PSD and ESD studies. Since our apparatus is only designed for UV-PLE experiments, a convenient system for study is crystalline tungsten, where there is a large body of PSD and ESD experimental results. In this paper, we present experimental results from the UVPLE of W(100) with the adsorbates H, O, and F.

EXPERIMENTAL

The experimental setup is shown in Fig. 1. A complete description of the data acquisition and analysis can be found in Ref. 1. Briefly, a UHV chamber is equipped with a 1.4 m time-of-flight (TOF) mass spectrometer and an Auger system for surface impurity analysis. A Nd:Yag pumped dye laser system with frequency summing capability generates (s)-polarized radiation at the wavelengths 355 and 318 nm. The laser beam is first passed through an external Raman cell, which removes temporal inhomogeneities and then is spatially apertured. The *unfocused* laser hits the tungsten target at a 45 degree angle, with the incident energy maintained near the threshold for ion species ejection (30 mJ/cm^2). Single shot TOF data is recorded by a transient digitizer (Transiac: 10 ns resolution; Analytek: 2 nsec resolution) along with the corresponding laser energy and the TOF voltages. The single shot data acquisition technique eliminates the possibility of a single shot, space-charge broadened KE distribution from being averaged in. The measured arrival times are converted to initial kinetic energy based on a model (Simion Ion Optics) of the TOF apparatus. In addition, the measured maximum kinetic energy as calculated by the Simion model is always compared with the results from a retarding potential experiment.

The tungsten (100) single crystal was treated in acetone solvent in accordance with the manufacturer's instructions and then transferred to the UHV chamber. The crystal was repeatedly baked though only to 500°C. The oxygen and hydrogen gases (research grade) were admitted into the chamber via a leak valve without additional purification. Prior to the introduction of adsorbates, the surface was dosed with nearly 100 K laser shots to remove impurities. The TOF spectra initially showed the desorbing products to be mostly Na$^+$ and K$^+$ ions. With continued surface "cleaning" at threshold laser fluences, we measured the simultaneous reduction in the desorbed alkali population along with the appearance of W$^+$ species.

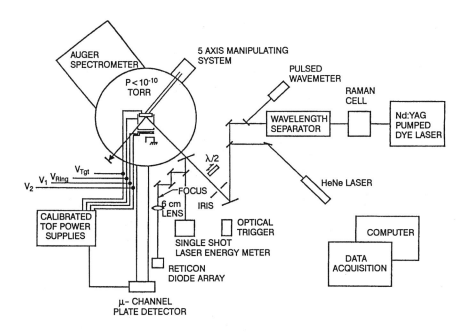

Figure 1. Schematic of the experimental apparatus

RESULTS

The UVPLE of adsorbate covered W(100) produces predominantly atomic ions. Some diatomic ions (NaH^+, KH^+) were also observed. We saw no change in the ejected KE or population distributions for the two wavelengths (355 nm, 318 nm) used. However, qualitative results place the ejection threshold lower for 318 nm laser excitation.

For the sake of clarity, we label the measured ejected species as follows: the adsorbate species (O^+, H^+, F^+), the impurity species (Na^+, K^+, NaH^+, KH^+); and the substrate species (W^+, WO^+). The other oxide compounds (e.g., WO_2^+, WO_3^+) were not detected. In addition, the ejected ion KE distributions cannot be fit to a Boltzmann. Figure 2a shows a TOF/KE distribution of photoejected W^+. Our TOF cannot fully resolve the four nearly equal tungsten isotopes. However, the $^{186}W^+$ is sufficiently resolved to identify the KE most probable (<KE>) as 2 eV. Similarly, Fig. 2b shows the results for the photoejected impurity species K^+ and KH^+. The upper KE axis correspond to the masses $^{39}K^+$ and $(^{39}K\,^1H)^+$. In comparison with the results shown in Fig. 2a, both the impurity and the substrate ions have the same <KE>. Although not shown, the photoejected Na^+ and NaH^+ products also have the same <KE>. This similarity in the <KE> between substrate and impurity compounds was also measured in our previous studies on crystalline Ag and Al targets [1, 2].

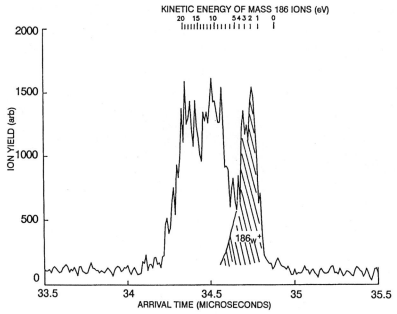

Figure 2a. TOF/KE distribution of photoejected W^+.

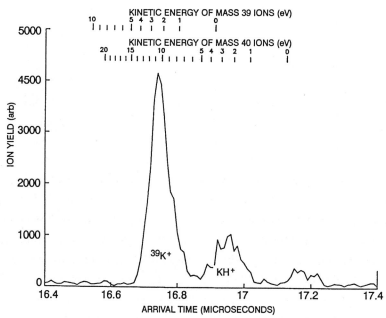

Figure 2b. TOF/KE distribution of photoejected K^+ and KH^+.

A surprising result in the UVLPE of W(100) is the measured adsorbate KE distributions. Figure 3a and 3b show the results for the photoejected ions F^+ and H^+. Two important features must be pointed out. First, the KE distributions of the adsorbates do not agree with that of the substrate and impurity species. Second, the KE distribution of the ejected F^+ is clearly split, though in the case of the H^+ ion, the splitting is not as definite.

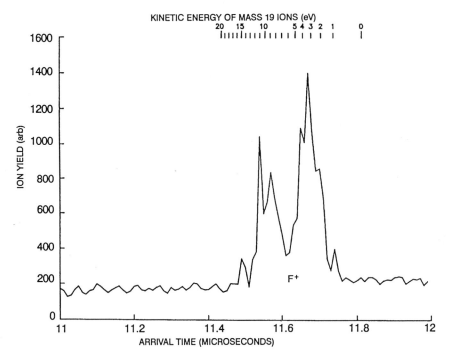

Figure 3a. TOF/KE distribution of photoejected F^+.

The addition of oxygen to W(100) clearly produces a photophysical reaction that lowers the threshold for removing W^+ and WO^+ ions. Figure 4 shows three TOF mass spectra taken at different O_2 gas pressures. The three spectra are taken with the laser fluence maintained near the threshold for removing W^+ species when no oxygen is in the chamber. The figure shows that the addition of oxygen enhances the removal of both WO^+ and W^+ ions. It is known that at room temperature, oxygen forms less than a monolayer coverage [7]. The observed pressure dependence in Fig. 4 reflects the competition between oxygen adsorption and photodesorption/reaction process. Our attempt to measure the photoejected O^+ KE distribution was unsuccessful. The signal to noise (one O^+ trace per 400 laser shots) was too poor to get an accurate TOF distribution. However, using the retarding potential technique, we could qualitatively state that the O^+ KE is in excess of 9 eV. We are currently refining our data acquisition method to allow accurate measurement of the

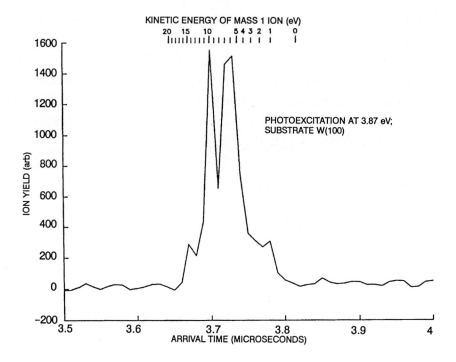

Figure 3b. TOF/KE distribution of photoejected H^+.

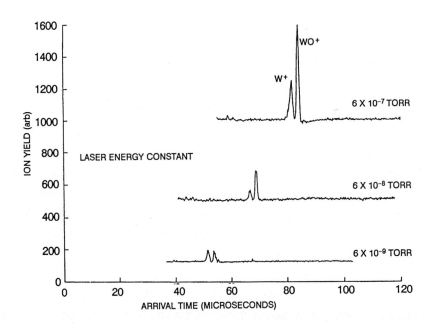

Figure 4. Time-of-flight mass spectra for various oxygen concentrations at constant laser energy.

O^+ KE distribution. The low O^+ desorption yield could be due to low cross section for desorption. This is also observed in ESD studies of O_2/W(100). Figure 5 shows a TOF/KE spectrum that includes both the W^+ and WO^+ products. The calculated KE axis is for the masses $^{186}W^+$ and $(^{186}W\ ^{16}O)^+$. The results show that for the oxygenated system, the <KE> for both the W^+ and the WO^+ ions is the same. Furthermore, the results in Fig. 5 also agree with the W^+ KE distribution measured prior to the addition of oxygen (Fig. 2a). Therefore, the adsorption reaction seems not to change the W^+ KE distribution.

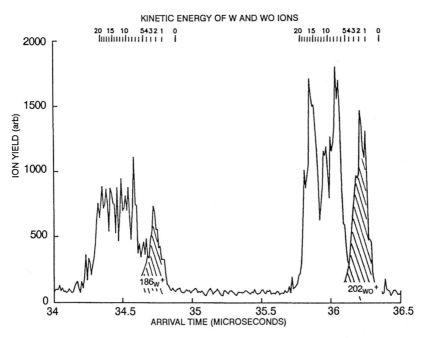

Figure 5. TOF/KE distribution of photoejected W^+ and WO^+.

DISCUSSION

Four main conclusions can be drawn from this experiment. First, we measure the ejection of highly electronegative species in positive charge states (i.e., O^+ IP 13.6 eV, F^+ IP 17.4 eV). This is more noteworthy considering that the photon energies used are 3.5 and 3.9 eV. Second, the KE distributions of the ejected adsorbate species (H^+, F^+) differ with those of the substrate (W^+, WO^+) and the impurity ions (Na^+, NaH^+, K^+, KH^+). Third, the adsorbate KE distributions show a bimodal profile, while those of the impurity species do not. Fourth, the addition of oxygen on W(100) has no effect on the ejected W^+ ion KE distribution. The oxygenation of the surface lowers the laser fluence threshold for removing the substrate species.

To our knowledge, no measurements of the desorbed KE exist for W^+ and WO^+ following PSD/ESD excitation. Therefore, a direct comparison with PSD/ESD data can only be made for the species F^+ and H^+. Our result for the F^+ KE distribution (peaks at 3 and 10 eV) partly agrees with that from ESD studies. The ESD results report a single peak at 3 eV [8]. Similarly, our measured H^+ ion KE distribution (IKD) also shows a bimodal shape (peaks at 5 and 10 eV), though the profile is less resolved. The ESD studies on H_2/W report the H^+ IKD with a peak at 1.7 eV [9]. However, for H^+ desorption from H_2O/W, the reported peak in the IKD is at 4.1 eV [9]. In our experiment, we could not bake the target above 500°C and therefore could not remove all surface oxygen. The adsorption of H_2, together with UV laser excitation, could have catalyzed the formation of H_2O on the tungsten. Given this reaction possibility, we could then explain our H^+ IKD peak at 5 eV as resulting from H_2O/W excitations. We also measure some H^+ ions with KE near 2 eV but cannot unambiguously attribute the signal with the desorbing of H^+ ion from an H_2 adsorbate. In ESD of H_2O/W and H_2/W, the H^+ ion yield desorbed from H_2O/W is 100 times larger [9]. In summary, our measured F^+ and H^+ low kinetic energy peaks in the IKD are in agreement with those of ESD studies. However, we are hard pressed to explain the bimodal shape in our F^+ and H^+ ion KE distributions. Based on the KF theoretical model, we would have to argue that both F and H have an additional binding site on tungsten. Furthermore, we wonder if it is just a fortuitous coincidence that both the F^+ and H^+ high KE peak values are nearly the same.

Our results also show that the alkali ion (Na^+ and K^+) distributions are nearly identical to those of the substrate species (W^+, WO^+). This equivalence was also seen in our previous studies on Ag and Al targets [1, 2]. Of interest is the difference in the IKD between the desorbed electropositive (Na^+, K^+) and the electronegative (F^+, H^+) products. The electropositive products have a lower $<KE>$ and a solitary IKD peak. The electropositive adsorbates tend to be covalently bonded, similar to the tungsten metal. The electronegative species, more likely, are bound ionically. Therefore, a larger energy release is potentially available for an ionically bound atom undergoing a transition to a repulsive Madelung well.

In summary, our results show for the first time that the UV pulsed laser excitation of adsorbates on W(100) remove ionic species that have $<KE>$ similar to that observed in ESD/PSD studies. Our measurements, not presented here, show that the ion desorption process occurs via a multiphoton absorption event. Based on our previous studies, we believe that the desorption occurs via an electronic induced process similar to that described by the Knotek-Feibelman theory.

REFERENCES

1. H. Helvajian and R. P. Welle, *J. Chem. Phys.* 91, 2616 (1989).
2. H. S. Kim and H. Helvajian, *J. Phys. Chem.* (accepted for publication August 1991).
3. L. Wiedeman and H. Helvajian, *Mater. Res. Soc. Proc.* 191, 217 (1990).
4. L. Wiedeman, H. S. Kim and H. Helvajian, *Laser Ablation-Mechanisms and Applications*, Springer Verlag (1991).
5. D. Menzel and R. Gomer, *J. Chem. Phys.* 41, 3311 (1964); P. A. Redhead, *Can. J. Phys.* 42, 886 (1964).
6. M. L. Knotek and P .J. Feibelman, *Phys. Rev. Lett.* 40, 964 (1978).
7. J. Kirschner, D. Menzel and P. Staib, *Surf. Sci.* 87, 1267 (1979).
8. Ch. Park, M. Kramer and E. Bauer, *Surf. Sci.* 109, 1533 (1981).
9. M. Nishijima and F. M. Propst, *Phys. Rev.* B2, 2368 (1970).

EXCIMER LASER ABLATION OF CdTe

P.D. Brewer, J.J. Zinck, and G.L. Olson
Hughes Research Laboratories
Malibu, CA 90265

ABSTRACT

KrF excimer laser irradiation (248 nm) of CdTe (100) at fluences below the melting threshold removes surface atoms and produces reversible changes in composition and structure that depend on the laser fluence and number of laser pulses. At fluences below 65 mJ/cm^2, the products desorbed from the surface consist of neutral Cd atoms and Te$_2$ molecules, and the velocity distributions of the desorbed products are well-described by a Maxwell-Boltzmann distribution. At higher fluences, Cd$^+$, Te$^+$, and Te$_2^+$ are also ejected, and the velocity distributions of the neutral and ion species become non-Maxwellian. Over the temperature range from 380-400°C the instantaneous laser-induced desorption rates are more than six orders of magnitude higher than the vacuum sublimation rates measured over the same temperature range. The composition of the CdTe surface can be reversibly controlled from stoichiometric to a Te-rich condition by varying the laser fluence over the range from 15 to 65 mJ/cm^2. The dynamics of the photo-stimulated desorption and the fluence dependent changes in surface composition are described in terms of the kinetic competition between the formation and desorption of Te$_2$ and the desorption of Cd atoms from the laser-irradiated surface.

INTRODUCTION

Pulsed laser irradiation of elemental and compound semiconductors results in desorption of surface atoms at a rate which depends on the photon energy and the density of absorbed photons.[1-5] In the case of compound semiconductors, the ablation process is generally accompanied by a fluence dependent change in surface composition.[6] The mechanism of laser-induced desorption has been investigated by numerous workers.[4,7,8] However, there is no comprehensive description which incorporates effects due to electronic excitation of atoms in the solid, formation of transient absorbing species, and lattice heating produced by non-radiative relaxation of the electronically excited states of the semiconductor. Since the excitation and relaxation processes that accompany pulsed laser irradiation of an absorbing, semiconductor surface are very fast (sub-picosecond), the establishment of the relative contributions made by electronic and thermal processes to the overall desorption mechanism has, in general, been difficult.

The recent work of Kumazaki and co-workers[9] and Nakai, et al.[10] showed that sub-bandgap excitation can be used to stimulate desorption of surface atoms without lattice heating. To explain their data they invoked a mechanism in which absorption occurs at localized surface defect sites causing bond breaking and ejection of atoms from the surface. Their results point strongly to a mechanism dominated by nonthermal processes. On the other hand, when above-bandgap radiation is employed, the absorption process is nonlocal and the distinction between thermal and nonthermal contributions is obscured. In this paper we suggest that when a highly absorbing semiconductor surface is exposed to above-bandgap laser radiation, both thermal and nonthermal components contribute to the ablation mechanism. We show that KrF excimer laser (248 nm) irradiation of CdTe surfaces produces reversible fluence dependent changes in surface composition and structure. These composition changes and the velocity distributions of particles ejected from the laser irradiated surface are well-described by a thermal mechanism. However, the laser-induced desorption rate at a given temperature is several orders of magnitude larger than the vacuum sublimation rate measured at the same temperature. This large difference in rates cannot be reconciled by a purely thermal model for the ablation process.

EXPERIMENTAL

The experimental apparatus and procedures used in this study are described in detail elsewhere;[3,4] therefore only the salient features of the experimental approach will be described here. CdTe(100) samples were chemically etched in a 1/16% Br_2/ethylene glycol solution and transferred to an ultrahigh vacuum chamber for the ablation experiments. In situ measurements of surface composition and structure were performed using Auger electron spectroscopy (AES) and reflection high energy electron diffraction (RHEED). KrF excimer laser pulses were directed into the chamber via a UV-grade viewport. The nominal laser pulsewidth was 20 nsec, and the repetition rate was maintained at 1 pps to avoid cumulative substrate heating effects. The identity and velocity of the desorption products were measured using time-of-flight (TOF) mass spectrometry. Ablation rates were deduced from post-irradiation stylus profilometry measurements.

RESULTS AND DISCUSSION

Desorption Products and TOF Distributions

Time-of-flight mass spectrometry was used to identify the desorption products and to measure the velocity distributions of the species ejected from the CdTe surface during KrF excimer laser irradiation. When the CdTe surface was irradiated in the fluence range from 15-65 mJ/cm^2 only neutral Cd atoms and Te$_2$ molecules were detected. Typical TOF spectra are shown in Figure 1 for both species detected at normal incidence to the surface at a fluence of 27 mJ/cm^2. At this fluence the material removal rate is ~0.1 monolayers/pulse. At this low ablation rate, neither collisions nor the formation of a Knudsen-layer affects the velocity distributions of the ejected species.[8,11] Over the laser fluence range from 27-65 mJ/cm^2, the measured velocity distributions of the neutral Cd atoms and Te$_2$ molecules are well-described by a single component Maxwell-Boltzmann distribution. TOF mass spectrometry measurements were also used to determine the dependence of the desorbed product velocity distribution as a function of bulk substrate temperature over the range 23-300°C. The most probable

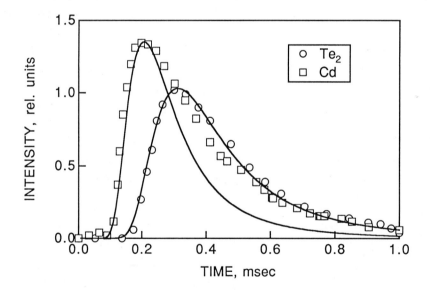

FIGURE 1. TOF distribution of Cd atoms and Te$_2$ molecules ejected from CdTe(100) surface by KrF excimer laser irradiation (27 mJ/cm^2). Intensities are uncorrected for mass spectrometer sensitivities. Solid lines are fits of a simulated Maxwell-Boltzmann velocity distribution to the experimental data.

velocities for both of the neutral products were in excellent agreement with those deduced from the Maxwell-Boltzmann velocity distribution when a temperature equal to the sum of the peak transient temperature and the substrate temperature was used in the calculation of the Cd velocity distribution. The peak transient temperature reached during laser irradiation was calculated using a one-dimensional heat flow model.[12] Input parameters to the model included the laser fluence and published values for the optical and thermal properties of CdTe.[13,14]

At fluences greater than 65 mJ/cm^2 ionized species (Cd^+, Te^+ and Te_2^+) are ejected from the surface. In this fluence regime the neutral products are no longer characterized by a single Maxwellian but, instead, have a bimodal velocity distribution.[4] Under these conditions, where the ablation rate is >10 Å/pulse, complications arising from surface melting and collisions of desorbed products in the near-surface region must be considered. The appearance of positive ions with increasing fluence is a behavior contrary to that observed by Nakai, et al.[10] and Hattori and co-workers[2] in their studies of the effects of sub-bandgap laser radiation on GaP surfaces. In that work, a decrease in the number of positive ions with increasing laser fluence was observed. These differing observations can be rationalized by considering the relative importance of localized electron excitation and thermal effects on the dynamics of laser-induced desorption. In bulk excitation of the near-surface region, which occurs during the irradiation of CdTe with 5 eV photons in a 30 nsec pulse, strong electron-phonon coupling will aid thermally-driven surface diffusion and desorption of neutral species. As the fluence is increased, multiphoton excitation results in the formation and ejection of positive ions from the surface. However, in sub-bandgap irradiation of GaP, the excitation is thought to be localized on surface defect sites. Hattori and co-workers[2,10] suggest that the defect state density is reduced by photon absorption. Since the defect states are the origin of the photo-generated ions, a decrease in the defect state density will result in an attendant decrease in the number of ions ejected from the surface.

<u>CdTe Ablation Rate</u>

The CdTe removal rate per pulse as a function of KrF excimer laser fluence is shown in Figure 2. Measurable desorption of Te_2 and Cd occurs at a threshold fluence of approximately 15 mJ/cm^2. The removal rate is observed to depend exponentially on laser fluence. TOF mass spectrometry measurements of the desorbed product flux indicated that the removal rate per pulse was constant for

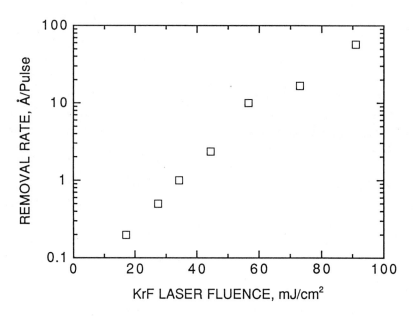

FIGURE 2. CdTe ablation rate as a function of KrF laser fluence.

multiple pulse exposure at a constant fluence.[4] This finding is in contrast to observations made in other compound semiconductors where multiple pulse exposure produces a continual change in removal rate due to a progressive depletion of the most volatile component in the surface layer.[6,15,16] It also differs from the result obtained when sub-bandgap radiation is employed.[2,10] In that case, removal of defect sites with increasing laser fluence results in a decreased flux of ablation products.

We point out that over the temperature range from 380-400°C the instantaneous laser-induced removal rates (averaged over the 20 nsec laser pulsewidth) are 6 to 8 orders of magnitude higher than the vacuum sublimation rates measured over the same temperature range.[17] This implies that excimer laser ablation of CdTe is not a purely thermal process, and that photo-induced bond dissociation may be facilitating the desorption of photoproducts from the surface. Taken together with the velocity distribution data presented in the previous section these results imply that the mechanism for above-bandgap laser ablation of CdTe consists of a nonthermal (bond-breaking) component which is initiated by absorption of photons by the bonding molecular orbitals. This process is followed by thermally driven diffusion and desorption of photoproducts having a velocity distribution characterized by the equilibrium surface temperature.

Fluence Dependent Changes In Composition And Structure

The composition of a CdTe (100) surface irradiated with a KrF excimer laser depends strongly on the laser fluence and the number of laser pulses. The fluence dependence of the Cd/Te ratio after multiple pulse exposure is shown in Figure 3. At fluences above 40 mJ/cm^2, preferential desorption of Cd atoms from a clean, well-ordered CdTe(100) surface results in the formation of a Te-rich surface which has no observable RHEED pattern. Auger depth profiling measurements indicate that the compositionally altered layer is $\leq$20 Å thick. Exposure of a stoichiometric surface to fluences <40 mJ/cm^2 results in the desorption of equal amounts of Cd and Te from the surface without altering the original composition.

We have found that a compositionally altered surface which does not exhibit a RHEED pattern can be returned to stoichiometry and to its original crystalline order by irradiating the surface at an appropriate fluence.[4,18] This behavior is illustrated by Figure 4 which shows the composition change produced by low fluence (35 mJ/cm^2) irradiation of a Te-rich surface (Cd/Te=0.4). The Te-rich surface was prepared by exposure of a stoichiometric surface to KrF laser pulses

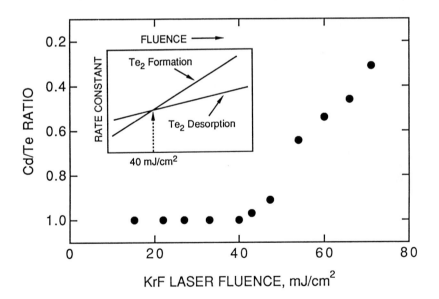

FIGURE 3. Cd/Te ratio at laser-irradiated surface vs. KrF laser fluence. Inset shows relationship between kinetics for Te$_2$ formation and desorption proposed to explain dependence of surface composition on laser fluence.

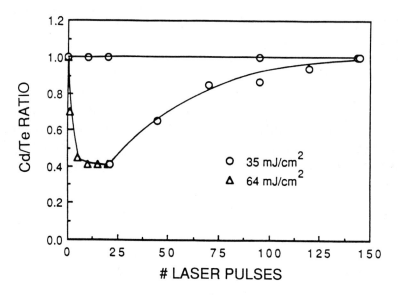

FIGURE 4. Changes in CdTe surface composition produced by multiple pulse, KrF excimer laser irradiation at two different fluences. Irradiation at 64 mJ/cm^2 results in a Te-rich surface; subsequent exposure at 35 mJ/cm^2 returns the surface to stoichiometry. Irradiation of a stoichiometric surface at a fluence of 35 mJ/cm^2 leaves surface composition unchanged.

at a fluence of 64 mJ/cm^2. In this example, multiple pulse, reduced fluence exposure returns the Te-rich surface to stoichiometry. It is also possible to return the surface to any desired composition between a highly Te-rich value and stoichiometry. Since the approach to a given composition depends upon the total energy incident on the surface, the number of pulses required to achieve that composition is greater for lower fluences.

To explain the fluence dependent changes in CdTe composition we propose the following elementary steps for the excimer laser desorption process:[4]

$$CdTe_{(c)} + h\upsilon \rightarrow Cd\cdots Te_{(x)} \quad \text{(Step 1)}$$
$$Cd\cdots Te_{(x)} \rightarrow Cd_{(g)} + Te_{2\,(ads)} \quad \text{(Step 2)}$$
$$Te_{2\,(ads)} \rightarrow Te_{2\,(g)} \quad \text{(Step 3)}$$

where $CdTe_{(c)}$ is the crystalline CdTe surface, $Cd\cdots Te_{(x)}$ is the disordered surface, $Te_{2(ads)}$ are adsorbed Te_2 molecules, and $Cd_{(g)}$ and $Te_{2(g)}$ are desorbed Cd and Te_2 molecules. We assume that the the activation energy for Te_2 formation is greater than that for Te_2 desorption, and that the two curves cross at an intermediate temperature (~600°C, 40 mJ/cm^2 -- see inset, Figure 3). The reason for selecting this particular temperature for the crossover point was that at fluences <40 mJ/cm^2 (600°C) the surface remains stoichiometric even though many surface atoms are desorbed, whereas at higher fluences the surface becomes increasingly Te-rich. Since the disruption of the Cd-Te bond thermodynamically favors the formation of the Te-Te bond and the liberation of Cd we also assume that the kinetics of Te$_2$ formation and Cd desorption are coupled. Thus, under high fluence (>40 mJ/cm^2) conditions Te$_2$ formation dominates over Te$_2$ desorption resulting in formation of a Te-rich layer at the surface. Irradiation of the Te-rich film at reduced fluence will produce a Te concentration at the surface that is controlled by the relative kinetics for formation and desorption. At all fluences less than 40 mJ/cm^2, Te$_2$ desorption dominates, and a stoichiometric film is obtained.

Although the simple picture given above is generally consistent with our experimental observations, a more accurate description requires a consideration of the fact that high fluence laser irradiation produces a compositionally graded layer ~20 Å thick (highest Te concentration at the surface). Under high fluence conditions each pulse removes >10 Å of material (and hence virtually all of the compositionally modified layer). Therefore, it requires only a few pulses to reach a steady state composition at the surface (see Figure 4). However, using low fluence conditions sufficient to restore stoichiometry, the thickness of the material that is removed during each pulse is much less than the thickness of the compositionally graded zone that was formed during the previous high fluence exposure. Thus, a larger number of pulses is required to establish the Cd/Te ratio at the new steady state value determined by the competitive rates for Te$_2$ formation and desorption at that fluence (see inset, Figure 3).

SUMMARY AND CONCLUSIONS

We have investigated fluence dependent changes in desorption rate, composition, and product velocity distributions which occur during KrF excimer laser irradiation of CdTe. We have shown that the velocity distribution of desorbed products at fluences less than 65 mJ/cm^2 can be accurately described by a Maxwell-Boltzmann distribution, and that there is an excellent correlation between the calculated peak surface temperatures and the observed translational temperatures of the ejected products. The ablation rate/pulse has been measured, and we observe that the ablation rate at a given temperature exceeds the vacuum sublimation rate by 6-8 orders of magnitude.

KrF laser irradiation produces fluence dependent changes in surface composition and structure. We find that the composition and structure can be reversibly controlled from Te-rich and amorphous to the stoichiometric value with surface order by varying the laser fluence over the range from 15-65 mJ/cm^2. The reversibility of the surface composition observed during irradiation at different fluences suggests that more than one competitive thermally activated process is operative during the generation and desorption of the products. A simple model is presented which treats the fluence dependent composition changes in terms of competition between formation and desorption of Te$_2$ and desorption of Cd during laser irradiation. These results strongly support the hypothesis that the mechanism for excimer laser ablation of CdTe is complex and consists of both a nonthermal component (photo-induced bond breaking) and a thermal component which we suggest controls diffusion and desorption of the photo-generated species from the thermally equilibrated surface.

ACKNOWLEDGEMENTS

The authors wish to thank J.E. Jensen and J.A. Roth for helpful discussions, and we gratefully acknowledge C.A. LeBeau for his expert technical assistance.

REFERENCES

1. T. Nakayama, N. Itoh, T. Kawai, K. Hashimoto, and T. Sakata, Radiat. Eff. Lett. 67, 129 (1982); T. Nakayama, M. Okigawa, and N. Itoh, Nucl. Instr. Meth. Phys. Res. B1, 301 (1984).

2. K. Hattori, Y. Nakai, and N. Itoh, Surf. Sci. Lett. 227, L115 (1990).

3. P.D. Brewer, J.J. Zinck, and G.L. Olson, Appl. Phys. Lett. 57, 2526 (1990).

4. P.D. Brewer, J.J. Zinck, and G.L. Olson, to appear in Mat. Res. Soc. Symp. Proc. 201, (1991).

5. K. Ichige, Y. Matsumoto, and A. Namiki, Nucl. Instr. Meth. Phys. Res. B33, 820 (1988).

6. V.M. Donnelly, V. McCrary, and D. Brasen, Mat. Res. Soc. Sym. Proc. 75, 567 (1987).

7. Y. Nakai, K. Hattori, A. Okano, N. Itoh, and R.F. Haglund, Jr., Nucl. Instr. Meth. Phys. Res., in press 1991.

8. R. Kelly, J. Chem. Phys. 92, 5047 (1990).

9. Y. Kumazaki, U. Nakai, and N. Itoh, Phys. Rev. Lett. 59, 2883 (1987).

10. Y. Nakai, K. Hattori, and N. Itoh, Appl. Phys. Lett. 56, 1980 (1990).

11. J.D. Cowin, D.J. Auerbach, C. Becker, and L. Wharton, Surf. Sci. 78, 545 (1978); I. NoorBatcha, R.R. Lucchhese, and Y. Zeiri, J. Chem. Phys. 86, 5816 (1987).

12. H.S. Carslaw and J.C. Jaeger, *Conduction of Heat in Solids* (oxford University, Oxford, 1959). For a review of heat flow calculations see: P. Baeri and S.V. Campisano, in *Laser Annealing of Semiconductors*, ed. by J.M. Poate and J.W. Mayer (Academic Press, New York, 1982) pp 75-109.

13. *Handbook of Optical Constants of Solids,* Ed. Edward D. Palik (Academic Press, Inc, New York, 1985) pp. 409-427.

14. G.A. Slack, and S. Galginaitis, Phys. Rev. 133, A253 (1964); C.M. Bhandari and D.M. Rowe, *Thermal Conduction in Semiconductors* (John Wiley & Sons, New York, 1988).

15. B. Strizker, A. Pospieszczyk, and J.A. Tagle, Phys. Rev. Lett. 47, 356 (1981), A. Pospieszczyk, M.A. Harith, and B. Strizker, J. Appl. Phys. 54, 3176 (1983).

16. A. Namiki, T. Kawai, Y. Yasuda, and T. Nakamura, Jpn. J. Appl. Phys. 24, 270 (1985).

17. J.M. Arias and G. Sullivan, J. Vac. Sci. Technol. A5, 3143 (1987); J.J. Dubowski, J.M. Wrobel, and D.F. Williams, Appl. Phys. Lett. 53, 660 (1988).

18. J.J. Zinck, P.D. Brewer, and G.L. Olson, to be published.

Part III

Polymers/Medicine

IR-LASER ABLATION IN MEDICINE:
MECHANISMS AND APPLICATIONS

Thomas F. Deutsch
Wellman Laboratory
Massachusetts General Hospital
Boston, MA 02114

Abstract: The use of pulsed infrared lasers for tissue ablation has led to increased control over thermal damage to tissue and to more precise control over material removal. Tissue removal is mediated by the explosive vaporization of water, which can lead to clinically significant tissue effects.

Tissue ablation forms the basis of conventional laser surgery. Early laser surgery experiments used primarily cw lasers, but recently there has been increased interest in pulsed-laser tissue ablation and in the physical mechanisms involved. Pulsed lasers can produce less thermal damage than continuous wave lasers by minimizing the effects of thermal diffusion, see below. In addition, pulsed lasers offer more precise control over the amount of tissue removed due to the pulse by pulse nature of the ablation. Some pulsed infrared (IR) lasers, such as the Er:YAG laser and the Co:MgF$_2$ laser have ability to cut bone with little thermal damage, from 5-15 μm in the case of Er:YAG and about 180 μm in the case of Co:MgF$_2$. These potential advantages of pulsed IR lasers for tissue ablation have been recognized and, coupled with the availability of a variety of compact pulsed mid-IR solid state lasers, have led to research which has advanced our understanding of pulsed IR laser tissue ablation.

That the thermal damage accompanying ablation could be spatially confined by using laser pulses short compared to the thermal relaxation time of the laser-heated volume was confirmed in a series of experiments using TEA CO$_2$, pulsed Er:YAG and cw CO$_2$ lasers for tissue ablation.[1,2] These lasers were chosen in order to be able to vary both pulsewidth and optical penetration depth. The extent of tissue damage was found to be in agreement with the prediction of a simple thermal model. A series of experiments to measure the fluence dependence of pulsed IR laser ablation showed that, to first order, pulsed laser ablation of tissue can be viewed as explosive vaporization of water, which constitutes approximately 85% of most soft tissues.

We have used the tunable Co:MgF$_2$ laser to measure the ablation threshold of hard and soft tissue as a function of wavelength. We varied the wavelength from 1.81 to 2.14 μm, a wavelength region in which the absorption length of water

---

This work was supported by the SDIO-MFEL program under Office of Naval Research Contract N00014-86-K-0017 and, in part, by a subcontract from Schwartz Electro-Optics under a National Institutes of Health SBIR program.

varies by a decade.[3] For soft tissues the ablation threshold tracked the optical absorption length of water; for bone there was little wavelength dependence, consistent with its low water content.

While simple blowoff models of ablation predict thresholds reasonably well, simple models of thermal damage to tissue are not completely successful. In particular, they would predict thermal damage only at the bottom of an ablation crater. In practice damage has a nearly uniform thickness over the interior surface of the ablation crater. Simple thermal models predict that the damage at the bottom of the crater will track the absorption length of water. However when thermal damage to cornea was studied using the $Co:MgF_2$ laser only a weak wavelength dependence was found. More recent models of thermal damage have suggested that a layer of liquid, probably molten collagen, at the walls of the crater may serve as a heat sink which mediates damage.[4] This model is consistent with some of the observed features of tissue damage, in particular with the fact that damage is found at the walls of the crater as well as the bottom.

The dynamics of tissue ablation have also been studied and have led to an awareness that explosive removal of tissue can lead to mechanical and acoustic effects that may be clinically significant. Flash photography, framing-cameras, and optical pump-probe techniques have all been used to study ablation dynamics. Flash photography was used to study both normal-spiking-mode and Q-switched Er:YAG laser ablation of skin and bone. The flash photographs showed that material was ejected at supersonic velocities, approximately 1400 m/s. A shock wave, propagating with a velocity of approximately 1600 m/s in the surrounding air, could frequently be seen in the photographs, ahead of the plume front. The presence of this shock wave indicates that high pressures are present in front of the target. Pump-probe measurements were also used to determine the motion of the particle front. These measurements demonstrated that each micropulse in the normal-spiking-mode pulse train was capable of ablating and rapidly ejecting tissue.[5]

Framing-camera pictures of $Co:MgF_2$ laser ablation of cornea and liver showed explosive removal of both materials, but also indicated differences in the nature of the explosions. The ablation of the cornea starts with the formation of a cavity which finally breaks; by contrast, chicken liver ablation shows what may be multiple bubbles forming. Such differences can be attributed to differences in the mechanical properties of the two tissues. Previous work has shown that tissue mechanical properties affect the efficiency of ablation by a pulsed CO_2 laser.[6]

The explosive nature of pulsed laser ablation can lead to damage in the unablated tissue. Tissue tearing has been observed in the ablation of cornea by a Q-switched Er:YAG laser at high fluences. Although 2.94 μm radiation from the Er:YAG laser has a penetration depth in water of approximately 1 μm, vaporization of surface water in tissue can result in dessication and thus a deeper penetration of radiation into the tissue. This can result in subsurface heated tissue which is confined by the tissue above. The confined, heated water vapor results in subsurface regions of high pressure which may cause tissue tearing. In addition, rapid removal of tissue from the surface of the target may induce shock waves in tissue and lead to its mechanical damage. Tissue damage several hundred microns below the ablated surface has been detected following the ablation of skin by a 193-nm ArF excimer laser. Since the UV radiation penetrates only a few microns in skin, the damage was attributed to acoustic effects.[7] Similar effects may occur in pulsed IR laser ablation.

In a clinical situation in which ablation occurs with a fluid, such as blood, surrounding the target, the mechanical effects may be enhanced because of the confinement of the water vapor formed. The effects of confinement by water have been examined in a somewhat different clinical context, the fragmentation of kidney stones using stress waves produced by a laser-induced plasma. These effects enhance the strength of the stress waves by up to a factor of 10.[8] Furthermore, delivery of laser energy by an optical fiber system in contact with tissue, a situation typical of many clinical applications, may also lead to confinement of the expanding water vapor and so to increased mechanical damage. Mechanical effects may be clinically significant in applications such as laser angioplasty, the ablation of fatty and calcific plaque from artery, where a damaged tissue surface may lead to clinically unacceptable effects. Ophthalmic applications of lasers, such as refractive surgery on the cornea using pulsed ultraviolet excimer lasers, are another example of a clinical application of laser ablation in which mechanical effects may be important.

REFERENCES

1. K.T. Schomacker, Y. Domankevitz, T.J. Flotte, T.F. Deutsch, Lasers Surg. Med. 11, 141, (1991)
2. J.T. Walsh, T.J. Flotte, R.R. Anderson and T.F. Deutsch, Lasers Surg. Med. 8, 108 (1988)
3. J.T. Walsh, T.J. Flotte, and T.F. Deutsch, Lasers Surg. Med. 9, 314 (1989)
4. A.D. Zweig, B. Meierhofer, O.M. Müller, C. Mischler, V. Romano, M. Frenz, and H.P. Weber, Lasers Surg. Med. 10, 262 (1990)
5. J.T. Walsh and T.F. Deutsch, Applied Physics B 52, 168, (1991)
6. J.T. Walsh and T.F. Deutsch, IEEE Trans. Biomed. Eng. 36, 1195 (1989)
7. S. Watanabe, T.J. Flotte, D.J. McAuliffe, and S.L. Jacques, J. Invest. Dermatol. 90, 761 (1988)
8. P. Teng, N.S. Nishioka, R.R. Anderson, and T.F. Deutsch, IEEE J. Quantum Electron. QE-23, 1845 (1987)

Pulsed Laser Ablation of Biological Tissue: Review of the Mechanisms

Alexander A. Oraevsky, Rinat O. Esenaliev, Vladilen S. Letokhov

Institute of Spectroscopy, USSR Academy of Sciences,
142092 Troitsk, Moscow Region, USSR.

1. History, Fundamental Phenomena and Problem Statement

The laser ablation of biotissue, i.e. the destruction and ejection of a certain volume of tissue under the effect of a high-power laser radiation, is finding ever increasing application in such areas of medicine as angioplasty [1], ophthalmology [2], dermatology [3], and cellular microsurgery [4]. The greatest promise is held by the pulsed ablation regime which has substantial advantages over continuous irradiation conditions [5, 6]. To select optimal pulsed ablation conditions (radiation wavelength and pulse duration and repetition rate) for a particular biological tissue, one needs an exact description of the mechanisms involved in the ablation process.

The first works on the ablation of soft tissues ranging from polymers [7] to atherosclerotic blood vessels (see review [8]) with various types of lasers have demonstrated the merits of pulsed UV lasers as concerns the ablation efficiency, extent of thermal injury of the adjacent layers, and correspondence between the shape of the crater formed as a result of ablation and the radiation intensity distribution over the cross section of the laser beam. The explanation of the phenomena observed has in these first works been based on the description of the pulsed UV ablation process by a mechanism involving a "cold" photochemical dissociation of molecular bonds [9]. But the absence of overheating of the tissue surrounding the crater resulting from pulsed laser ablation can be explained within the framework of the thermal explosion mechanism [10] as well. The photothermal model of the pulsed laser ablation of biological tissue has been verified by many experiments. Such facts as the low quantum yield of photochemical reactions for tissue biomolecules ($\simeq 1\%$) [11] and high radiationless relaxation rate of excited electronic-vibrational states in the condensed phase ($\simeq 10^9$-10^{10} s^{-1}) [12], and also the presence of a plume of steam and other hot gaseous products issuing with supersonic velocities [13, 14] point to a thermal character of the pulsed ablation of biotissue in the UV region of the spectrum.

The thermal nature of ablation by IR and visible laser pulses is practically beyond a doubt [5, 13]. Nevertheless, paradoxes requiring an explanation seem indispensable here too. The principal paradox is the fact that the maximum temperature reached at the threshold incident energy fluence calculated for the ablation of soft water-bearing tissue proceeding from known absorption and scattering coefficients is below 100°C, the boiling point of water. This paradox can be resolved if consideration is given to the optical inhomogeneity of biotissue [14].

Of great interest, to our view, is the introduction of photothermomechanical stress in the description of pulsed ablation [15, 16]. The estimates presented below show that the shock compression caused by the sudden thermal expansion of biotissue in the course of its rapid heating can explain quite well a number of experimental results.

Note that the fundamental phenomena experimentally observed to occur in the process of pulsed laser ablation of biological tissue, namely, (1) the presence of a threshold incident pulse energy fluence for ablation, (2) the formation of shock waves, (3) the gasdynamic sputtering of the ablation products with supersonic velocities, and (4) the dependence of the ablation efficiency and threshold energy fluence on the attenuation coefficient of the tissue, can, in principle, be explained within the framework of any of the mechanisms mentioned above: the photochemical, photothermal, or photomechanical one. Up to now, there has been no discussion of the relationship between the contributions from each of these mechanisms for particular irradiation conditions and types of tissue. The present work is aimed at considering the possible biotissue ablation mechanisms and conditions conducive to their realization.

2. Photochemical Mechanism

The photochemical pulsed ablation mechanism presumes that the laser energy absorbed by tissue biomolecules excites electronic states in biopolymers (proteins) that lie above the dissociation and ionization limits of these molecules. The direct dissociation of molecular bonds, or their mediate dissociation through reactions with ions or radicals formed from the highly excited electronic states, leads to the splitting of long polymer chains into short fragments (see Fig. 1a). The numerous bond breaks cause pressure to rise materially inside the iradiated tissue volume, which makes the molecular fragmets escape from the tissue. The threshold dissociation and ionization energies $\hbar\omega_{d,i}$ range between some 6 and 8 eV. The corresponding laser wavelengths fall within the UV region of the spectrum. Therefore, any appreciable quantum yields of photochemical reactions in proteins and other biopolymers can only be expected at irradiation wavelengths λ shorter than 200 nm. Unfortunately, no quantitative quantum yield measurements have been made for photochemical reactions in biomolecules in this region of the spectrum.

High-lying electronic states with an energy above 6 eV in biomolecules can also be populated as a result of two-quantum (two-step) excitation by radiation in the near UV region. For example, the energy fluence of 308-nm XeCl laser radiation ($\hbar\omega_L$ = 4eV) necessary to realize an efficient two-step excitation, which requires an efficient population of an intermediate electronic state, is given by

$$\Phi^* \simeq (\tau_p / \tau_{rel} \sigma_1) \hbar\omega_L \quad [J/cm^2] \qquad (1)$$

where τ_p is the laser pulse duration, τ_{rel} is the relaxation time of the excited electronic states, σ_1 is the absorption cross section, and $\hbar\omega_L$ is the laser quantum energy. For τ_p = 20 ns, $\tau_{rel} \simeq$ 1 ns, $\sigma_1 \simeq 10^{-17}$ cm^2 and $\hbar\omega_L = 6.45\times10^{-19}$ J, the fluence Φ^* corresponds to the experimentally observed threshold energy fluence values for ablation of biological tissue in the near UV region, $\Phi_{thr} \simeq 1$ J/cm^2. But there is no data in the literature on the square-law dependence of the ablation efficiency on the irradiation intensity.

The quantum yield φ of molecular bond dissociation drops sharply with the increasing irradiation wavelength as

$$\varphi \sim \exp{-(\hbar\omega_{d,i} / \hbar\omega_L)} \qquad (2)$$

and becomes negligible even in the near UV region of the spectrum [11]. At the same time, the probability of radiationless relaxation of excitation to heat grows rapidly (Fig. 1b). The role of photochemical processes in the ablation of biotissue by pulses with a wavelength of λ > 250 nm can be considered negligible.

3. Photothermal Explosion Mechanism

The necessary condition for explosive boiling is to make the tissue substance in the irradiated tissue volume go to an overheated metastable state within a time interval shorter than that required for heat to diffuse from this volume, the temperature of the tissue rising high enough for an intense formation therein of gas-phase bubbles in a fluctuation manner. Within the framework of this model, the pulsed laser ablation process can be subdivided into three stages: opticophysical, thermophysical, and gasdynamic.

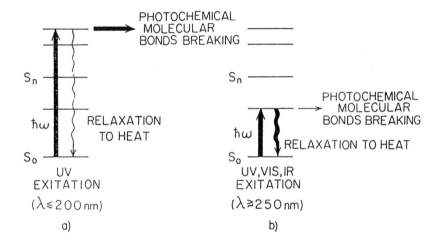

Fig. 1. Illustrating the opticophysical stage of biotissue ablation in the case of predominance of (a) photochemical mechanism and (b) photothermal mechanism.

During the first stage, there take place the absorption of laser radiation by the tissue and the subsequent conversion of the excitation energy into heat. Depending on the irradiation wavelength, the absorbing chromophores in the tissue are either biomolecules (mainly proteins) or water. In both cases, water serves as a reservoir for the thermal energy into which the laser radiation energy is converted. As the irradiated tissue volume is heated so that the tissue temperature rises above the threshold value, the opticophysical stage of the ablation process comes to an end.

The second, thermophysical ablation stage consists in the formation of gas-phase bubbles in the overheated tissue volume. The process is accompanied by the rise of pressure in this volume, the resultant shock wave propagating over the heated tissue layer faster than the metastable overheated liquid phase is formed. The pulsed heating of the tissue, the rate of which substantially exceeds the rate of heat consumption in evaporation in the existing nucleation centers, satisfies the explosive boiling conditions. Under such conditions, the number of the gas-phase bubble formation centers reaches saturation, and the nucleation rate $J(T)$ depends very strongly on the extent of overheating relative to the threshold explosive boiling temperature [17]:

$$J(T) = n(T)\{6\sigma(T)/[(2+p'/p'')\pi m]\}^{1/2} \times \exp[-\Omega(T)/kT]\exp[-W_c(T)\psi(\theta)/kT], \quad (3)$$

where n is the number of tissue molecules per unit volume, $\sigma(T)$ is the surface tension of the metastable liquid at the temperature T, p' is the external pressure, p'' is the pressure of the gaseous ablation products formed, m is the mass of the overheated liquid molecule, $\Omega(T)$ is the specific vaporization heat per molecule, k is the Boltzmann constant, $W_c(T) = 16\pi\sigma^3(T)/3(p'' - p')^2$ is the work of formation of a critical-size gas-phase bubble at the temperature T, and $\psi(\theta) < 1$ is the coefficient of the angle of contact, which allows for the decrease in the work of formation of a gas-phase bubble on the nonwettable surface of a nucleation center.

Calculation by formula (3) yields the following threshold temperature for the explosive boiling of a soft water-bearing tissue: $T_{thr} = 280°C$.

The thermophysical stage is followed by the gasdynamic stage. In the course of this last stage of the pulsed laser ablation process, there takes place the ejection of the overheated tissue volume and the gasdynamic sputtering of the ablation products with supersonic velocities. Collisions between microscopic tissue particles, molecules, atoms, and radicals in the hot gas cloud formed as a result of ablation in air cause the thermochemiluminescence of the ablation products [14]. The ablation products begin to luminesce at the start of the gasdynamic stage of the ablation process. At energy fluences close to the threshold value, the luminescence pulse lags behind the exciting laser pulse, which bears out the existence of the thermophysical stage involving the formation of the metastable phase of the overheated biological tissue. The experimental detection of the delayed luminescence of the hot ablation products is an important evidence in favor of the photothermal ablation mechanism.

Within the framework of the above thermal model, the minimum energy required for the ablation of a unit tissue mass is expressed as

$$E_{thr} \equiv E/\Delta M \, [J/g] = C(T_{thr} - T_0) \approx 1000 \text{ J/g}, \qquad (4)$$

where $C = 3.5$ J/g deg is the specific heat of tissue containing 75% water, T_{thr} is the threshold ablation temperature, and T_0 is the initial tissue temperature.

If we compare the threshold specific ablation energy value obtained from (4) with the experimental energy values measured for various irradiation wavelengths ranging from the IR to the UV region, we will bring out the main paradox of the theory of explosive boiling of tissue. The paradox is that the experimentally measured specific energy deposited by the laser pulse of the threshold energy fluence Φ_{thr} in the most heated, surface tissue layer turns out to be a fraction of the energy E_{thr} calculated by formula (4). Table 1 lists the threshold expansion energy values found from the expression:

$$E_{thr}^{exp} = \delta \cdot \Phi_{thr} \cdot \mu_a / \rho \qquad (5)$$

using tissue density $\rho = 1.08$ g/cm^3, experimental threshold fluence Φ_{thr}, absorption coefficient μ_a and numerical coefficient δ, measured at the laser wavelength used. The coefficient δ allows for the fact that total effective pulse energy fluence inside the tissue is several times the incident energy fluence as a consequence of summation of the incident energy fluence and the diffuse back-scattered energy fluence [18]. This numerical coefficient comes about its maximum value when the laser beam radius becomes commensurable with the radiation penetration depth l_{eff}. The coefficient δ was computed by means of a program calculating the propagation of laser radiation in biological tissue by the Monte-Carlo simulation [19].

Table 1 lists the values of the parameters entering into expression (5) for the 1st-4th harmonic frequencies of the Nd:YAG laser. As can be seen from the Table 1, the E_{thr}^{exp} values are by a factor of 5÷12 lower, than the E_{thr} values obtained from expression (4).

This paradox can possibly be explained by a spatially nonuniform heating of biotissue by laser pulse. At wavelengths shorter than 1.4 μm laser rdiation can be absorbed by microchromophore centers in the tissue whose volume is materially smaller than the total irradiated tissue volume. The explosive boiling of such overheated microcenters and the ensuing mechanical removal of the entire irradiated tissue bulk sharply reduces the volume energy density E_{thr} necessary for the ablation process to occur [14]

Another approach to the resolution of the paradox consists in the explanation of the experimentally observed pulsed laser ablation parameters by the action of a photomechanical tensile stress mechanism.

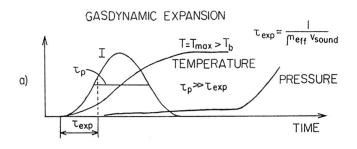

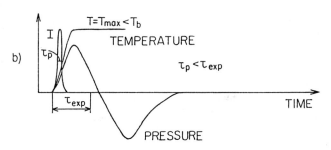

Fig.2. Illustrating temperature and pressure rise in biological tissue irradiated by (a) long and (b) short laser pulses.

4. Photomechanical Mechanism

To gain an understanding of this mechanism, let us compare the pressures produced in the heated tissue volume by a long and a short laser pulses (Fig. 2). We will compare the pulse duration τ_p with the characteristic hydrodynamic expansion time τ_{exp} of the heated tissue [16]. Let the pulse duration τ_p be long enough in comparison with the time τ_{exp} equal to the time it takes for the sound to propagate to the depth l_{eff} to which the the incident energy fluence Φ_0 penetrates the tissue. In that case, while a given tissue layer is being heated by the laser pulse, an increased pressure (compression) pulse resulting from the thermal expansion of the irradiated tissue propagates over the irradiated tissue volume, followed by a rarefaction (negative pressure) pulse (Fig. 2a). Thus, the rise of pressure in the irradiated tissue volume due to a gradual heating of the tissue is compensated for by the rarefaction produced behind the sound wavefront.

Now, let the tissue heating rate be substantially higher than the hydrodynamic tissue expansion rate, with the laser pulse duration τ_p being much shorter than the characteristic time

$$\tau_{exp} = l_{eff} / c_s \qquad (6)$$

where c_s is the sound velocity in the tissue. In that case, by the moment the heating of the irradiated tissue comes to an end, a valuable pressure builds up therein, determined by the heat expansion coefficient β and isothermal compressibility ξ_T of the tissue. To illustrate, at the threshold pulse energy fluence,

$$P_{thr} = (A^*/2)(\beta / \xi_T \rho C)(\Phi_{thr} \mu_a \delta), \qquad (7)$$

where $A^* = [1 - \exp(-\tau_p/\tau_{exp})] \tau_{exp}/\tau_p$ is a factor allowing for the attenuation of pressure in the tissue. The build-up in the irradiated volume of a pressure in excess of the critical rupture pressure of tissue causes the escape (spallation) of the heated layer through the free surface, like the escape of a compressed spring.

Table 1 lists the experimental values of the threshold energy fluence, measured for nanosecond laser pulses, along with the ablation depth values for atherosclerotic plaque l_{abl}, calculated for the maximum ablation efficiency η from the expression:

$$l_{abl} = \Delta M_{abl}/\rho S = \eta E/\rho S = (\eta/\rho) \cdot \Phi^{max} \qquad (8)$$

for irradiation wavelengths of 266, 355, 532 and 1064 nm. For comparison Table 1 also presents the values of the spallation depth l_{sp}, calculated by the formula:

$$l_{sp} \approx l_{eff} \ln (P^{max}/P_{sp}) \qquad (9)$$

using the values of the pressure P^{max} in the tissue heated in a pulsed manner at the maximum ablation efficiency and the critical rupture pressure (the pressure P_{sp} for water was taken at 100 bar [20]). Our rough estimates show a fairly good agreement between the experimental tissue ablation depths at λ = 266 nm and the theoretical values. Thus the ablation of biological tissue by short (τ_p < 10^{-8} s) UV laser pulses can be described by means of the photothermomechanical stress mechanism. But in the visible and IR regions of the spectrum, the model involving the thermal explosion of the nonuniformly overheated tissue seems much more plausible.

Table 1.

Laser pulse parameters, optical properties of atherosclerotic plaque and characteristics of ablation by Nd:YAG laser harmonics

λ, nm	266	355	532	1064
μ_a, cm^{-1}	180	15	3.2	0.5
l_{eff}, μm	25	182	430	2800
Φ_{thr}, J/cm^2	0.8	4	24	45
δ	1.2	2.15	2.75	3.7
E_{thr}^{exp}, J/g	160	120	200	80
η, mg/J	1.2	0.21	0.13	0.06
l_{abl}, μm	48	27	52	27
Φ_{max}, J/cm^2	4	13	40	55
τ_p, ns	10	10	15	20
A^*	0.74	0.96	0.97	1
P_{thr}, bar	192	187	308	125
l_{sp}, μm	55	330	700	1180

5. Optimum Ablation Conditions

Based on the experiments performed and comparison drawn between the possible pulsed laser ablation mechanisms, one can summarize conditions conducive to the most efficient ablation of biological tissue as follows.

1. The wavelength of the laser pulse should be selected so as to ensure its maximum absorption by the tissue being treated, and the laser radiation should be delivered to the spot of interest via a light guide.

2. To decrease the thermal injury of the tissue surrounding the ablation crater, the laser pulse duration should be made shorter than the time it takes for heat to diffuse out of the irradiated tissue volume. On the other hand, the pulse should not be too short in order to avoid the development of shock stresses that may inflict mechanical damage to the surrounding tissue.

3. The photochemical mechanism of ablation of biological tissue can hardly be realized in the optical region of the spectrum, very short irradiation wavelength being required to achieve this. To realize the photothermal ablation mechanism, the irradiated layer should be heated to such a temperature as makes its lifetime till explosive boiling shorter than the time of heat diffusion from the irradiated volume.

References

1. R. Marcruz, J.R.M. Martins, A.S. Tupinambas, E.A. Lopes, A. Pena, V.B. Carvalho, E. Armelin, L.V. Delcourt: Possibilidades terapeuticas do raio laser em ateromas, *Arq. Bras. Cardiol.* **34**, 9-13, 1980.
2. S.L. Trokel, R. Srivanasan, B. Braren: Excimer laser surgery of the cornea, *Am. J. Ophthalmol.* **96**, 710-715, 1983
3. R.J. Lane, R. Linkster, J.J. Wynne, A. Torres, R.G. Geronemous: Ultraviolet-laser ablation of skin, *Lasers Surg. Med.* **4**, 201-206, 1984.
4. T. Kasuya, M. Tsukakoshi: Laser microirradiation of cells, *Laser Science and Technology*, ed. by V.S. Letokhov, C.V. Shank, Y.R. Shen, H. Walther, vol. 1, Harwood Academic Publishers, Chur-London-Paris-New York-Melbourne, 1988.
5. N.P. Furzikov: Different lasers for angioplasty: Thermooptical comparison, *IEEE J. Quant. Electron.* **QE-23(10)**, 1751-1755, 1987.
6. R.O. Esenaliev, A.A. Oraevsky, V.S. Letokhov: Laser ablation of atherosclerotic blood vessel tissue under various irradiation conditions, *IEEE Trans. Biomed. Eng.* **36(12)**, 1188-1194, 1989.

7. R. Srinivasan, V. Mayne-Banton: Self developing photoetching of polyethylene terephthalate films by far ultraviolet excimer laser radiation, *Appl. Phys. Lett.* 41(6), 576-578, 1982.

8. J.M. Isner, R.H. Clarke: The current status of lasers in the treatment of cardiovascular diseases, *IEEE J. Quant. Electron.* QE-20, 1406--1420, 1984.

9. R.Srinivasan: Ablation of polymers and biological tissue, *Science* **234**, 559-563, 1986.

10. Ya.B. Zel'dovich, Yu.P. Raizer: Physics of shock waves and high temperature hydrodynamics phenomena, Academic Press, New York, 1966.

11. A.A. Oraevsky, S.L. Jacques, R.A. Sauerbrey, G.H. Pettit, I.S. Saidi, F.K. Tittel, P.D. Henry: XeCl ablation of atherosclerotic tissue: Optical properties and energy pathways, *Lasers Surg. Med.*, **in press**, 1991.

12. *Radiationless processes in molecules and condensed phases*, ed. by F.K. Fong, Springer-Verlag, Berlin-Heidelberg-New York, vol. 15, p. 360, 1976.

13. A.K. Dmitriev, N.P. Furzikov: Mechanism of laser ablation for biotissues, *Sov. Phys. Izvestiya* 53(6), 1105-1110, 1989 (in Russian).

14. A.A. Oraevsky, R.O. Esenaliev, V.S. Letokhov: Kinetics and mechanism of laser ablation of atherosclerotic tissue under nanosecond and picosecond pulsed irradiation, *Lasers Life Sci.*, **to be published**, 1991.

15. M.S. Kitai, V.L. Popkov, V.A. Semchishen: Dynamics of UV excimer laser ablation of PMMA caused by mechanical stresses: Theory and experiment, *Macromol. Chem. Macromol. Symp.* 37, 257-267, 1990.

16. R.S. Dingus, R.J. Scammon: Grüneisen stress induced ablation of biological tissue, *SPIE*, in press, 1991.

17. V.P. Skripov, E.N. Sinitsin, P.A. Pavlov, G.V. Ermakov, G.N. Muratov, N.V. Bulanov, V.G. Baidakov: Thermophysical properties of liquids in metastable phase, Atomizdat, Moscow, pp. 9-24, 1980 (in Russian).

18. R.R. Anderson, H. Beck, U. Bruggeman, W. Farinelli, S.L. Jacques, J.A. Parrish: Pulsed photothermal radiometry in turbid media: Internal reflection of back - scattered radiation strongly influence the optical dosimetry, *Appl. Opt.* 28(2), 2256-2262, 1989.

19. M. Keijzer, S.L. Jacques, S.A. Prall, A.J. Welch: Light distributionin artery tissue: Monte-Carlo simulations for finite diameter laser beams, *Lasers Surg. Med.* 9, 148-154, 1989.

20. D.E. Grady: The spall strength of condensed matter, J. Mech. Phys. Solids, 36(3), 353-384, 1988.

ETCHING POLYMER FILMS WITH CONTINUOUS WAVE ULTRAVIOLET LASERS - THE PHOTOKINETIC EFFECT

R. Srinivasan

UVTech Associates, 2508 Dunning Drive

Yorktown Heights, NY 10598, U. S. A.

INTRODUCTION

The intense research activity over the past several years on the interaction of pulsed, ultraviolet laser radiation from an excimer laser (193nm, 248nm, 308nm or 351nm) with organic polymer surfaces [1] has been carried out exclusively at power densities greater than 1 MW/cm^2 using laser pulses of < 1μs pulse width (FWHM). The matrix shown in Figure 1 places these investigations in perspective. The width of a pulse from an excimer laser cannot be changed by more than a factor of ten by simple adjustments to the operating conditions. Considerable effort has been devoted to the design of novel circuits for excimer lasers which would produce pulses longer than those available from standard, commercial excimer lasers which typically deliver pulses of duration from 5ns to 35ns. Such attempts have resulted in the construction of an excimer laser with a pulse width of 1 μs at 308nm [2]. In its interactions with polymer surfaces, it produced results which were not exceptional when compared to results that were obtained with pulses in the 5-35ns range [3].

It is possible to produce pulsed, ultraviolet laser radiation with pulse widths > 1μs by chopping the beam from a continuous wave (cw) ion laser. Since these lasers are available with ultraviolet output from 0.1 watt to several watts, such pulses can be focussed down to a spot such that the power density approaches 1 MW/cm^2. In our investigations, we have used power densities which range from 10 - 100 kW/cm^2 and the

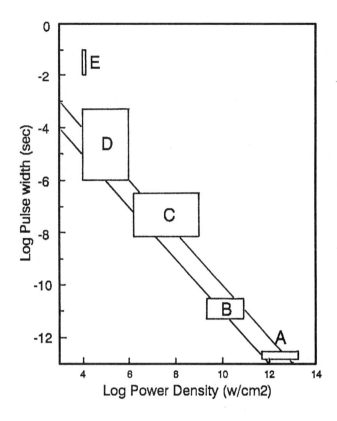

Figure 1. Matrix of power densities and pulse widths at which UV laser:polymer interactions have been studied. A, femtosecond pulses; B, picosecond pulses; C, nanosecond pulses; D, present work; E, Ref. 9. All of the possible conditions within each box have not been investigated. The diagonal lines correspond to constant fluences of 1 J/cm^2 (upper) and 0.1 J/cm^2 (lower) respectively.

pulse widths ranged from 10 μs to 1000 μs. The chopping of the beam was achieved by merely translating the surface past the laser spot as shown in Figure 2. The turntable on which a polymer film was placed was rotated at a controlled speed by means of an electric motor and a speed control. It is evident that the transit time that it takes the laser spot to sweep past an area equal to its own diameter constitutes a pulse width. Chopping the laser beam in this manner produced a novel and unexpected **Photokinetic Effect** which led to the etching of lines on a polymer surface with good definition and very little lateral thermal damage [4]. The relation of the power density - pulse width domain over which those results were obtained to past work using excimer lasers of various pulse widths is indicated in Figure 1.

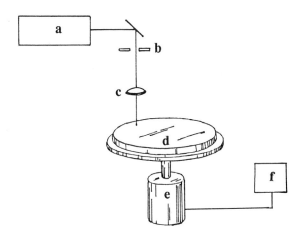

Figure 2. Schematic view of the apparatus used. a, Argon-ion laser; b, Shutter; c, Focussing lens; d, turntable; e, Drive motor; f, Motor control. The polymer film samples were mounted on the turntable and held in place with adhesive tape.

ETCHING POLYIMIDE (KAPTON™) FILMS [4]

Commercial argon ion lasers have output in the ultraviolet region from 275nm to 380nm. By the use of special mirrors it is possible to isolate the output at a single line (e.g., 275nm) or use the combined output of several lines over a narrow wavelength range (e.g., 300-330nm). It was possible to produce pulses which ranged in duration from 8 ms to 500ms by chopping this output with a camera shutter. When single pulses of these durations (350-380nm) at a power density of 1.9×10^4 W/cm^2 interacted with a **stationary** Kapton film, there was a black spot on the surface which was raised <u>above</u> the surface and whose diameter was greater than that of the laser spot. Even at 8ms exposure time, the blackening considerably exceeded the diameter of the laser spot. When the exposure time was >100ms the swelling of the surface of the film was large enough to distort the surrounding unexposed, unblackened surface. The film did not show any sign of being etched. At the same power density of the laser spot, when the turntable was rotated, the polymer was etched smoothly (i.e., without any distortion of the surface) and to a depth

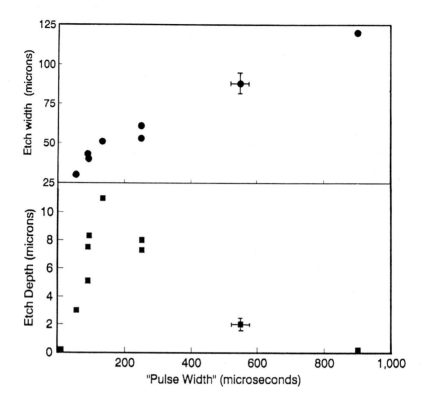

Figure 3. Etch depth and etch width of line etched on Kapton film by laser spot as a function of "pulse width". Wavelength, 350-380nm; Power density, 1.9×10^4 watts/cm^2. Nominal dia. of spot, 100 μm. Uncertainties in individual measurements are shown in data points at 550 μs.

of several microns. Figure 3 shows a plot of the etch depth and the width of the cut at the polymer surface (1 minute exposure) as a function of the "pulse width", the latter being varied by varying the speed of rotation of the turntable. The term "pulse width" will hereafter be placed within quotation marks in this article whenever it really means a transit time over a spot on a surface and does not pertain to a light pulse of a certain duration. The etched circles were all 5.0 (± 7%) cm in dia. There is clearly an optimum "pulse width" at which the etching action is a maximum. Since the total duration of the exposure for all of the circles was constant (i.e., 1 min), the total number of joules of light that was deposited on the surface was constant. At the short end of the "pulse width" range it is not surprising that as the "pulse width" tends to 1 μs, the etch depth starts to fall off from the optimum value because the regime of photoablation effectively begins at this "pulse width" and the power density that is used here is ~100-fold smaller than the threshold for photoablation [3, 5-7]. At the other limit, as the "pulse width" tends to 1 ms, the radiation merely pyrolyzes the film and the black product seems to limit the depth to which the radiation can penetrate the surface. This situation is similar to the static experiment that was mentioned at the beginning of this section. Even at a "pulse width" of 500µs, the bottom of the cut that is etched is irregular (steep hills and valleys) in contrast to the results at shorter "pulse widths" which lead to a tapered cross-section. The width of the cut at the surface is seen (Figure 3) to narrow progressively as the "pulse width" narrows from 500µs to < 100µs and is never wider than the laser spot itself which had a nominal diameter of 100µm. A moving spot evidently does not give rise to the thermal damage that is seen when the same total dose of light energy in joules is delivered to the surface in a single pulse. The extreme sensitivity of the etching process to the speed at which the laser spot moved over the surface of the polymer has been pointed out [4]. A 30% change in the "pulse width", under appropriate conditions, resulted in >250% change in the etch depth. A further point of contrast between the photoablation process that is brought about in this material by excimer laser pulses and the present process is that the latter does not cause ablation, i.e., the explosive ejection of the products - there is no audible report that accompanies it and there is little evidence of debris that is scattered with force.

ETCHING POLYMETHYL METHACRYLATE [8]

The etching of polymethyl methacrylate (=PMMA) surfaces by a moving spot of ultraviolet laser light offers some special insight into this etching phenomenon. In a study that was carried out in 1985, films of PMMA were etched by Bjorkholm and his co-workers [9] using a focussed, moving spot of laser radiation of 257.2 nm wavelength. They used a power density of ~10 kW/cm^2 at the work surface and translated the beam at speeds such that the "pulse widths" ranged from ~10ms to ~400ms. The conditions under which they operated are shown in Figure 1. The maximum thickness of the material that was etched was only 600nm. No energy density or intensity threshold for the etching

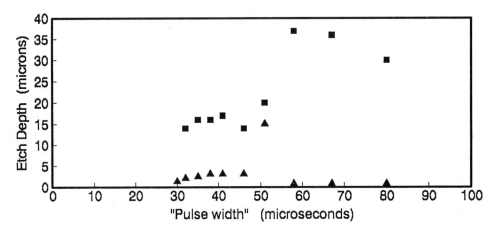

Figure 4. Etch depth as a function of "pulse width" in the etching of doped PMMA. Wavelength, 300-330 nm; number of pulses, 330 for all points; triangles, before solvent treatment; squares, after solvent treatment.

process (which the authors refer to as "ablation") was observed. The etch depth depended only on the deposited energy density and was independent of the rate of deposition, i.e., the process obeyed a reciprocity relationship. The data given above on the etching of Kapton show that there is a striking difference between etching with a moving ultraviolet laser spot in the "pulse width" range from 1 - 1000 μs (as used in photokinetic etching) and <1000 μs as used by these workers. Photokinetic etching shows that there is a failure of the reciprocity relationship and the etch depth is a maximum at one "pulse width". It was of interest to extend these observations to PMMA itself in order that a direct comparison between photokinetic etching and photoablation can be made on one hand and also compare the results that were obtained at "pulse widths" > 1ms to results that were obtained at shorter "pulse widths".

Figure 4 shows a plot of the etch depth as a function of the "pulse width" for a constant number of pulses when the surface of a sample of doped PMMA is etched by a moving beam of 300-330nm laser radiation. The spot size was nominally 60μm and the power density was 1.1 x 10^5 W/cm^2. The sample used was PMMA of molecular weight 140,000 and it was doped with 2% of Tinuvin^{TM}, a commercial material that is added to PMMA to stablize it towards degradation by ultraviolet light. The behavior of this exact material under pulsed, ultraviolet radiation from an excimer laser has been reported before [10].

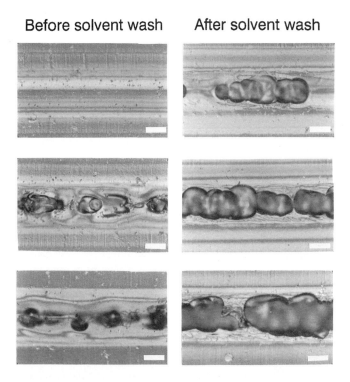

Figure 5. Optical microphotographs of surface of doped PMMA which was etched with cw ultraviolet laser light. These are three of the samples whose etch depths are recorded in Figure 4. The bar in every photo corresponds to 50 μm. The "pulse widths" correspond to **top**, 46 μs; **middle**, 51 μs; **bottom**, 67 μs.

The data in Figure 4 show that there is an optimum "pulse width" at which the etch depth is considerably greater than at "pulse widths" that are either shorter or longer. The abrupt transition from this optimum value to an absence of etching at a slightly longer pulse width was puzzling at first sight. However, when the PMMA sample was washed with a solvent mixture of methyl isobutyl ketone and isopropyl alcohol (3:1) which selectively dissolved out the photoproducts, the situation became clearer. Figure 5 shows photomicrographs of the surfaces in which lines have been etched at three different "pulse widths". The etching process appears to proceed by the unzipping of the polymer chain. The resulting evolution of volatile products pushes the partly molten material out of the cut. This molten material is readily dissolved by the solvent treatment. The "pulse width" has to be long enough to cause sufficient decomposition of the polymer so that the etching proceeds at a reasonable rate. At greater "pulse widths" so much molten material appears to be formed that it blocks the passage of light into the channel that is cut. When the etching is terminated, this molten material solidifies in and above the channel and is registered as a failure to cut. After treatment with a solvent, the actual depth to which the polymer surface has been etched can be measured. This is seen (Figure 4) to increase steadily with increasing "pulse width". A scanning electron microphotograph of a cut channel (Figure 6) shows that the bubbling of the material causes the bottom of the trench that is cut to be considerably distorted.

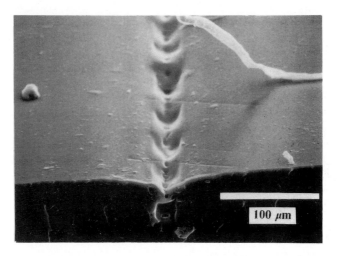

Figure 6. Scanning electron microphotograph of surface of doped-PMMA etched with cw ultraviolet laser (350-380nm) radiation. The white strip of material lying on the surface represents polymer that was removed by etching.

ETCHING OTHER POLYMERS

Photokinetic etching has been found to be effective on a variety of polymeric solids including synthetic polymers such as polyethylene terephthalate, epoxy polymers, and polycarbonates as well as natural polymers such as paper and animal muscle tissue [8]. In all these examples cw ultraviolet radiation in the wavelength range from 275nm to 380nm was used at a power density of 10 - 100 kW/cm^2.

CONCLUSIONS

The photokinetic etching effect which is seen in the interaction of a moving spot of cw ultraviolet laser radiation with the surface of an organic solid offers a new modality for the use of UV laser photons. It may develop into a practical method to cut a variety of polymer films to a precision of a few tens of microns with no detectable lateral thermal damage. Its potential use as a technology will complement the use of pulsed, excimer laser radiation for the same purpose.

It is too early to propose a mechanism for the photokinetic effect. There appears to be a role in the process for the photothermal effects of the kind that have been proposed in the past [5] for the action of pulsed, excimer lasers on the same substrates. The chemistry of the process seems to depend strongly on the composition of the material.

ACKNOWLEDGMENT

The author wishes to thank Coherent, Inc. of Palo Alto, California for their support of this research.

REFERENCES

1. R. Srinivasan and B. Braren, Chem. Rev., **89**, 1303 (1989).
2. R. S. Taylor and K. E. Leopold, J. Appl. Phys., **65**, 22 (1989).
3. R. S. Taylor, D. L. Singleton and G. Paraskevopoulos, Appl. Phys. Lett., **50**, 1779 (1987).
4. R. Srinivasan, Appl. Phys. Lett., **58**, (in press) 1991.
5. J. H. Brannon, J. R. Lankard, A. I. Baise, F. Burns and F. Kaufman, J. Appl. Phys., **58**, 2036 (1985).
6. G. Koren and J. T. C. Yeh, Appl. Phys. Lett., **44**, 1112 (1984).

7. R. Srinivasan, B. Braren and R. W. Dreyfus, J. Appl. Phys., **61**, 372 (1987).

8. R. Srinivasan, (to be published).

9. J. E. Bjorkholm, L. Eichner, J. C. White, R. E. Howard and H. G. Craighead, J. Appl. Phys., **58**, 2098 (1985).

10. R. Srinivasan and B. Braren, Appl. Phys., **A45**, 289 (1988).

MECHANISTIC INSIGHT IN THE LASER-PULSE SPUTTERING OF POLYMERS BY COMBINED PHOTOGRAPHY AND GAS-DYNAMIC ANALYSIS

Roger Kelly, Bodil Braren, and K.G. Casey
IBM Research Division, T.J. Watson Research Center, Yorktown Heights,
NY 10598, U.S.A.

ABSTRACT

The gas-dynamic effects which are so prominent in laser-pulse sputtering have three limiting cases. The one limit could be described as an unsteady adiabatic expansion (UAE) due to gas which escapes from a finite reservoir ("outflow" model). The second limit is that the particles are released from target surface, form a Knudsen layer (KL), and then enter a UAE ("KL-UAE" model). Closely related is the third limiting case, which allows for condensation at the surface ("condensation" model). Explicit photographs suggest that either the KL-UAE or the condensation model is relevant to laser-pulse sputtering of the polymer polymethylmethacrylate (PMMA).

GAS-DYNAMIC RESULTS

In connection with laser-pulse sputtering, especially of polymers, there is now an abundance of new information of two types: (a) real-time photographs have been made of those sputtered particles able to absorb or scatter light [1,2], (b) gas-dynamic descriptions are finally available for pulsed situations [3,4].

We would like to regard the sputtering mechanisms as residing in two "families". The one family describes the basic release process, in particular whether it is predominantly (i) electronic (i.e. due to short-lived excitation), (ii) chemical (i.e. due to long-lived excitation, for example broken bonds), or (iii) thermal (i.e. due to the vaporization of otherwise stable fragments). The other family describes the gas-dynamic interactions that take place consequent to the basic release process. There are again 3 possibilities. (i) There could be a UAE like that which occurs when the membrane confining a finite reservoir of gas is broken, a well-known problem since it describes some aspects of guns ("outflow" model). (ii) There could be the somewhat more complicated process in which particles are released, effectively layer-by-layer, from the surface and form first a KL and only then enter a UAE ("KL-UAE" model). A KL is the region extending about 3 mean free paths beyond a desorbing surface where, provided the particle density is high enough, the independently emitted particles come to equilibrium. In practice it is manifested by the system developing a positive flow velocity (u_K) corresponding to Mach number $M = u_K/a_K = 1$, where "a_K" is the sound speed at the KL boundary. When the pulse of release ends, u_K falls abruptly to zero. (iii) The third possibility is like the KL-UAE model but when the release pulse ends there is simultaneously flow towards the surface, leading to condensation, as well as flow away from the surface ("condensation" model). This is taken into account by having u_K take on a negative value, $-u_o$, at the KL boundary [5].

In some respects these three possibilities give similar results. (a) The flow pattern at first involves a particle density which decreases rapidly with distance from the target surface up to an expansion front (fig. 1, curves "0.5" and "1"). (b) At a critical time there is an abrupt or near-abrupt decrease in density at the target surface (fig. 1, curves "2" and "4"). But they also give different results in two respects. (c) The critical time for the density decrease at the target surface is bounded differently for the different models. For the outflow model the time is $t = \tau_s$, when the disturbance triggered by the outflow is reflected from the back of the reservoir and finally reaches the surface. Approximately, τ_s is given by

$$\tau_s = \text{(crater depth)/(sound speed)} \approx 150 \text{ ns}. \tag{1}$$

For the KL-UAE and condensation models the critical time is $t = \tau_r$, when the release process terminates, a time which could be as short as the laser pulse or as long as the time for excitation or heat to be retained. τ_r is not restricted as in eq. (1). (d) In the outflow model the flow pattern for $t > \tau_s$ involves two regions, with a low density near the target surface and then a region in which the density decreases rapidly to zero (like curves "2" and "4" of fig. 1 but without the part between the lines of contact (LOC)). The KL-UAE and condensation models have an additional region where the density shows a remarkable increase with distance (fig. 1, curves "2" and "4").

PHOTOGRAPHY

Considering UV laser pulse bombardment of PMMA, a fairly definite result relates to the question of whether the mechanism more nearly resembles the outflow, the KL-UAE, or the condensation model. We find that the experimental value of the time for density decrease is ~6 µs compared with the laser pulses having a length of ~20 ns (figs. 2). This immediately excludes the outflow model, where the characteristic time should be very much shorter (eq. (1)). In addition, a region in which the density increases with distance is a prominent feature in the photography even if some of the details suggest spurious imaging effects in which the axis is imaged preferentially (fig. 2b). Again the outflow model is excluded and the overall conclusion is that the laser-pulse sputtering of PMMA resembles more nearly what we term the KL-UAE or condensation model. (These cannot be distinguished with the information available.) We suggest that the reason for the outflow model not being relevant is that the absorption depth for PMMA is rather high (6-18 µm), so the density of energy deposition is low and the necessary massive bond breakage that would make outflow possible does not happen. Indeed, there is good evidence that for low fluences many pulses are needed to initiate laser sputtering with PMMA[1].

With a knowledge that the critical time is τ_r rather than τ_s and that it is of order 6 µs, some progress can be made with the question of whether the release process is better described as electronic, chemical, or thermal. It follows, from the knowledge that the thermal diffusivity of PMMA is $\kappa \approx 0.0007$ cm^2/s, that the heat diffusion distance, $2(\kappa t)^{1/2}$, for 6 µs is about 1 µm. Since this is smaller than the absorption depth (6-18 µm), heat retention will be an important effect. By contrast there is probably no role at all for energy retention in broken bonds, as the relevant time constant is macroscopic, perhaps even infinite. In our terminology the release mechanism is thus better described as thermal than chemical; whether or not there is an electronic component due to short-lived excitation cannot, on the other hand, be addressed at all with the information available.

REFERENCES

1. R. Srinivasan, B. Braren, and K.G. Casey, J. Appl. Phys. 68, 1842 (1990).
2. B. Braren, K.G. Casey, and R. Kelly, Appl. Phys. B (submitted).
3. R. Kelly, Nucl. Instr. Meth. B46, 441 (1990) and J. Chem. Phys. 92, 5047 (1990).
4. R. Kelly, Phys. Rev. (submitted).
5. T. Bergstrøm and T. Ytrehus, Phys. Fluids 27, 583 (1984).

FIGURES

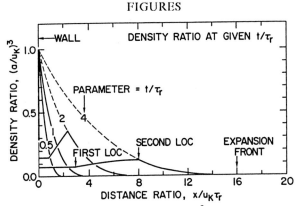

Fig. 1. Calculated values of density ratio, $\rho/\rho_K = (a/u_K)^3$, vs distance ratio, $x/u_K\tau_r$, appropriate to laser-pulse sputtering for the KL-UAE model. Here ρ is density, $\underline{a}$ is sound speed, and $\underline{x}$ is distance. For $t < \tau_r$ the particle density decreases rapidly with distance from the surface. At $t = \tau_r$ the previously porous wall at $x = 0$ is suddenly sealed and as a result the boundary condition changes abruptly from $u = u_K$ to $u = 0$. The flow breaks up into three regions separated by two lines of contact (LOC). The curves are for atoms with $\gamma = C_p/C_V = 5/3$.

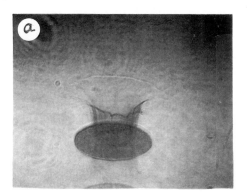 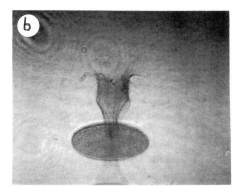

Fig. 2. PMMA targets with a thickness of 1 mm were exposed to single laser pulses (0.4 J/cm², 193 nm, ~20 ns, diameter = 780 µm, normal incidence) in air. The ejecta were photographed by firing above the target a second laser pulse (596 nm, ~1 ns, incidence at 30° from the target surface) with a known delay and thereby imaging both the ejecta and the shock wave (not visible in the photographs shown). All particles that either absorb or scatter are seen, although there appears to be an imaging threshold which causes the axis to be imaged preferentially (see (b)).

(a) <u>Delay of 3.4 µs.</u> The ejecta are seen at an oblique angle of incidence as they escape from the target surface. Note how the expansion of the sputtered particles is essentially one-dimensional, as assumed in the gas-dynamics of fig. 1. The surface of the target shows the darkening characteristic of surface damage.

(b) <u>Delay of 7.0 µs.</u> The outstanding feature here is the tendency for preferential imaging near the axis of the column of ejecta, a feature that we take to indicate an abrupt fall in density near the target surface combined with a slight radial density gradient. If there is an imaging threshold then the radial gradient will lead to loss of image for large radius. Important for the comparison with theory is that the region with imaging near the axis has progressed only part way up the column of ejecta, a situation resembling curves "2" and "4" of fig. 1.

Part IV

Biological Molecules

LASER DESORPTION AND MULTIPHOTON IONIZATION OF SOME SMALLER BIOMOLECULES: RECENT RESULTS AND PROSPECTS.

Claus Köster, Josef Lindner, Gary R. Kinsel, Jürgen Grotemeyer[*]
Institut für Physikalische und Theoretische Chemie
Technische Universität München
Lichtenbergstr. 4, D-8046 Garching, Germany

ABSTRACT

Some results are presented concerning investigations in laser desorption of neutral organic molecules by various laser wavelength into a supersonic jet followed by either single or multiphoton ionization. The results demonstrate that resonant absorption of photons in the desorbing materials must take place in order to observe high quality mass spectra. Non resonant laser desorption requires significantly higher laser powers than the resonant desorption. It is shown that in case of smaller biomolecules multiphoton ionization is superior to single photon ionization with 118 nm photons. Some applications of the multiphoton ionization on small protected nucleotides are shown and disscussed.

INTRODUCTION

The adoption of lasers with different wavelength, such as in the infra-red (IR), visible (VIS) and ultraviolett (UV) regime, to mass spectrometry yielded in a large number of applications. Beside the use in ion activation (photodissociation) [1] lasers have been introduced throughout the years also to ion production. First applications were dealing with simultaneous desorption and ionization [2]. In recent years this technique has found large interest in the desorption of large bioorganic ions, such as intact proteins, enzymes and nucleotides, from a matrix [3].

Alternative methods for ionization by laser radiation use either multiphoton absorption [4] or single photon ionization by vacuum ultraviolett photons [5] of already gas phase based molecules. Through its features **mu**ltiphoton **i**onization (MUPI) added some further opportunities to mass spectrometry. These are the possibilities of *soft ionization* with the formation of mainly molecular ions, *tuneable fragmentation* with an easy adjustment of the degree of fragmentation and *wavelength dependence* of the mass spectrum. Especially these superiour features of multiphoton ionization are demonstrated in Figure 2 by comparing the multiphoton and single photon ionization mass spectra of the tetrapeptide Leu-Trp-Met-Arg-AcOH.

For the successful investigations of non volatile biomolecules by light it is necessary to separate the desorption from the ionization in space and time. To reach this requirement different techniques have successfully been examined [6], which incorporate laser desorption [7] and liquid introduction of neutrals [8].

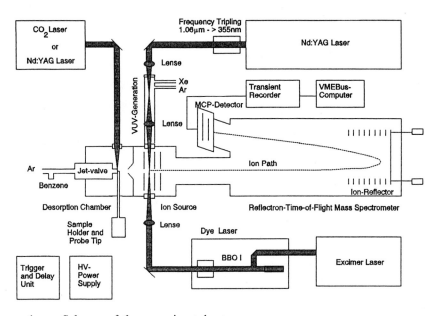

Figure 1   Scheme of the experimetal setup.

EXPERIMENTAL SETUP

The mass spectrometer system used for the investigations is a Bruker TOF 1 prototype with some home build modifications. It consists of three different parts: the laser desorption unit, the laser ionization source and the **R**eflectron-**T**ime-**o**f-**F**light (RETOF) mass spectrometer (shown in Figure 1).

In our experimental setup neutral molecules are transformed into gas phase through either laser desorption or other means separately from the ionization step. The laser desorption can be induced by either 10.6 μm radiation from a CO_2-laser with different pulse length ($22\mu s$ or 80 ns) or by the four wavelength of a small Nd:YAG laser (1064, 532, 355, 266 nm; puse length 5-10 ns). The ionization can occur either by multiphoton absorption of photons delivered from an excimer pumped dye laser system or by single VUV-photon absorption of photons produced by a conversion of 355 nm photons from a Spectra Physics Nd:YAG laser (DCR 3-D) [9]. The complete experimental setup is described in larger detail elsewhere [10].

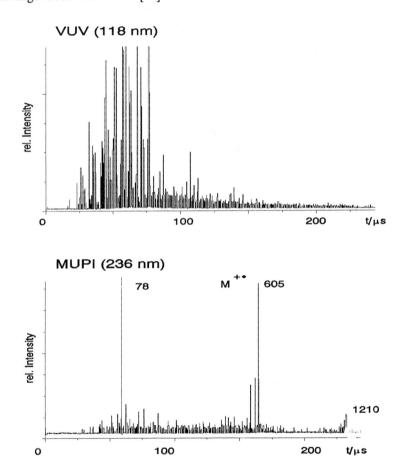

Figure 2 Comparison of VUV and MUPI-mass spectra of the tetrapeptide Leu-Trp-Met-Arg-AcOH.

INVESTIGATIONS OF THE LASER DESORPTION OF NEUTRALS

Investigations in the laser desorption of neutrals and the mechanisms leading to gas phase based molecules can be divided into several questions. High importance gains the question of the influence of laser pulse length, laser irridiance, laser wavelength and the influence of matrix assistence to the desorption of neutrals [11]. In this short paper no complete answers to these question nor all facettes of the results can be given or discussed.

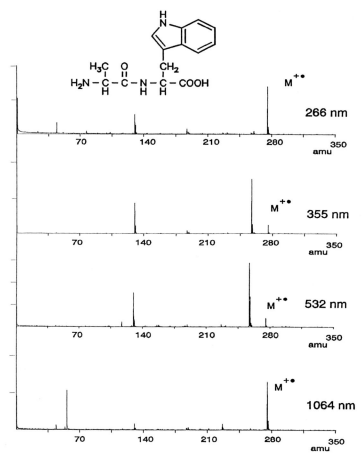

Figure 3 MUPI-mass spectra of the dipeptide Alanyl-tryptophan produced by various desorption laser wavelength.

A further problem is dealing with the question whether mainly intact neutrals are desorbed or the impact of the laser beam on the surface is leading to photo products and therefore to fragmentation of the sample before ionization. Due to the fact that the experimental setup used allows a nearly simultaneous measurement of VUV and MUPI mass spectra of laser desorbed molecules, one can compare the effects of both ionization techniques with respect to the desorption of neutral molecules. Figure 2 demonstrates that the desorption of larger molecules by a CO_2-laser is accompanied by a major fraction of photoproduced neutrals in the desorption step. But it should be noted that the strong fragmentation in the VUV-mass spectrum of Leu-Trp-Met-Arg-AcOH can not be explained alone by the desorption process, but must also be accounted to the ionization procedure. This is based on the fact of the large excess energy delivered to the molecular ion by the VUV-photon. But one can clearly state that the desorption of neutrals from a surface is accompanied by a heavy fragmentation of neutral molecules. It should be noted that the amount of fragmentation is linked to the thermal and photochemical stability of the sample desorbed [12].

Clearly the comparison of both mass spectra shows the superiority of multiphoton ionization. In the MUPI mass spectrum only two major ion signals are found which can be identified as the molecular ion of the tetrapeptide (m/z 605) and that of benzene (m/z 78) used as an internal mass marker.

Generally accepted in literature is the conception that the use of photons with high energy (e.g. UV-photons) in laser desorption leads to ions and to a less intense appearance of neutral molecules, while longer wavelength (e.g. IR-photons) produce more neutrals instead of ions from a surface. We have therefore investigated the wavelength dependence of the desorption of various organic molecules.

In Figure 3 the MUPI mass spectra of simple dipeptide, alanyl-tryptophan, are shown which are obtained by the four wavelength of the Nd:YAG-laser. Obviously this compound can be desorbed as neutral from the surface by all wavelength. Here it should be noted that the useage of all wavelength ranging from the near-infra-red to the ultraviolett is closely linked to the stability of the sample. Alanyl-Tryptophan shows a high stability to thermal as well as photochemical stress. But turning to even longer wavelength, i.e. 532 nm and 1064 nm, as shown in Figure 3 and in Table 1, yields not only in the production of pure molecular ions, but give also rise to intense signals of metal-atoms desorbed from the probe tip. As seen from Table 1 desorption of molecules at these wavelength is only possible by a high intense, focussed laser beam on the probe tip. Contrary to the other wavelength, desorption occurs here after an induction time of a couple of desorption laser shots. This behaviour can be explained through the fact that first the sample is removed from the metal surface without producing intact molecules. After this induction time the substrate is heated by the laser leading to the measureable desorption of metal atoms and intact neutral molecules. This process can be described

Compound	Wavelength of desorption laser			
	266 nm	355 nm	532 nm	1064 nm
Pentacen	3 (21000)	18 (1400)	10 (300)	14 (0)
Estradiol	9 (1000)	20 (0)	>250^f(Fe+) (0)	>320^f(Fe+) (0)
ALA-TRP	7 (4400)	61^f (0)	>250^f(Fe+) (0)	>320^f(Fe+) (0)
LEU-PHE	14 (150)	220^f(Fe+) (0)	n.s. (0)	n.s. (0)

^{a} Absorptivities measured in methanol (L mol^{-1} cm^{-1}). ^{f} Desorption thresholds measured using focussed LD conditions. (Fe+): Appearance of Fe$^+$ ions. n.s. Signals of post ionized neutrals not observed.

Table 1 Measured threshold irridiances of neutral molecules (values in MW/cm^2).

as a thermal desorption or, since here the laser radiation is not absorbed by the sample, as a non resonant desorption process. At laser wavelength, where only intact molecules without addition of metal atoms are desorbed, a resonant absorption process of photons in the molecules on the surface must been occured as already discussed for the desorption of ions by laser radiation [13].

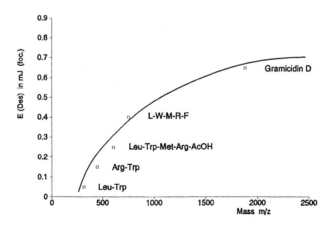

Figure 4 Desorption irridiance for different peptides at 266nm desorption wavelength versus the mass (ionization wavelength 255 nm).

This idea of resonant desorption versus non-resonant desorption is supported by the measured desorption thresholds and the absorptivities of the molecule in solution given also in Table 1. In cases where the absorptivity is neglectable and also the thermal stress of the molecules is very low, no desorption of intact molecules could be detected. This demonstrates that a thermal desorption model is not appropriate to describe the volatilization of molecules from a surface by laser radiation.

Furthermore the results demonstrate that other than the IR-wavelength can be effectively used for the desorption of neutrals, notably the 266 nm wavelength. Comparing the different desorption wavelength it should be noted, that the amount of desorbed material at 266 nm is in the same order than by the desorption with the CO_2-laser.

Figure 4 lists the results of the threshold measurements at 266 nm laser desorption for a series of tryptophan-containing peptides with increasing mass. Again it was possible to obtain mass spectra of post-ionized molecules without creating any abundant Fe^+ signals. Apparently the irridiance necessary for desorption is increasing with the mass for smaller and medium sized molecules, but stays even for larger molecules in the regime of resonant absorption. It should be noted that the desorption irridiance is not increasing linear, but raise to an upper limit thus again in accordance to theoretical considerations of resonant absorption of photons in bulk layers.

MULTIPHOTON IONIZATION OF SMALL NUCLEOTIDES

Details of the multiphoton ionization and/or fragmentation procedure have been published throughout the years in greater detail [14]. During the last years MUPI-mass spectra have been presented for a large number of substances, such as drugs, peptides [15], nucleosides, nucleotides [16], aromatic compounds and others [17].

As a result from investigations of protected nucleotides used in the nucleotide synthesis the mass spectra of two fully protected, isomeric deoxyribonucleoside 3'-phosphates, d(CpAp) and d(ApCp) are shown in Figure 5. The ionizing wavelength was tuned to the maximum of the molecular ion yield, which was found to be at 242.5 nm. In both mass spectra of these isomeric compounds the molecular ion is formed with high intensity at mass 1405. As seen from the shorthand structures the intense signals found in the mass spectra can easily be used for the unambiguous determination of the sequence in these dinucleotides. Breakdown of the bond between the sugar moeity and the intermediate phosphate group is leading to the signals at mass m/z 615 in case of the

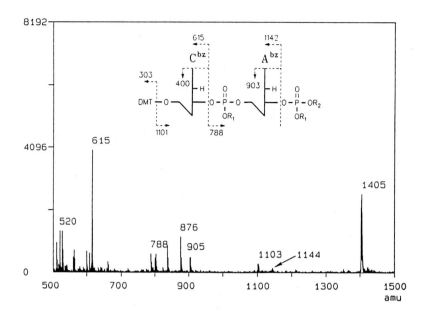

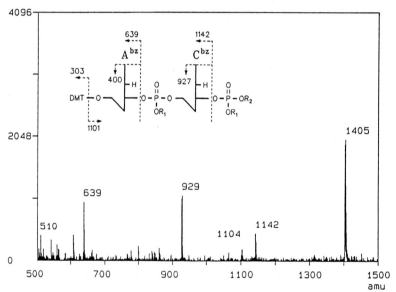

Figure 5 MUPI-mass spectra of the protected dinucleotides d(CpAp) and d(ApCp) (MW 1405 Da). Inserts show the main fragmentation reactions.

d(CpAp) nucleotide, while the isomeric structure d(ApCp) yield in the signal of m/z 639. These signals identify the 5'-end of the chain. The counter ion to this fragmentation reaction is only observed in case of the d(CpAp) and found at mass m/z 788. It should be noted that these structural important fragmentations are observed in all possible isomeric 16 nucleotides [16].

CONCLUSION

It has been shown that the combination of laser desorption and multiphoton or single VUV-photon ionization yields in results which can lead to a better understanding of the desorption process of materials from bulk layers. The results presented allow to distinguish between different processes in the laser desorption of neutrals. It has also been shown that beside IR-radiation also other wavelength of the spectrum can be used for the desorption of materials as long as the sample absorbs the radiation. As a result, the most efficient method for desorbing and detecting neutral intact molecules of organic and in particular, of larger, thermally labile compounds, is using resonant laser desorption and resonant laser ionization.

ACKNOWLEDGEMENT

This work was supported by grants from the Deutsche Forschungsgemeinschaft (GR 917/1-2 und GR 917/6-1) and the Bundesministerium für Forschung und Technologie (13N5307). Dr. G.R. Kinsel gratefully acknowledge a Alexander von Humbolt fellowship.

REFERENCES

1. For a recent overview see : Busch K.L., Glish G.L., McLuckey S.A., *Mass Spectrometry/Mass Spectrometry*, VCH Publisher Inc., Weinheim, New York, 1988.

2. Hillenkamp F., Unsöld E., Kaufmann R., Nitsche R.; *Appl.Phys.* **8,** 341 (1975).

3. a) Karas M., Hillenkamp F.; *Anal. Chem.* **60,** 2299 (1989).
   b) Beavis R.C., Chait B.T.; *Rapid Commun.Mass Spectrom.* **3,** 233 (1989).
   c) Salehpour M., Perera I., Kjellberg J., Islamian M.A., Hakansson P., Sundqvist, B.U.R.; *Rapid Commun.Mass Spectrom.* **3,** 259 (1989).

4. a) Lubman D.M.; *Mass Spectrom.Rev.* **7,** 535, 559 (1988).
   b) Grotemeyer J., Schlag E.W.; *Angew.Chem.Int.Ed.Engl.* **27** 447 (1988).

5. a) Schühle U., Pallix J.B., Becker C.H., *J.Vac.Sci.Technol.* **A6,** 936 (1988);
   b) Becker C.H., *J.Vac.Sci.Technol.* **A5,** 1181 (1987).
   c) Schühle U., Pallix J.B., Becker C.H., *J.Am.Chem.Soc.* **110,** 2323 (1988).

6. Köster C., Dey M., Lindner J., Grotemeyer J.; *Proc. 2nd International Meeting "Spectroscopy Across The Spectrum"* A.M.C. Davis (Ed.), The Royal Society of Chemistry, London, (1990) in press.

7. a) Boesl U., Grotemeyer J., Walter K., Schlag E.W.; Org.Mass Spectrom. **21,** 645 (1986).
   b) Kinsel G.R., Lindner J., Grotemeyer J., Schlag E.W.; J.Chem.Phys. (1991) in press.

8. Köster C., Grotemeyer J., Schlag E.W.; *Z.Naturforsch.* **45a,** 1285 (1990).

9. a) Van Bramer S.E., Johnston M.V.; *J.Am.Mass Spectrom.Soc.* **1,** 419 (1990).
   b) Kung A.H., Young J.F., Harris S.E.; *Appl.Phys.Lett.* **22,** 301 (1973).

10. Boesl U., Grotemeyer J., Walter K. and Schlag E.W.; *Anal. Instrum.* **16** 151 (1987).

11. Beavis R.C., Lindner J., Grotemeyer J., Schlag E.W.; *Chem.Phys.Lett.* **146,** 310 (1988).

12. C. Köster, J. Grotemeyer; unpublished results.

13. a) Hillenkamp F., Karas M.; *Int.J.Mass Spectrom.Ion Proc.* **69,** 265 (1986) and literature cited there.
    b) Karas M., Bachmann D., Bahr U., Hillenkamp F.; *Int.J.Mass Spectrom.Ion Proc.* **78,** 53 (1987).

14. see ref. [4]; Grotemeyer J., Schlag E.W.; *Acc.Chem.Res.* **22,** 399 (1989).

15. Grotemeyer J., Schlag E.W.; *Org.Mass Spectrom.* **23,** 388 (1988).

16. a) Lindner J., Grotemeyer J., Schlag E.W.; *Int.J.Mass Spectrom.* **100,** 267 (1990).
    b) Lindner J., Grotemeyer J.; *J.Mol.Struc.* (1991) in press.

17. Grotemeyer J., Lindner J., Köster C., Schlag E.W.; J.Mol.Struc., **217,** 51 (1990).

MATRIX-ASSISTED LASER DESORPTION AND IONIZATION OF BIOMOLECULES

Brian T. Chait and Ronald C. Beavis
The Rockefeller University
New York, N.Y. USA 10021

Recent developments in the volatilization and ionization of large molecules using matrix-assisted ultraviolet laser desorption have made it possible to produce intact protonated molecule ions from proteins with molecular masses ranging from a few thousand to greater than 100,000 mass units[1].

In the paper presented at the workshop, we described the construction and performance of a linear time-of-flight mass spectrometer with improved performance for the measurement of proteins and other biomolecules. Details on this instrument and its performance can be found in references 2-5. The specifications of the present instrument are summarized below:

mass range	1 to greater than 300,000 u
resolution	300 - 500 Full width half maximum
mass accuracy	0.01% for resolved peaks
sensitivity	< 1 pmol
universality	Spectra have been obtained from more than 200 different proteins including heavily glycosylated proteins and membrane proteins (see example given in Fig. 1).

During experiments designed to elucidate the detailed role of the matrix and its effect upon the mass spectrum, we discovered a number of new matrix materials[2-5] with improved properties compared with the nicotinic acid, which was the matrix used by the originators of the matrix-assisted laser desorption technique[1]. One class of these new matrix materials - the cinnamic acid derivatives sinapic acid, ferulic acid, and caffeic acid - was found to have especially favourable properties:

(i) Spectra obtained from nicotinic acid were found to exhibit ion peaks showing adduction to the protein of a large number of photochemically generated products from the matrix[2]. This adduction caused a considerable lowering of the quality of the spectra and the information that could be extracted therefrom. The cinnamic acid derivatives produced spectra exhibiting much lower levels of photochemically generated adduction products and were therefore of much higher quality[3,5].

(ii) The spectra from nicotinic acid were obtained with laser irradiation from a frequency quadrupoled Nd(YAG) laser giving a wavelength of 266 nm. Because the cinnamic acid derivatives absorb strongly at longer wavelengths (to greater than 350 nm)[4], it is also possible to obtain spectra with a much less expensive, low power nitrogen laser.

(iii) The cinnamic acid derivatives showed an unprecedented ability to produce high quality spectra from complex mixtures of peptides and proteins in the face of large concentrations of non-proteinaceous impurities such as salts, buffers, and lipids[6]. It has even proved possible to obtain informative mass spectra from unpurified or partially purified biological fluids[6,7].

A number of other fundamental features of matrix-assisted laser desorption were described. These include:

(a) The finding that a significant proportion of the fully accelerated intact protein ions appear to undergo metastable decomposition in the flight-tube of the mass spectrometer[8], indicating that the proteins are vibrationally excited during the desorption/ ionization and ion acceleration process.

(b) The finding that the initial velocity of polypeptide ions (prior to ion acceleration by externally applied electric fields) is the same (750 m/s) for polypeptide ions ranging in mass from 1,000 to 16,000 u[9], indicating that the polypeptide ions are entrained in a supersonic expansion of the matrix molecules after the laser ablation event.

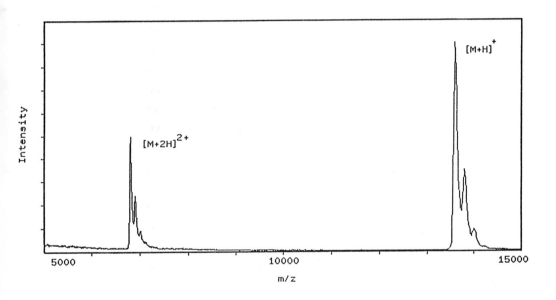

Fig. 1. Positive matrix-assisted laser desorption mass spectrum of phospholipase A_2 from the venom of *Crotalus attrox*. 1 pmol of protein inserted into the mass spectrometer .
Measured molecular mass = 13,582
Calculated molecular mass = 13,581.

REFERENCES

1. M. Karas and F. Hillenkamp, "Laser Desorption Ionization of Proteins with Molecular Masses Exceeding 10,000 Daltons" Anal. Chem. __60__ 2299 (1988).

2. R.C. Beavis and B.T. Chait, "Factors Affecting the Ultraviolet Laser Desorption of Proteins" Rapid Commun. Mass Spectrom. __3__ 233 (1989).

3. R.C. Beavis and B.T. Chait, "Cinnamic Acid Derivatives as Matrices for Ultraviolet Laser Desorption Mass Spectrometry of Proteins" Rapid Commun. Mass Spectrom. __3__ 432 (1989).

4. R.C. Beavis and B.T. Chait, "Matrix-Assisted Laser-Desorption Mass Spectrometry Using 355nm Radiation" Rapid Commun. Mass Spectrom. __3__ 436 (1989).

5. R.C. Beavis and B.T. Chait "High-Accuracy Molecular Mass Determination of Proteins Using Matrix-Assisted Laser Desorption Mass Spectrometry" Anal. Chem. __62__ 1836 (1990).

6. R.C. Beavis and B.T. Chait, "Rapid, Sensitive Analysis of Protein Mixtures by Mass Spectrometry" Proc. Natl. Acad. Sci. [USA] __87__ 6873 (1990).

7. D.E. Goldberg, A.F.G. Slater, R. Beavis, B.T. Chait, A. Cerami and G.B. Henderson, "Hemoglobin Degradation in the Human Malaria Pathogen *Plasmodium Falciparum*: A Catabolic Pathway Initiated by a Specific Aspartic Protease" J. Exp. Med. __174__ 961 (1991).

8. R.C. Beavis and B.T. Chait, "Investigation of the Matrix Isolated (UV) Laser Induced Polymer Sublimation Using a Time-of-Flight Mass Spectrometer" NATO ASI Series, W. Ens and K.G. Standing (Eds.), 1991. In press.

9. R.C. Beavis and B.T. Chait, "Velocity Distributions of Intact High Mass Polypeptide Molecule Ions Produced by Matrix Assisted Laser Desorption" Chem. Phys. Lett. 1991. In press

LASER ABLATION OF INTACT MASSIVE BIOMOLECULES

Peter Williams, David Schieltz, Cong-Wen Luo, Robert M. Thomas,
and Randall W. Nelson[1]

Department of Chemistry, Arizona State University, Tempe, AZ 85287-1604

Overview. The processes of separation, mass measurement and sensitive detection of large biomolecules (proteins, oligonucleotides, oligosaccharides) are of great importance in modern biochemical research, but the techniques in use at the present time, mainly based on the hindered electromigration of these molecules in gel materials, are relatively slow and have limited resolution. These functions - separation, mass measurement and detection -- are just those for which modern mass spectrometry has great power and speed. It has therefore been of great interest to develop ways to characterize large biomolecules mass spectrometrically. The fundamental difficulty that is encountered is that mass spectrometry requires *gas-phase* ions. The term "gas-phase" somewhat misleadingly suggests thermally-induced reversible transitions across a phase boundary; for large biomolecules, decomposition is the result of such thermal excursions. In fact, mass spectrometry requires simply a molecule launched on a free ballistic trajectory in space. As particle size increases, it is increasingly possible to achieve such trajectories mechanically. The most familiar analog to the processes now under development for the production of massive gas-phase ions is a sneeze, which through a high-velocity gas flow projects into the atmosphere an aerosol containing intact massive virus particles. The volatilization techniques currently exciting great interest in the mass spectrometry community work in analogous fashion. We discuss here one such technique,

---

[1] Present address: Vestec Corp., 9299 Kirby Dr., Houston, TX 77054

in which pulsed laser ablation produces a high-velocity gas jet to propel intact large biomolecules into the gas phase.

Pulsed laser ablation techniques. There are several major difficulties with pulsed laser ablation of intact large molecules. Large molecules may be bound to substrates, or neighboring molecules, by numerous hydrogen bonds or extensive Van der Waals interactions, the aggregate strength of which can be comparable with, or even much greater than, covalent bond strengths within the molecule. The difficulty of breaking all of these bonds simultaneously without breaking a single covalent bond in the molecule is extreme. Further, simply heating the molecule produces incoherent, random motion, not the required coherent motion of the molecule as a whole away from the surface. If ions are required to be produced in the ablating laser pulse rather than in a subsequent ionization process following volatilization, then ionization mechanisms are an additional concern.

Potentially general solutions to these problems are now available. The key feature is the embedding of the target molecules in a matrix which will act as a propellant, and may also modify the heating of the molecules and play a role in ionization. When the matrix material is chosen correctly, its explosive decomposition or volatilization produces a high-velocity gas jet directed away from the target surface which propels the embedded biomolecules on the desired ballistic trajectories. The problem of target species degradation is addressed by incorporating into the system a chromophore into which the laser energy is initially coupled. This can take the form of a fine colloidal metal suspension /1/, a strong UV-absorbing compound such as nicotinic acid /2/ or sinapic acid /3/, or, in our work, a metal substrate underlying a thin, transparent target film /4/. Thermal or photolytic decomposition of the chromophore material itself provides the high-velocity gas jet in the work of Hillenkamp et al /2/ and Chait and Beavis /3/. In the work of Tanaka et al /1/ and our work /4/, explosive volatilization of a solvent -- glycerol /1/ or water /4/ -- heated by contact with the irradiated chromophore is operative.

The problem of selectively breaking the bonds binding solvent or chromophore adducts without breaking structural bonds in the target molecule has not been addressed directly, but such breakage appears to be a fortunate consequence of the overall process.

Inevitably, it seems, the target molecules must be heated to some extent through contact with the heated matrix material and/or direct excitation by the laser, and this seems to be sufficient to cause the molecules to shed excess solvent molecules. These heated molecules must then survive flight through a mass spectrometer without undergoing unimolecular decomposition. Two effects diminish the probability of decomposition. First, the sheer complexity of a large molecule offers a large number of normal modes into which energy is partitioned, and the statistical probability of concentrating enough energy into one particular bond to rupture it is low. Second, and probably more important, a consequence of the explosive volatilization process is that the resulting gas plume acts as a supersonic jet, producing significant cooling of the initially excited molecules /4/.

Ionization and detection. In each of the three processes mentioned above, a fraction of the ablated large molecules are found to be ionized. There is evidence in the UV matrix studies that ions are formed by attachment of ionized fragments of the matrix molecules /2/. In our work, ablating with a tunable dye laser operating in the visible region of the spectrum, ionization efficiency increased when the laser was tuned into resonance with the $3s^2S - 3p^2P$ transition in atomic sodium /5/, and we speculated that the observed molecular ions of oligonucleotides and proteins were sodium ion adducts. Sodium ions could have been produced from sodium atoms in the ablated vapor plume by multiphoton ionization, or through resonance-enhanced collisional ionization processes as suggested by Yeung /6/. By analogy, similar gas-phase photoionization processes might be occurring in the UV-matrix approaches.

Detection of massive ions is limited because extremely massive ions move too slowly, even accelerated to several tens of kiloelectron volts, to eject secondary electrons from an electron multiplier surface. It appears that molecular ions more massive than ~ 30 kDa are detected by secondary *ion* rather than secondary electron emission /7/. Because of the twin complications of ionization and detection of large molecules, results obtained using mass spectrometry can be ambiguous: failure to observe molecular ions under a given set of conditions may not signal a failure of ablation, but simply that the molecules were not ionized or that the ions formed were not detectable. We have made extensive use of a simple collection technique to assess the efficiency of ablation separate from mass spectrometer concerns /4/. With this approach, we can evaluate the supersonic jet character of the ablated plume, by determining the angular distribution of ablated material on the collector, and can assess the survival or destruction of the collected sample molecules using gel electrophoresis. We discuss here briefly some examples of this approach.

Upper mass limit. It is interesting to inquire how heavy a molecule could be volatilized using these laser techniques. It seems possible that the only limits to ablation are set by the mechanical strength of the large target molecules. The upper limit of ablation of intact DNA molecules was assessed using a frozen thin film of an aqueous solution of a restriction enzyme digest (*Hin*dIII digest of *lambda* phage DNA). This digest contained DNA molecules ranging in molecular weight from ~ 80 kDa up to ~ 15 MDa. The ^{32}P-labelled DNA was ablated from a frozen thin solution film on a copper substrate, and analysed using gel electrophoresis. Fig. 1 shows the size distribution of ablated material superimposed on that of the original mixture. Molecules

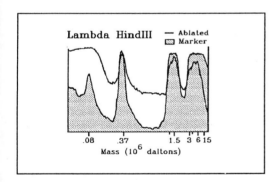

Figure 1. Line scans of electrophoresis autoradiograms of HindIII restriction enzyme digest of Lambda phage DNA. Shaded region: starting material; Line: material ablated from thin layer of frozen aqueous buffer solution on a copper substrate

up to ~ 6 MDa (~ 9,400 kilobase-pairs) were detectable in the ablated deposit, but there is no evidence that the largest molecules (~ 15 MDa, ~ 25 kilobase-pairs) survived. This is understandable from a mechanical viewpoint: DNA molecules of this size are about 8 μm long, and are highly prone to breakage simply by hydrodynamic shear forces. This work indicates that, if the twin problems of ionization and efficient detection could be overcome, mass spectrometric characterization of nucleic acid molecules in the megadalton range might be possible.

Other chromophores. The ablation process outlined above (energy coupling via highly absorbent chromophore, ablation via matrix explosion) should work quite generally. We have begun to look into possible chromophores in the visible region of the spectrum. These should be compatible with our frozen solution approach which we feel gives the best compatibility with biochemical processes, and the possibility to tailor molecular conformation (via pH or added solutes) and ionization efficiency (via resonant ablative ionization of a chosen solute). Incorporating the chromophore in the propellant matrix would be useful in minimizing the ablated sample thickness, and thus the instantaneous gas load following ablation, and in obviating the need to move the sample repeatedly, since numerous laser shots should be possible on a given area given a chromophore with sufficiently high extinction coefficient. We have looked initially at

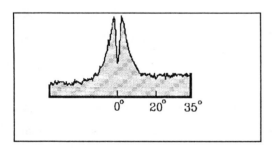

Figure 2. Densitometer line scan through deposit from laser-ablated frozen black ink droplet. The angular scale is approximate. The dip in the center is due to a hole punched to allow laser access to the sample.

dyes, and in particular at black inks. For ink, white paper makes a simple collector on which to visualize the angular distribution of the ablated material. Fig. 2 shows an image of an ablated ink deposit on a paper collector. The deposit is clearly strongly forward-peaked, reminiscent of our earlier thin film target results /4/; 80-90% of the ablated material falls within a half-angle of ~ 10°, which corresponds to an angular distribution varying roughly as $\cos^{50}\theta$. Following Kelly /8/, this corresponds to a Mach number of ~ 5 in the expansion.

High Mach numbers suggest that cooling should be efficient and that ablated large molecules should have a good chance of survival. We have evaluated the latter probability using the small protein lysozyme (molecular weight ~ 14,500 Da), which is somewhat easier to work with than nucleic acid samples. Fig. 3 shows a

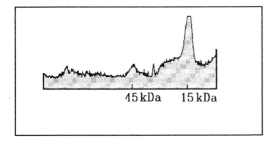

Figure 3. Line scan through electrophoresis assay for lysozyme laser-ablated from frozen buffer solution with black ink chromophore.

videodensitometer line scan of a gel electrophoresis assay of lysozyme ablated from a frozen solution using a black ink chromophore. Not only does the lysozyme survive intact, but there is evidence that a 45 kDa protein contaminant in the starting mixture also was ablated intact. These results are a strong indication that the matrix-assisted ablation technique is remarkably tolerant of variations in the explosive matrix and the chromophore, and suggest that it should be possible to tailor combinations of these two to achieve optimum compatibility with the chemistry of the material to be studied, with the operational demands of the mass spectrometer, and with the available laser.

Acknowledgement. Our work is supported by the Department of Energy Human Genome Program under Grant DE-FG02-91ER61127, and by NSF Grant CHE 9006939.

References

1. K. Tanaka, H. Waki, Y. Ido, S. Akita, Y. Yoshida and T. Yoshida, Rapid Commun. Mass Spectrom. **2**, 151 (1988)
2. M. Karas, D. Bachman, U. Bahr and F. Hillenkamp, Int. J. Mass Spectrom. Ion Processes **78** 53 (1987)
3. R. Beavis and B. Chait, Rapid Commun. Mass Spectrom. **3**, 233 (1989)
4. R.W. Nelson, M.J. Rainbow, D.E. Lohr and P. Williams, Science **246**, 1585 (1989)
5. R.W. Nelson, R.M. Thomas and P. Williams, Rapid Commun. Mass Spectrom. **4**, 348 (1990)
6. H-M Ping and E.S. Yeung, Anal. Chem. **61**, 2546 (1989)
7. B. Spengler, D. Kirsch, R. Kaufmann, M. Karas, F. Hillenkamp and U. Giessmann, Rapid Commun. Mass Spectrom. **4**, 301 (1990)
8. R. Kelly and R.W. Dreyfus, Nucl. Instrum. Methods Phys. Res. B. **32**, 341, (1988)

APPLICATIONS OF MATRIX-ASSISTED LASER DESORPTION FOURIER TRANSFORM MASS SPECTROMETRY FOR BIOMOLECULES[*]

R. L. Hettich and M. V. Buchanan
Analytical Chemistry Division, Oak Ridge National Laboratory
Oak Ridge, TN 37831-6120

INTRODUCTION

The recent advances in matrix-assisted laser desorption (LD) mass spectrometry have indicated tremendous potential for the analysis of large biomolecules (MW > 100,000 Daltons) such as peptides and oligonucleotides.[1,2] To achieve soft ionization for these large molecules, a matrix compound such as nicotinic acid is mixed with a small amount of the biological analyte. This matrix helps absorb the laser radiation and subsequently reduce the fragmentation of the analyte ions. Although most of the matrix-assisted laser desorption development has employed time-of-flight mass spectrometers, the objective of our research is to develop and refine laser desorption Fourier transform mass spectrometry (FTMS) for the detailed structural characterization of biomolecules at the isomeric level. This paper summarizes the development of matrix-assisted laser desorption FTMS for the examination of normal and modified oligonucleotides, including investigation of experimental parameters which influence the useable FTMS mass range.

EXPERIMENTAL

All experiments were performed with an Extrel FTMS-2000 Fourier transform mass spectrometry equipped with a Quanta Ray DCR-11 pulsed Nd:YAG laser. The 266 nm laser radiation was focussed onto the sample disk to provide power densities of

---

[*] Sponsored by the U.S. Department of Energy, Office of Health and Environmental Research, contract DE-AC05-84OR21400 with Martin Marietta Energy Systems, Inc.

approximately 10^6-10^7 W/cm^2. The experiment was initiated by firing the laser to desorb and simultaneously ionize the sample. The resulting positive or negative ions were then trapped in the FTMS cell where they could be manipulated and examined in detail. A complete mass spectrum could be obtained from one laser shot, although in most cases several laser shots were signal averaged to provide better spectra. Following ejection of matrix ions, high resolution and accurate mass measurements were used to determine the empirical formulas of the analyte ions. Additional structural information was obtained by collisional dissociation and ion-molecule reactions and in most cases provided the information necessary to differentiate isomers. High resolution ion excitation and ejection can be accomplished with the "Stored Waveform Inverse Fourier Transform" (SWIFT^R) ion manipulation technique.

RESULTS

Following protocols developed previously for matrix-assisted laser desorption FTMS of small peptides [3], we have examined the potential of this technique for the characterization of normal and modified nucleic acid constituents, such as oligonucleotides. These compounds are quite polar and are difficult to examine by mass spectrometry due to their non-volatility and ease of fragmentation. Matrix-assisted laser desorption FTMS using 266 nm radiation and a nicotinic acid matrix could be used to generate abundant molecular (M-H)$^-$ ions as well as informative fragment ions for these compounds. The negative ion spectra provided detailed structural information about the sequence and degree of modification of these oligonucleotides. Nucleotide adducts such as (benzo(a)pyrene tetrol)-N^2-deoxyguanosine monophosphate can also be characterized at the picomole level with this technique.

Experimental parameters for the matrix-assisted LD technique were optimized with oligonucleotide dimers. The negative ion spectra for these compounds revealed not only molecular weight information from the (M-H)$^-$ ions but also sequence information from the fragment ions. For example, the isomeric dimers d(5'-TC-3') and d(5'-CT-3') each revealed an (M-H)$^-$ ion at m/z 530, although their fragment ions were significantly different and could be used to differentiate the two isomers. The most abundant sequence ion formed by cleavage of the phosphate linkage was m/z 306 for d(TC) whereas the most abundant sequence ion for d(CT) was m/z 321. The observation of these sequence ions verifies that the major fragmentation occurs from the 5' end of the oligonucleotide. Thus, these fragment ions identify the nucleic acid residue at the 3' end of the oligomer and provide isomeric differentiation. Modified oligonucleotide dimers such as alkylated $d(T)_2$ could also be characterized by this technique. For these compounds, the fragment ions provided not only sequence information for the oligomers but also identification and location of the modifications.

Matrix-assisted laser desorption FTMS proved to be a powerful method for the structural characterization of oligonucleotide trimers and tetramers as well. In addition to formation of molecular (M-H)⁻ ions for these oligomers, abundant fragment ions were also observed, resulting primarily from cleavage of nucleosides from the 5' end. This fragmentation has been outlined above for dimers and provides sequence information as well as isomeric differentiation for these larger oligomers. The nicotinic acid matrix was essential for the success of these experiments. For example, matrix-assisted LDFTMS of the trimer d(GGT) indicated an (M-H)⁻ ion at m/z 899 and fragment ions at m/z 650 and 321. Examination of this dimer under identical experimental conditions except that the nicotinic acid matrix was not used, revealed complete loss of the (M-H)⁻ ion as well as the larger fragment ions. Figure 1 shown below illustrates the negative ion spectrum for the tetramer d(AGCT). Note the (M-H)⁻ ion at m/z 1172 and the fragment ions at m/z 939, 739, 610, and 321. The ions at m/z 1172, 939, 610, and 321 provide molecular weight and sequence information for this tetramer and verify that the structure is d(5'-AGCT-3'). To verify this fragmentation pathway, an isomeric tetramer d(TGCA) was also examined. This compound also revealed an (M-H)⁻ ion at m/z 1172, but had fragment ions at m/z 948, 619, and 330 which were sufficient to differentiate these two isomers. Isomeric tetranucleotides could be unambiguously identified by examination of their fragment ions, indicating that compounds which differ in their sequences, such as d(5'-GGCC-3') and d(5'-GCGC-3'), can be differentiated as well as compounds which differ only the location of the 3' end, such as d(5'-GGCC-3') and d(5'-CCGG-3'). The presence of modifications such as alkyl adducts on these tetramers can also be identified.

Examination of larger oligonucleotides failed to reveal molecular ions, implying that additional experimental parameters for the FTMS experiments need to be examined.

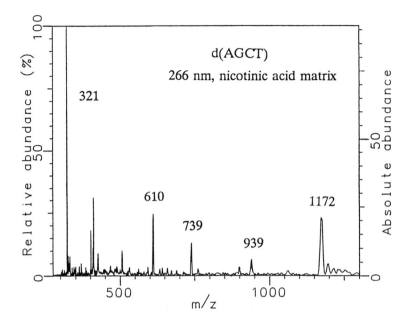

Figure 1. Negative ion matrix-assisted laser desorption FTMS spectrum of d(AGCT).

Because ions in the m/z 10,000 - 100,000 range have been observed in time-of-flight experiments, it appears that the matrix-assisted laser desorption technique is successful at desorbing these large ions. The major factors which need to be further addressed for the FTMS experiments involve the trapping and detection of these large ions. Other researchers have indicated that these large laser-desorbed ions may have significant kinetic energy distributions.[4] While this factor does not adversely affect the TOF experiments, it would have substantial implications for the FTMS experiments. Ions which have kinetic energies significantly larger than the trapping well of the FTMS (typically 1-5 V) will be lost from the trap if they cannot be sufficiently cooled or decelerated. We are attempting to address the ion kinetic energy issue by various experiments designed to examine the kinetic energy distributions and trapping efficiencies for large ions. Retarding grid experiments have been evaluated as a means of trapping ions with kinetic energies up to approximately 15 eV. For this experiment, the potential energy difference between the sample disk and the trapping plate of the FTMS cell was varied to either retard or accelerate the laser-desorbed ions. Preliminary investigations indicate that retarding grid experiments are not successful in slowing the ions significantly presumably due to the fact that the space-charged ion cloud generated by laser desorption is probably not dramatically affected by the relatively small potentials placed on these plates. Further experiments are in progress to examine collisional cooling and electrostatic deceleration as methods of more effectively removing the excess kinetic energies of the large ions.

CONCLUSIONS

Matrix-assisted laser desorption FTMS provides a powerful technique for the generation and structural characterization of normal and modified oligonucleotides. Examination of these compounds using 266 nm radiation and a nicotinic acid matrix revealed (M-H)⁻ ions as well as fragment ions which provided structural information. The fragmentation observed resulted primarily from elimination of nucleosides from the 5' end of the oligomers and provided information which could be used to differentiate isomers. Experiments are in progress to extend the application of this ionization technique for larger biomolecules. Preliminary results indicate that ion trapping considerations need to be examined, due to the fact that large laser-generated ions may have significant kinetic energy distributions and may be poorly trapped in the FTMS cell. Current research is directed at determining efficient methods of cooling or decelerating the laser-generated ions.

REFERENCES

1. M. Karas and F. Hillenkamp, Anal. Chem. 60, 2299 (1988).
2. R.C. Beavis and B. T. Chait, Rapid Comm. Mass Spec. 3, 233 (1989).
3. R.L. Hettich and M.V. Buchanan, J. Amer. Soc. Mass Spec. 2, 22 (1991).
4. B. Spengler, Y. Pan, R.J. Cotter, and L.-S. Kan, Rapid Comm. Mass Spec. 4, 99 (1990).

Comparison of Atomization Processes: Trace Element Analysis Using RIS of Laser-Irradiated and Ion-Bombarded Biological and Metal Surfaces

H.F. Arlinghaus and N. Thonnard

Atom Sciences, Inc., 114 Ridgeway Center, Oak Ridge, TN, 37830

ABSTRACT

Resonance Ionization Spectroscopy (RIS) is becoming recognized as an emerging field with wide applications. RIS is an analytical technique providing extreme element sensitivity and selectivity by detecting ionization produced from lasers tuned to excited states of the analyte. The sensitivity and selectivity of the RIS process is especially valuable for ultra-trace element analysis in samples where the complexity of the matrix can be a serious source of interference. Using either Sputter-Initiated RIS (SIRIS) or Laser Atomization RIS (LARIS), it is possible to localize with high spatial resolution ultra-trace concentrations of a selected element to the parts per trillion level. We have compared both the SIRIS and LARIS techniques to determine their characteristics to localize and quantitate trace elements on biological and metal surfaces, and their different response as a function of atomization parameter, substrate and analyte. Using 266 nm, 10 ns atomization pulses on organic matrices, quantitation with the LARIS technique was identical to SIRIS, but yielded a factor of 100 greater signal. Quantitation on metallic matrices was not as good with LARIS (at 532 nm) than with SIRIS.

I. INTRODUCTION

The extremely high ionization efficiency and element specificity of Sputter-Initiated Resonance Ionization Spectroscopy (SIRIS) and Laser Atomization RIS (LARIS) enables measurements that frequently are very difficult or impossible with other techniques. In the SIRIS/LARIS technique, the abundant neutral atoms released by the sputtering process are ionized by precisely-tuned laser beams and counted in a mass spectrometer. Using either SIRIS or LARIS, it is possible to localize and quantitate with high spatial resolution ultra-trace concentrations of a selected element at or near the surface of solid samples. The sensitivity and selectivity of the RIS process is especially valuable for trace element analysis in materials where the complexity of the matrix is frequently a serious source of interference. In the last several years, SIRIS was successfully applied in a variety of fields such as semiconductors[1-5], geo- and cosmochemistry[6-7], environmental[8] and biomedical sciences[9], and DNA sequencing[10-12] while LARIS applications are more limited[13,14]. In this paper we will present data showing differences between SIRIS and LARIS response as a function of atomization parameter, substrate and analyte, and discuss the strength and weaknesses of both atomization processes in light of the sample matrix, RIS detection and sensitivity.

II. EXPERIMENT

A schematic of the SIRIS/LARIS instrument is shown in Figure 1. The system consists of a duoplasmatron microbeam ion gun (incident angle = 60^0 from normal), a pulsed flood electron gun, a RIS laser system consisting of a Quanta Ray DCR-2A pulsed Nd:YAG laser (repetition rate 30 Hz) pumping two Quanta Ray PDL-1 pulsed dye lasers, a Quanta Ray Nd:YAG laser and a Lambda Physik LPX 105e excimer laser for laser atomization, a computer-controlled (x, y, z, ϕ) sample holder, two HeNe lasers for exact target positioning, a video imaging system for sample observation, a sample interlock system, and a mass spectrometer detection system (for more details, see[2]). In the SIRIS experiments, a 9 keV pulsed Ar^+ ion beam of typically ~2 μA current and 500 ns pulse width in a 50-200 μm diameter spot size is used to bombard samples, while for the LARIS experiments, the second (532 nm) and fourth (266 nm) harmonic of a Nd:YAG laser (pulse width = 8 ns) and ArF (193 nm) of a excimer laser (pulse width = 10 ns) are used to irradiate samples. The atomization laser output can be adjusted with a variable attenuator with typical energies in the range of 0.01-1000 μJ/pulse on a spot size of 75-150 μm in diameter. Suppression of secondary ions produced by the ion sputtering or laser atomization process is achieved by a combination of the relative timing between the ion sputtering or laser atomization pulse and the firing of the RIS lasers, timed extraction voltage switching, and electrostatic energy analysis. The selectively ionized atoms are extracted and directed through an electrostatic and magnetic-sector, or time-of-flight, mass spectrometer onto an ion detector with measurements made in the analog or single ion counting modes. The high selectivity of the RIS process (a selectivity of more than 10^9 has been shown[13]) also helps maintain linearity in the mass spectrometer ion extraction region by reducing space charge effects that would otherwise be present due to ionization of the major constituents of the sample. In the SIRIS experiments, for insulating samples, charge compensation is achieved by flooding the samples with low energy electrons (20-70 eV) between the ion pulses, thereby permitting self-adjustment of the surface potential without damaging the sample. For the LARIS experiments, no charge compensation is needed with insulating samples. Imaging is achieved by changing the x or y target position while the atomization laser beam or ion beam position remains fixed.

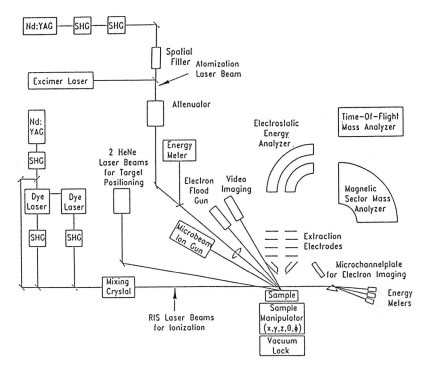

Figure 1. Schematic illustration of the SIRIS/LARIS instrumental set-up.

III. RESULTS AND DISCUSSION

As good correlation between analyte concentration and SIRIS signal for semiconductor and metallic matrices had already been shown by Parks, et al.[15], our first attempts at LARIS measurements were made on the identical samples[13]. At the time these measurements were made, the complete system shown in Figure 1 was not yet available and sample atomization was only possible at 532 nm. After removal of surface impurities, reasonably good reproducibility and correlation with analyte correlation was obtained with gallium doped silicon samples (Figure 2a) using 2×10^7 W cm^{-2} 532 nm, 10 ns laser pulses. Measurements of aluminum in steel NBS Standard Reference Materials (Figure 2b), were rather poor under similar irradiation conditions (4×10^7 W cm^{-2}). Examination of the steel samples after analysis indicated that significant sample modification occurred. Even though the matrix was not very different in all four of these samples, significant differences in atomization behavior was seen.

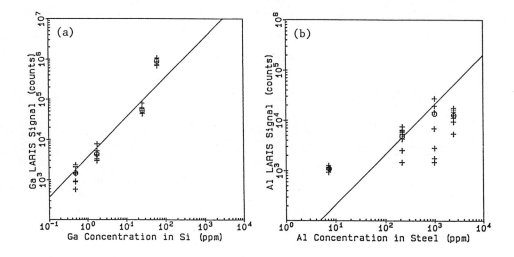

Figure 2. LARIS signal correlation with concentration of gallium in silicon (a) and aluminum in steel (b). Crosses are the individual sums of 500 laser shots; the mean is denoted with circles.

To compare the SIRIS and LARIS response, 5% gelatin in water was uniformly doped with ^{99}Tc at concentrations between 10 ppb and 100 ppm. Gelatin was chosen as the matrix because it closely resembles the physical and chemical form of a range of biological soft-tissues in which analysis is often limited by the presence of molecular interferences. Figure 3 shows the correlation of ^{99}Tc concentration in gelatin with the ^{99}Tc SIRIS and LARIS (266 nm) signal. Technetium was ionized by a uv photon (313.3 nm), a visible photon (554.2 nm) and a 1064 nm infrared photon for the ionization step. The correlation of signal with concentration both with SIRIS and LARIS is linear over the entire range of Tc concentrations used, down to 10 ppb. The detection limit for SIRIS and LARIS was much lower than 10 ppb because no background could be detected, as the signal virtually disappeared by detuning the laser wavelength. The results demonstrate that identical quantitative results with either SIRIS or LARIS can be achieved in biological materials, but with a factor of 100 more signal in the LARIS mode. Thus, trace element measurements in the 100 ppt level or lower are possible with LARIS. Since the sputtering yield of gelatin is similar to that of biological tissues, uniformly doped gelatin standards can be used to calibrate trace element analyses in biological materials using either SIRIS or LARIS.

A promising new DNA sequencing technique is to label DNA with stable metal isotopes[10-12]. Four separate isotopes from an element are attached to the A, G, C, and T terminated fragments of the DNA, allowing these fragments to be combined in one electrophoresis gel lane. The isotope-labeled DNAs can be analyzed directly with SIRIS or LARIS in the gel. To test the

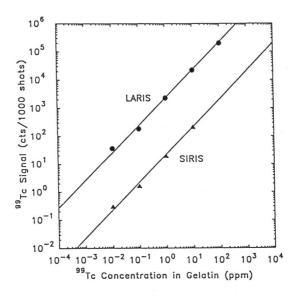

Figure 3. Correlation of ^{99}Tc concentration gelatin with the SIRIS and LARIS signal.

SIRIS/LARIS response to metal-labeled DNA, oligonucleotides were labeled with ^{116}Sn and mixed with a 6% solution of gelatin (final concentration = 550 pmoles/ml). Figure 4 shows the ^{116}Sn yield as a function of reciprocal laser energy and wavelength for various samples. RIS of tin was done with a three color scheme. 266 nm (solid line) and 532 nm (dotted line) laser light was used to measure the ^{116}Sn yield in the same gelatin sample. The Sn yield increases by several orders of magnitude with increasing laser energy, with a higher increase observed for shorter atomization laser wavelength. The higher Sn yield for 266 nm is mainly due to a higher absorption coefficient which increases the desorption yield and the fragmentation rate. The optimum atomization laser energy at 266 nm is 50-150 μJ/pulse; at 532 nm, 300-600 μJ/pulse. At higher laser energies, significant background signal from isobaric molecular species, presumably due to space charge effects and/or delayed emission, started to occur. The dashed line shows the ^{116}Sn yield measured in a NBS steel sample (^{116}Sn concentration ~130 ppm) as a function of laser energy at 532 nm. The ^{116}Sn signal increased more than six orders of magnitude before reaching a maximum at ~70 μJ/pulse, followed by a decrease. Since the direct ionization signal did not increase above 70 μJ/pulse, we conclude that space charge effects and plasma formation led to a decrease of the signal. Also, it may be seen that the ^{116}Sn signal increase observed for steel is much higher than the ^{116}Sn signals in gelatins. This is due to the steel surface having a much higher absorption coefficient than gelatin, and to no molecular bonds needing to be fragmented to obtain atomic Sn.

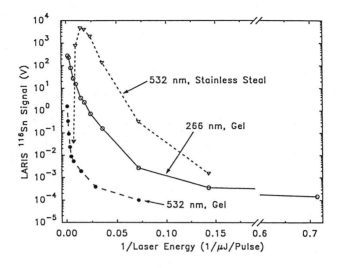

Figure 4. ^{116}Sn yield as a function of reciprocal laser energy at 532 nm and 266 nm atomization laser wavelength for a ^{116}Sn-17-mer DNA primer in gelatin and a stainless steel sample (NBS 1263 SRM, ^{116}Sn concentration = 130 ppm).

The shape of the curves in this Arrhenius diagram indicates that the desorption process is not of thermal nature over the entire atomization laser energy distribution. Furthermore, thresholds for particle desorption could not be defined because particles emission was still observed at laser powers of 5×10^5 W cm^{-2} and lower. The measured yield as a function of atomization laser wavelength indicates that at shorter laser wavelengths, e.g., 193 nm, even higher Sn yields should be obtainable. Absorption of photons by tin compound attached to DNA starts at 266 nm and increase dramatically towards shorter wavelengths. Absorption coefficients for acrylamide gel at various wavelengths are high below 250 nm, with the tail of the absorption band going to 266 nm, and virtually no absorption at wavelengths longer than 355 nm. Therefore, wavelengths shorter than 250 nm for laser atomization of Sn-labeled DNA in polyacrylamide gel should lead to higher surface desorption yields.

Figure 5 shows SIRIS and LARIS position scans for different atomization laser wavelengths obtained from Sn-labeled DNA loaded on polyacrylamide gel and subjected to electrophoresis (a-d), and from Sn-labeled DNA hybridized to a complementary oligonucleotide which was bound to a single 25 μm fiber (e, f). For each SIRIS scan, a prominent band of ^{116}Sn was observed at the uv absorption site of the DNA band in polyacrylamide gel and on the fiber. To be able to directly compare the SIRIS signal with the LARIS signal, the same area scanned with SIRIS was scanned again with LARIS. Figure 5b shows the first LARIS scan using 532 nm

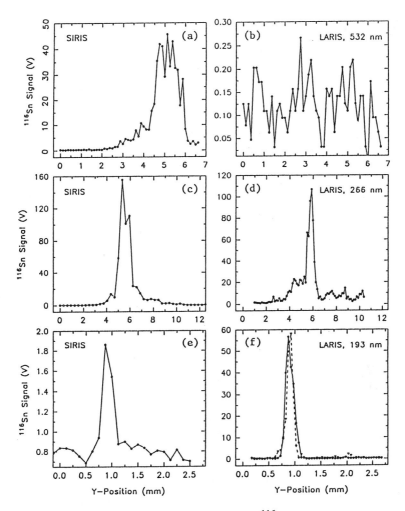

Figure 5. SIRIS (a,c) and LARIS (b,d) analysis of an ^{116}Sn-labeled 17-mer DNA band in polyacrylamide gel (total amount = ~ 500 pmol). The ion and laser beam direction was parallel to the electrophoresis direction. SIRIS (e) and LARIS (f) scans over a single fiber that has been incubated in 1 μM ^{116}Sn-labeled DNA solution. Solid and dashed lines are scans through two different y positions showing the reproducibility of the signal.

atomization laser wavelength and five shots per data point. In comparison to the SIRIS data (Figure 5a) only a very low ^{116}Sn signal was measured over the entire band area. No prominent band could be detected. The same area was scanned again using 30 laser shots per data point (not shown). This time, a strong signal was observed at the site of the DNA band. In the next scan, the strong signal observed in the previous scan again disappeared. We repeated this scan several times, observing widely varying random signals at the site of the DNA band. Since acrylamide and DNA are totally transparent for 532 nm laser light, we concluded that the high fluctuations of the observed ^{116}Sn signal is presumably caused by modifying physical parameters such as absorption coefficient and thermal conductivity in micron sized areas with consecutive laser shots. The fragmentation of the DNA and release of atomic tin is a variable and unreliable process at 532 nm, and occurs only when sufficient energy is deposited to modify the material as suggested above. Figure 4d shows a LARIS scan using 266 nm atomization laser wavelength. This time a prominent DNA band of ^{116}Sn was observed. The signal strength obtained with LARIS is similar to that of SIRIS. In Figure 4f two LARIS scans over a 25 μm fiber using 193 nm atomization laser wavelength are depicted. The two curves represent data obtained from different positions on the fiber. These data show well defined peaks of ^{116}Sn with very large signal-to-noise ratios and essentially identical results at the two different locations. Since the diameter of the fibers are obviously much smaller than that of the atomizing laser beam, the peak width implies that the beam diameter was approximately 130 μm. For this short atomization laser wavelength a much higher signal was obtained with LARIS than for SIRIS. The comparison between SIRIS and LARIS for different atomization laser wavelength demonstrates how important it is to find the right laser wavelengths and fluence for optimization of the LARIS yield. Further analysis using other wavelengths and matrices will be obtained in the near future.

VI. CONCLUSION

We conclude that both SIRIS are LARIS can provide very sensitive and selective element analyses in biological matrices. A higher signal can be obtained with LARIS than with SIRIS, but its matrix dependence makes it more difficult to implement as the atomization protocol needs to be carefully optimized for each matrix. The high selectivity of the SIRIS and LARIS techniques allows interference-free trace element analysis even in materials where the complexity of the matrix is frequently a serious source of interference. The efficiency of the ionization process reduces the sample size required for analysis by orders of magnitude compared to other techniques. In combination with a small focused ion or laser beam, both SIRIS and LARIS will be able to image trace elements in biological tissues with sub-cell spatial resolution and with extremely high sensitivity and selectivity. Additional measurements over a wider atomization power and wavelength range will be required to optimize the LARIS techniques to semiconductor and metallic samples.

V. ACKNOWLEDGEMENTS

We wish to acknowledge K.B. Jacobson, R.A. Sachleben, and their co-workers at the Oak Ridge National Laboratory for their encouragement and the preparation of the Sn-labeled DNA samples; A.W. McMahon of the Harwell Laboratory in the U.K. for the Tc-doped gelatin samples;

and D.W. Beekman (now at Martin Marietta Research Center) and M.T. Spaar of Atom Sciences for their invaluable support in the experiments. This work was supported in part by the Air Force Office of Scientific Research under Contract No. F49620-86-C-0073, the United States Department of Energy under Sub-contract No. 90X-SF316C and Contract No. DE-AC-05-89ER80735, and by the following U.K. organizations: the Department of Energy, the Department of the Environment, the Department of Health, the Ministry of Agriculture, Fisheries and Food, and the United Kingdom Atomic Energy Authority.

REFERENCES

1. J.E. Parks, M.T. Spaar, and P.J. Cressman, J. Crystal Growth 89, 4, 1988.
2. H.F. Arlinghaus, M.T. Spaar, and N. Thonnard, J. Vac. Sci. Technol. A8, 2318, 1990.
3. F.M. Kimock, J.P. Baxter and N. Winograd, Nucl. Instr. and Meth., 218, 287, 1983.
4. C.E. Young, M.J. Pellin, W.F. Calay, J.W. Burnett, D.M. Gruen, H.J. Stein, and S.S. Tsao, in Resonance Ionization Spectroscopy 1988, T.B. Lucatorto and J.E. Parks, ed., Inst. Phys. Conf. Ser. 94, 205, 1988.
5. S.W. Downey, A.B. Emerson, and R.F. Kopf, Resonance Ionization Spectroscopy 1990, J.E. Parks and N. Omenetto, eds., Inst. Phys. Conf. Series 114, 401, (Bristol: The Institute of Physics), (1991).
6. R.D. Willis, N. Thonnard, O. Eugster, Th. Michel, and B.E. Lehmann, Resonance Ionization Spectroscopy 1990, J.E. Parks and N. Omenetto, eds., Inst. Phys. Conf. Series 114, 275, (Bristol: The Institute of Physics), (1991).
7. J.D. Blum, M.J. Pellin, W.F. Calaway, C.E. Young, D.M. Gruen, I.D. Hutcheon, and G.J. Wasserburg, Anal. Chem. 62, 209, 1990.
8. A.W. McMahon, S.A. Mountfort, H.F. Arlinghaus, and M.T. Spaar, Resonance Ionization Spectroscopy 1990, J.E. Parks and N. Omenetto, eds., Inst. Phys. Conf. Series 114, 467, (Bristol: The Institute of Physics), (1991).
9. T.E. Quinn, T.J. Murphy, L.J. Moore, H.F. Arlinghaus, M.T. Spaar, E.H. Taylor, and N. Thonnard, Resonance Ionization Spectroscopy 1990, J.E. Parks and N. Omenetto, eds., Inst. Phys. Conf. Series 114, 333, (Bristol: The Institute of Physics), (1991).
10. K.B. Jacobson, H.F. Arlinghaus, H.W. Schmitt, R.A. Sachleben, G.M. Brown, N. Thonnard, F.V. Sloop, R.S. Foote, F.W. Larimer, R.P. Woychik, M.W. England, K.L. Burchett, and D.A. Jacobson, Genomics, 9, 51, (1991).
11. H.F. Arlinghaus, N. Thonnard, M.T. Spaar, R.A. Sachleben, F.W. Larimer, R.S. Foote, R.P. Woychik, G.M. Brown, F.V. Sloop, and K.B. Jacobson, Anal. Chem., 63, 402, (1991).
12. H.F. Arlinghaus, N. Thonnard, M.T. Spaar, R.A. Sachleben, G.M. Brown, R.S. Foote, F.V. Sloop, and K.B. Jacobson, J. Vac. Sci. Technol. A9, (3), 1991, in press.
13. D.W. Beekman, and N. Thonnard, in Resonance Ionization Spectroscopy 1988, T.B. Lucatorto and J.E. Parks, ed., Inst. Phys. Conf. Ser. 94, 163, 1989.
14. N.S. Nogar, R.C. Estler, M.W. Rowe, B.L. Fearey and C.M. Miller, in Resonance Ionization Spectroscopy 1988, T.B. Lucatorto and J.E. Parks, ed., Inst. Phys. Conf. Ser. 94, 147, 1989.
15. J.E. Parks, H.W. Schmitt, G.S. Hurst and W.M. Fairbank, Jr., in Analytical Spectroscopy, W.S. Lyon, ed., Anal. Chem. Symp. Ser. 19, 149, 1984.

Laser Desorption of Peptide Molecules and Ions
Using 193 nm Radiation

J. Paul Speir, Greg S. Gorman, and I. Jonathan Amster
Department of Chemistry, University of Georgia
Athens, Georgia 30602

Introduction.

Several investigators have studied both experimental and theoretical aspects of the laser desorption of solids. The laser desorption of nonvolatile molecules dispersed in a volatile matrix is now well established.[1-3] However, an understanding of the fundamental processes that would allow experimentalists to consistently and reproducibly carry out laser desorption experiments on pure samples of nonvolatile compounds is still lacking. Research in this laboratory is directed toward studies of the gas phase ion chemistry of peptide molecules and peptide ions. In particular, ultraviolet laser desorption is used to generate gas phase peptide neutral molecules from thin films of a pure solid. The gas phase reactions of peptide molecules with ions stored in a Fourier transform mass spectrometer can be studied using laser irradiation to desorb neutral peptide molecules. Such experiments have required the development of specialized experimental techniques, which are reported here.

Experimental.

All laser desorption experiments are carried out using the 193 nm output of an argon fluoride excimer laser (Questek, Model 2110.) Neutral and ionic products of laser desorption are analyzed using a Fourier transform mass spectrometer with a 1-3/4" cubic analyzer cell and a 1 Tesla electromagnet. The upper mass limit of this instrument is approximately 2000 Daltons. Some details of the analyzer cell are shown in Figure 1. Peptide coated sample stubs are transferred from a pump down vacuum chamber into the jaws of a sample holder mounted onto the analyzer cell, in the main vacuum chamber. The two stage sample introduction allows the analysis to be performed under ultrahigh vacuum (2-5 x 10^{-10} Torr), a pressure 2-3 orders of magnitude lower than can be achieved in experiments utilizing a direct insertion probe for sample introduction. A metal target carousel is mounted beneath the analyzer cell. One of five transition metal targets are positioned beneath an opening in the bottom cell plate. A rotary motion feedthrough is attached to the carousel and allows selection of the desired target without opening the vacuum chamber. Gas phase transition metal ions can be formed by focusing the output of the laser onto the selected metal target, as has been described by Freiser.[4]

Neutral desorption products are ionized by reactive collisions with ions stored in the analyzer cell of the Fourier transform mass spectrometer. The details of this ionization procedure, called laser desorption/chemical ionization (LD/CI), have been reported previously.[5,6] For neutral desorption experiments, the laser irradiance is approximately 10^6 W/cm^2. At this irradiance, only neutral desorption occurs without direct formation of any ions by the laser. At higher irradiance, ions are formed directly by the laser, and can be trapped and mass analyzed.

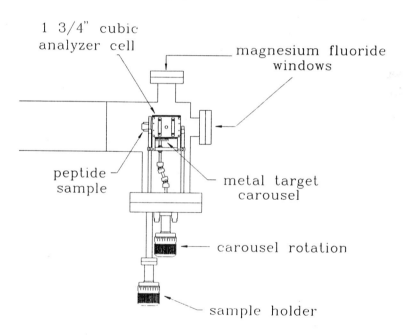

Figure 1. Details of the FTMS analyzer cell used for LD/CI experiments.

A machineable glass (Macor) stub with a 75 nm overlayer of gold and palladium is used as the sample substrate for laser desorption, shown schematically in Figure 2. Peptide samples with a thickness equivalent to 100-1000 monolayers are applied to the substrate using the electrospray method.[7] Peptides are obtained commercially or synthesized using the "tea-bag" version of the Merrifield synthesis.[8] The peptides are desalted using reverse phase chromatography on a 1 cm C18 Sep-Pak column (Waters Associates.)

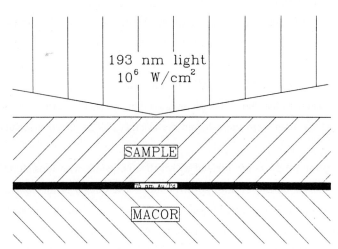

Figure 2. Cross sectional view of the sample and substrate used for LD/CI experiments, showing a film of a neat peptide sample corresponding to 750 monolayers on a 75 nm thick metal film, supported by a Macor substrate.

Results and Discussion.

The laser desorption of peptide neutral molecules without the direct formation of ions depends critically on the experimental conditions used. The substrate plays an important role in the desorption process. Figure 2 shows the relative thickness of a typical peptide sample (750 monolayers) and metal film overlayer used in these experiments. The ultraviolet light used for neutral desorption is strongly absorbed by the metal film. The 75 nm thickness of the metal film corresponds to the penetration depth of the 193 nm light into the metal,[9] so that the laser energy is completely absorbed in the metal film, and is not transmitted to the Macor stub. The energy is rapidly converted to heat, which is not readily conducted away from the film, due to the thermal insulating properties of the glass stub and the organic sample. Concentration of the laser light into this narrow region causes a rapid temperature jump, promoting the desorption of the peptide sample. Experiments in which the metal film thickness is varied from 25 nm to 150 nm show that 75 nm is optimum for the desorption of the small peptides (2-4 amino acid residues) investigated to date. Thicker metal films would be expected to depress the temperature jump by allowing heat to be conducted from the surface of the metal film into lower layers. Consistent with this prediction, thicker metal films (greater than 100 nm) are found to require higher laser irradiance to induce peptide desorption. At higher irradiance, ions are formed directly by the laser, which interfere with the observation of ion-molecule reactions of the desorbed neutrals. When the metal film is thinner than 50 nm, desorption requires higher laser irradiance, again causing the formation of ions directly by the laser. For such thin metal films, the laser radiation may be penetrating into the Macor underlayer, reducing the temperature jump at the surface. Under these conditions, sodium and potassium ions are formed by the laser desorption process, indicating that the radiation has penetrated into the Macor, which contains high concentrations of such ions.

The exclusive desorption of peptide neutrals requires some of the laser energy to be dissipated in the sample itself. The 193 nm light used for desorption experiments is strongly absorbed by most peptides, which have molar extinction coefficients on the order of 1000-10,000 $M^{-1}cm^{-1}$ at this wavelength. 1000 monolayers of a peptide is expected to absorb 10-90% of the incident radiation. This process is found to be critical to the observation of ion-molecule reactions between laser desorbed molecules and the ions stored in the Fourier transform mass spectrometer. Desorption experiments with peptides at 248 nm or 351 nm are unsuccessful in the production of neutrals that can react with stored ions. Experimentally, it is observed that desorption at the longer wavelengths causes stored ions to be ejected from the analyzer cell. This suggests that highly energetic collisions have occurred between the laser desorbed material and the trapped ions. The longer wavelengths are not strongly absorbed by the peptides investigated, and so the incident desorption beam is transmitted entirely to the metal film. One interpretation of these results is that the temperature jump at the metal-peptide interface volatilizes the sample in a narrow zone under the bulk of the sample, causing chunks of the sample to be launched into the stored ions, causing their ejection. Consistent with this interpretation is the observation that all of the irradiated sample is removed in a single laser shot at the longer wavelengths, while several laser pulses are required to remove all the sample at a spot when using 193 nm light. The desorption of chunks of material (spallation) has been observed for laser desorption experiments of DNA suspended in an ice matrix when the wavelength of the desorbing laser is not absorbed by the matrix.[10]

Using the experimental parameters described above, the desorption of intact peptide molecular neutrals is observed with a variety of small peptides. Chemical ionization of the laser desorbed neutrals using NH_4^+ as a reagent ion produces a protonated molecule exclusively for a variety of peptides.[5] Additional evidence for the desorption of intact molecules has been obtained by collecting the desorbed neutrals, and examining them using direct laser vaporization/ionization. A copper target on the metal target carousel, Figure 1, is positioned beneath the opening in the bottom cell plate. After firing the desorption laser at a peptide sample (pro-leu-val) approximately twenty times, the laser is directed at the copper target. A copper attached molecular ion is formed exclusively, as shown in Figure 3. Interestingly, the desorption of the metal attached ion was found to occur at lower irradiance than is observed when the peptide is applied directly to a metal target. Condensation of a peptide vapor on the metal target produces a sample film that is more finely divided and more uniform compared to the films produced by solution evaporation. The morphology of the deposition may play a role in the efficient desorption of peptide neutrals. Such effects will be the subject of future studies.

The LD/CI technique can be used to study reactions of peptide molecules with a variety of reagent ions. We have observed proton transfer reactions over a broad range of exothermicities (ΔH = 0 to -5 eV),[5] as well as charge exchange ionization (ΔH = 0 to -15 eV) and transition metal ion reactions. To date, this technique has been successful for peptides containing up to four amino acids. We hope to extend such studies to larger peptides as we develop a better understanding of the fundamentals of the laser desorption process.

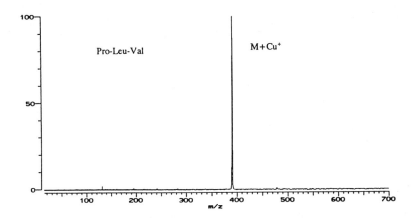

Figure 3. A mass spectrum showing copper attachment to a molecule of pro-leu-val, formed by direct laser ionization of the tripeptide on a copper substrate. The peptide was transferred to the substrate by laser desorption, as described in the text.

Acknowledgements. We gratefully acknowledge financial support from the National Science Foundation (CHE-9024922). Acknowledgement is made to the donors of the Petroleum Research Fund, administered by the American Chemical Society, for partial support of this work.

References.

1. Karas, M.; Hillenkamp, F. Anal. Chem. **1988**, *60*, 2290-2301.
2. Beavis, R. C.; Chait, B. T. Rapid Commun. Mass Spectrom. **1989**, *3*, 432-435.

3. Salehpour, M.; Perera, I.; Kfellberg, J.; Hedin, A.; Islamian, M. A; Hakansson, P.; Sundqvist, B. U. R. Rapid Commun. Mass Spectrom. **1989**, 3, 259-263.

4. Burnier, R. C.; Byrd, G. D.; Carlin, T. J.; Wiese, M. C.; Cody, R. B.; Freiser, B. S. In "Ion Cyclotron Resonance Spectrometry"; Hartmann, H.; Wanczek, K.-P., Eds.; Springer-Verlag: Berlin, 1982; 98-118.

5. Speir, J. P.; Gorman, G. S.; Cornett, D. S.; Amster, I. J. Anal. Chem. **1991**, 63, 65-69.

6. Amster, I. J.; Land, D. P.; Hemminger, J. C.; McIver, R. T., Jr. Anal. Chem. **1989**, 61, 184-186.

7. McNeal, C. J.; Macfarlane, R. D.; Thurston, E. L. Anal. Chem. **1979**, 51, 2036-2039.

8. Houghten, R. A. Proc. Nat. Acad. Sci. U.S.A. **1985**, 82, 5131-5135.

9. Ready, J. F. "Effects of High Power Laser Radiation"; Academic Press: New York, 1971; Chapter 3.

10. Nelson, R. W.; Rainbow, M. J.; Lohr, D. E.; Williams, P. Science **1989**, 246, 1585-1587.

ABLATION OF MATERIAL BY FRONT SURFACE SPALLATION

R. S. Dingus and R. J. Scammon
Los Alamos National Laboratory, Los Alamos, NM 87545

ABSTRACT

Laser irradiation can be utilized to remove (i.e., ablate) material in a controlled manner by a hydrodynamic process, referred to as front surface spallation. In this process, a thin layer next to a free surface is heated to a level (below vaporization) so rapidly that it cannot undergo thermal expansion during laser heating. This generates a stress pulse, which propagates both inward and toward the free surface, with an initial amplitude that can be calculated using the Grüneisen coefficient. As the pulse reflects from the free surface, a tensile tail can develop of sufficient amplitude, exceeding the material strength, that a layer will be spalled off, taking much of the laser-deposited energy with it. To achieve spallation conditions, the laser wavelength, pulselength and fluence must be tailored to the absorption depth, Grüneisen coefficient, and spall strength. Hydrodynamic calculations and analytical modeling are presented to explain the process and illustrate conditions under which it should be expected to occur. Under some conditions, front surface spallation can have advantages over ablation by thermal vaporization, where residual temperatures are generally higher.

1. INTRODUCTION

The information in this paper has been derived from efforts to develop computational techniques to model x-ray- and laser-target interaction for defense applications. The effort has included the performance of detailed experiments to gain confidence in the modeling. The knowledge gained from these efforts appears to have potential benefit to medical and other industrial applications. The computational modeling efforts have involved development of a basic understanding of the processes along with simple analytical codes; development of large, numerical, radiation-hydrodynamic-computer codes to analyze complex conditions; and development of the necessary materials properties data bases for these calculations. The discussion in this paper will emphasis semi-transparent materials, with a vacuum or gas in front, exposed to short pulselength lasers with fluences just above the ablation threshold. The laser wavelength will be assumed to be long enough that breakage of molecular bonds by photon absorption is not a dominant process. More general discussions of pulsed laser effects phenomenology are presented in Ref. 1-5.

A reasonable criterion for being semi-transparent is that a significant fraction, if not all, of the laser beam is absorbed in the material, but the laser pulselength is short enough that the laser penetration depth is larger than the depth to which energy is transported by thermal diffusion during the time of the laser pulse. For rapid heating, when the front surface dose is above the complete vaporization energy (i.e., ΔH_c, the energy per unit mass to take the material from its ambient temperature into the vapor state), material will be ablated by vaporization to a depth at which the dose equals the complete vaporization energy. A fraction of the material with a dose above the incipient vaporization energy (i.e., ΔH_i, the energy to take the material to the vaporization temperature, but excluding the latent heat of vaporization, ΔH_v; $\Delta H_c = \Delta H_i + \Delta H_v$; generally $\Delta H_i \ll \Delta H_v$) will also be vaporized. This vapor can entrain nonvaporized material from this region so that the precise amount of material ablated from this region is dependent on rates including those associated with both thermodynamics and hydrodynamics. As the heating pulselength gets longer (for semi-transparent materials), the situation becomes more complicated because there is time for cooling by ablation to peg the front surface temperature at the vaporization temperature, while in depth laser heating can cause superheating until instabilities result in rapid vaporization at some depth, which will drive off cooler material at lesser depths. For opaque materials, this will not occur because the heating source remains at the surface.

2. FRONT SURFACE SPALLATION PROCESS

For semi-transparent materials, if the laser pulselength and the time for conversion of absorbed laser energy to thermal energy are sufficiently short, such that the transit distance of an acoustic signal during the laser pulse is less than or of the order of one laser absorption depth, then the heating from laser absorption will induce a stress pulse whose amplitude is proportional to the absorbed fluence. At fluences below that for thermal vaporization, this stress pulse can cause front surface spallation, which is a form of ablation. This type of ablation generally requires less fluence per unit thickness ablated than is required by ablation from thermal vaporization or photochemical decomposition because the material is ejected in solid or liquid fragments rather than as molecules or atoms, which requires more energy to break more bonds. For short laser pulselengths (i.e., large laser flux), large compressive stress pulses can be imparted to the material by thermal vaporization or photochemical decomposition due to the rapidity with which the momentum of the ablated material is carried away.

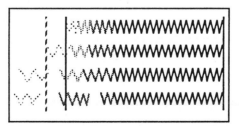

Figure 1. Illustration of front surface spallation process.

Figure 1 illustrates the basic process relating to front surface spallation. Imagine that the material of interest is a spring with a certain compressibility; Fig. 1 shows four configurations of this spring. Suppose that the front surface of the material (spring) is heated to some depth during a short time as illustrated by the hatched portion of the spring at the top in the figure. If the material had been heated slowly, thermal expansion would have caused it to expand to the vertical dotted line to the left in Fig. 1, as shown in the next spring down. However, because it was heated so rapidly, it acquires a significant velocity during the expansion (provided it is not restrained by another material in contact), which must be followed by deceleration to bring it back to rest. If the material has insufficient strength, the force associated with the deceleration will cause a fracture and a layer will keep going as shown by the next case. This fracture is referred to as spallation, or spall for short. Depending on the circumstances, the velocity of the residual material can cause one or more additional layers to spall as shown for the bottom spring.

Front surface spallation can also be described in more quantitative, but equivalent terms. Heating a (front surface) layer of material before it can thermally expand causes a (positive) compressive stress pulse in that region whose amplitude can be calculated using the Grüneisen coefficient. This stress pulse will cause dynamic expansion, both toward the front surface and in the opposite direction. The stress pulse will be reflected at the front surface by the shock impedance mismatch producing a reflected (negative) tensile pulse, if the impedance of the material in front is less than that of the material. To have spall, it is generally necessary that the material in front of the material be a gas or vacuum and not a liquid or solid; otherwise, the shock impedance of the material in front will prevent a tensile wave of sufficient amplitude from developing. The tensile pulse trails the compressive pulse into the material, thus causing the characteristic bipolar stress pulse seen in stress measurements at low fluences[6]. If at some depth, the tensile stress exceeds the tensile spall strength of the material, then the material will spall at that depth, creating a new boundary surface at which the stress becomes zero. The rest of the compressive pulse that is propagating toward the front surface will then be reflected from this new boundary. If the reflected tensile stress builds up to the spall stress again, then spall will occur again. Eventually, the criterion will no longer be met for another spall layer to develop and a residual tensile tail will be left in the material with an amplitude that is less than the spall strength of the material. If the back of the material is also a free surface, then the compressive pulse will reflect from there as a tensile pulse, which can cause back surface spall.

3. EQUATIONS FOR THERMOELASTIC STRESS

To understand the basic relations governing the spallation process, a purely thermoelastic response of the material is assumed along with an exponential laser deposition profile and constant absorption coefficient, μ. For the sake of discussion, it is assumed that the deposited laser energy is instantly converted into heat. The time scale is assumed to be short enough that thermal conduction can be ignored; however, it is assumed that the heat is distributed locally uniformly where it is deposited. The laser beam is assumed to expose a large diameter of material compared to the absorption depth.

For an incident fluence, F_o, into the material (after correction for reflection at the material surface), the fluence, F, at a depth x in the material is given by $F = F_o \exp(-\mu x)$. The energy density (energy per unit mass), E, deposited by the laser in the material at depth x is $E = -(1/\rho) \, dF/dx = (\mu/\rho) F_o \exp(-\mu x)$. A rectangular step function laser temporal profile is assumed with a pulselength, t_L, because it is most illustrative. Other pulse shapes can also be treated analytically[6] and calculations for arbitrary pulse shapes can be done numerically with hydrodynamic codes.

The material properties important to the photospall process are: the absorption coefficient, μ; the sound speed, c; the Grüneisen coefficient, Γ; and the spall strength, σ_s. It is convenient to define a characteristic time, t_o, where $t_o = 1/\mu c$, which is the time for an acoustic signal to propagate a distance of one laser absorption depth ($1/\mu$). Also, let τ_L be the ratio of the laser pulselength to this characteristic time, so $\tau_L = t_L / t_o = \mu c \, t_L$. It is instructive to discuss how the stress waveform develops in the material. For simplicity, the sound speed is assumed to be constant (actually at higher compressive stress, the speed is generally slightly higher, but this change is probably negligible for the low stresses and short transit distances of interest in the present discussion). If the laser energy was deposited instantaneously in the material, the stress, σ, in the material, just after deposition, would be compressive with an amplitude at a distance x into the material of

$$\sigma = \Gamma \mu F_o \exp(-\mu x) \quad . \tag{1}$$

Actually, because the front surface is a free surface (it is assumed that a gas, which has a shock impedance of essentially zero, or a vacuum is outside the front surface), the stress must be zero for all time at x = 0; this boundary condition is insured by the ensuing reflection at this surface. The stress pulse immediately begins to propagate both toward the surface and away from the surface with a velocity c. This bifurcation in direction, leads to a factor of two reduction in amplitude. For illustrative purposes, one can think of the stress given in Eq. 1 as consisting of two pulses, each with half the amplitude of Eq. 1; one moving inward, the other outward. At any given time, the stress is the superposition of these two pulses. The inward moving pulse simply translates with velocity c. Each element of the outward moving pulse reverses sign and direction as it reflects from the front surface. This results in a bipolar stress pulse, which, at any given time t, makes an abrupt transition at the location ct in the material from the largest tensile value to the largest compressive value. For times of the order of or less than the characteristic time, t_o, the peak tensile stress is building asymptotically from zero to its maximum negative value of $-\Gamma \mu F_o/2$, while the peak compressive stress is decreasing asymptotically from $+\Gamma \mu F_o$ to $+\Gamma \mu F_o/2$. During times for which $t >> t_o$, the stress amplitude decreases from the peak stress value, σ_p, to zero exponentially as $\exp(-\mu|x'|)$ in both directions, where |x'| is the distance from the location ct. Because of the assumed linear elastic behavior, similar statements could be made regarding the stress as a function of time at given positions in the material. It is interesting to note that nowhere in this paper are the processes under discussion dependent on a shock front[5] developing in the material.

If the laser has a finite pulselength, t_L, then stress propagation occurs during the pulse causing a number of differences from instantaneous energy deposition. For a rectangular step function laser temporal profile (as is being assumed), the tensile stress does not begin to develop until the laser pulse ends. During times for which $t \gg t_o$, and $t \gg t_L$, the tensile and compressive stress peaks become separated by a distance ct_L and the peak stress value, σ_p, is reduced by a factor A, which accounts for the stress relief that occurs during the laser pulse, where $A = (1-\exp(-\tau_L))/\tau_L$. This attenuation factor approaches 1 as τ_L goes to zero and approaches $1/\tau_L$ for large values of τ_L. The factor A is plotted versus τ_L in Fig. 2. As illustrated in Fig. 2, the peak stress is reduced substantially when τ_L is greater than one.

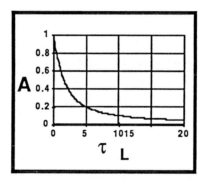

Figure 2. Stress attenuation versus ratio of laser pulselength to characteristic time

Thus, the peak tensile stress, σ_p, developed in the material occurs (assuming no spall) at times late compared to the laser pulse duration, t_L, and late compared to the characteristic time, t_o, and has an amplitude of

$$\sigma_p = (A/2) \Gamma \mu F_o \quad \text{for } t \gg t_L \text{ and } t \gg t_o . \tag{2}$$

It is interesting to note that, when $\tau_L \gg 1$, then σ_p is inversely proportional to t_L and independent of μ; that is,

$$\sigma_p \approx (1/2) \Gamma F_o / (ct_L) \quad \text{for } t \gg t_L, \quad t \gg t_o \text{ and } \tau_L \gg 1 . \tag{3}$$

To better understand the origin of the stress, it is helpful to examine the terms in Eq. 2. The Grüneisen coefficient can be written as $\Gamma = \beta / (\rho C_v \kappa_T)$ where β is the thermal expansion coefficient, ρ is the density, C_v is the specific heat at constant volume and κ_T is the isothermal compressibility of the material. Putting this into Eq. 2 gives $\sigma_p = (A/2)(1/\kappa_T)\beta(\mu/\rho)F_o / C_v$.

The product $(\mu/\rho)F_o$ gives the peak laser dose in energy per unit mass, which occurs at the front surface of the material; dividing this by C_v gives the temperature rise; multiplying this by β gives the thermal expansion; multiplying this by $1/\kappa_T$ gives the stress caused by this thermal expansion; multiplying this by 1/2 accounts for the stress propagating both inward and outward as discussed above; and multiplying this by A corrects for stress propagation during the laser pulse.

Equation 2 assumes a Mie-Grüneisen equation of state, which is applicable for the solid and liquid phases. At high fluences where vaporization occurs (i.e., $(\mu/\rho)F_o > \Delta H_i$), as the vapor expands, the equation of state more nearly approximates that of an ideal gas.

In reality it takes a finite time for materials to fracture when subjected to a tensile stress; fracture models have been developed to account for this complication[7]. Neglecting this complication for now, if the spall strength of the material is less than the peak tensile stress, then the material will spall at the location and time in the material where and when the peak stress reaches the spall strength. Thus, spall occurs if $\sigma_p > \sigma_s$. After the spallation occurs, depending on the circumstances, the tensile stress may build again to the spall strength amplitude, spalling off another layer. The total number of spall layers is approximately one to two times the ratio σ_p / σ_s and the total spall depth is approximately one to a few times the laser absorption depth $(1/\mu)$. These relations for the number of spall layers and the spall depth are not exact, principally because part of the tensile stress pulse generally will have propagated beyond the spall location by the time of each spall and because of the discrete nature of the spallation process in association with an exponentially increasing stress with distance into the material.

4. ANALYTIC THERMOELASTIC STRESSWAVE CALCULATIONS

Let $\tau = \mu c t$, $\tau_L = \mu c t_L$, and $\xi = \mu x$. Then, assuming that the laser flux is constant for a duration of t_L and with a total fluence of F_o, evaluating the integrals presented in the paper by Bushnell[8] gives:

$$\sigma = (1/(2\tau_L)) \Gamma\mu F_o [\ \exp\text{-}(\tau - \xi) \quad - \exp\text{-}(\tau + \xi) \quad] \quad \text{for } \tau \leq \tau_L,\ \xi \leq \tau \tag{4}$$

$$\sigma = (1/(2\tau_L)) \Gamma\mu F_o [\ \exp\text{-}(\xi - \tau) \quad - \exp\text{-}(\xi + \tau) \quad] \quad \text{for } \tau \leq \tau_L,\ \xi \geq \tau \tag{5}$$

$$\sigma = -(A/2) \Gamma\mu F_o [\ \exp\text{-}(\tau - \tau_L - \xi) - \exp\text{-}(\tau - \tau_L + \xi)] \quad \text{for } \tau \geq \tau_L,\ \xi \leq \tau - \tau_L \tag{6}$$

$$\sigma = -(A/2) \Gamma\mu F_o [B\exp\text{-}(\tau - \tau_L - \xi) - \exp\text{-}(\tau - \tau_L + \xi)] \quad \text{for } \tau \geq \tau_L,\ \tau - \tau_L \leq \xi \leq \tau \tag{7}$$

$$\sigma = (A/2) \Gamma\mu F_o [\ \exp\text{-}(\xi - \tau) \quad + \exp\text{-}(\xi + \tau - \tau_L)] \quad \text{for } \tau \geq \tau_L,\ \xi \geq \tau \tag{8}$$

where $A = (1 - \exp(-\tau_L))/\tau_L$ and $B = (\exp 2(\tau - \tau_L - \xi) - \exp(-\tau_L))/(1 - \exp(-\tau_L))$ (9)

Figures 3 and 4 present the results of calculations using Eqs. 4-9, which assume that the

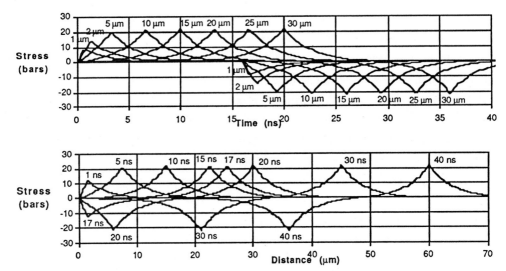

Figure 3. Thermoelastic stresswave development for long laser pulselength compared to characteristic time. Parameters used were: $\mu = 2700$ cm^{-1}, $c = 1.5$ μm/ns, $\Gamma = 0.15$, $F_0 = 0.07$ J/cm^2, $t_L = 16$ ns. Resultant values are: $\tau_L = 6.5$, $\sigma_p = 22$ bar.

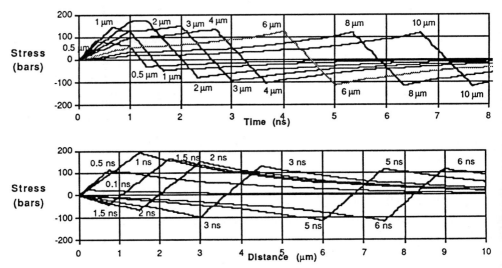

Figure 4. Thermoelastic stresswave development for short laser pulselength compared to characteristic time. Parameters used were: $\mu = 2700$ cm^{-1}, $c = 1.5$ μm/ns, $\Gamma = 0.15$, $F_0 = 0.07$ J/cm^2, $t_L = 1$ ns. Resultant values are: $\tau_L = 0.41$, $\sigma_p = 117$ bar.

spall strength is large enough that spall does not occur. In Fig. 3, the material properties used are those estimated for corneal tissue exposed to a typical ArF laser pulse[5,6]. Figure 4 is for the same conditions except that the laser pulselength is 1 ns, instead of 16 ns. In the top part of Figs. 3 and 4, the stress in the material is plotted as a function of time at different positions in the material. The bottom part of Figs. 3 and 4 gives the stress as a function of position in the material at different times. In these figures, it may be noted that the separation between the positive and negative peaks is equal to the pulse length, that no negative stress develops during the laser pulse and that the peak stress amplitude is much larger for the 1 ns pulse than for the 16 ns pulse as indicated by Eq. 2.

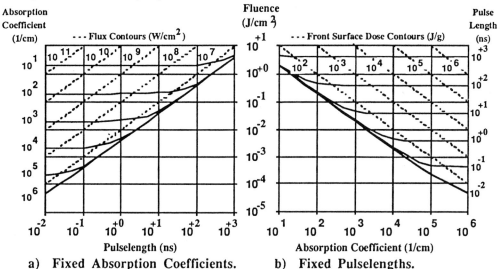

a) Fixed Absorption Coefficients.   b) Fixed Pulselengths.

Figure 5. Fluence threshold for front surface spallation calculated from Eq. 2 using $\sigma_s / \Gamma = 100$ bar, $c = 2$ μm/ns, and $\rho = 1$ g/cm^2.

Figure 5 gives threshold fluence curves for front surface spallation from Eq. 2 for the specific parameter values indicated (typical of a liquid), plotted as a function of pulselength for various absorption coefficients in Fig. 5 a) and as a function of the absorption coefficient for various pulselengths in Fig. 5 b). The curves can be easily scaled to different parameter values using Eq. 2 For liquids and solids, typical order of magnitude values of parameters are: 0.3 to 3 for the Grüneisen coefficient, 5 to 50 kbar for the spall strength of solids[9], 10 to 100 bars for the spall strength of liquids[9], 1 to 8 μm/ns for the sound speed, and 1 to 20 g/cm^3 for the density. Because the spall strength for liquids is small, front surface spallation will frequently tend to cause ablation of exposed solid targets down to the melt depth. The dashed lines in Figs. 5 a) and b) are respectively contour lines for constant flux (F_0/t_L) and constant front surface energy deposition ($\mu F_0/\rho$). The threshold fluence curves in Fig. 5 merge in the vicinity of $\mu c t_L = 1$ such that for much larger t_L or μ, the fluence is determined by Eq. 3 (which is independent of μ) and for much smaller t_L or μ, the fluence is determined by Eq. 2 with A = 1 (which is independent of t_L). For example, Fig. 5 shows that (for the parameters used in Fig. 5) for a fixed absorption coefficient of 10^4 cm^{-1}, the threshold fluence decreases as the pulselength is decreased to about 1 ns, but does not decrease much for shorter pulselengths. Similarly, for a fixed pulselength of 1 ns, the threshold fluence decreases as the absorption coefficient is increased to about 10^4 cm^{-1}, but does not decrease much for larger absorption coefficients. Figures 5 a) and b) respectively show that, for the parameters used, a flux of at least 4×10^6 W/cm^2 and a front surface dose of at least 20 J/g are required for front surface spall. This flux is not far from that at which aerosol breakdown in air[1] may become a problem so that some exposures of interest might need to be performed in vacuum.

For medical applications where tissue apparently has a reasonably high Grüneisen coefficient and a reasonably small spall strength, front surface spallation should be a possible means of ablation[5]. For aluminum, as an example of a metallic solid, the Grüneisen coefficient is about 1, the melt energy is about 1000 J/g (desirable to drive to melt to reduce spall strength), the penetration depth ($1/\mu$) for optical photons is about 10^{-6} cm, the sound speed is about 5 μm/ns, which suggests a pulselength of about 2 ps or less (to keep $\mu c t_L < 1$). The thermal diffusivity, k, is about 1 cm^2/s so that in 2 ps, the thermal diffusion depth ($\delta = \sqrt{4kt}$) is about 3×10^{-6} cm, which exceeds the laser penetration depth, so that Eq. 2 would have to be corrected for thermal diffusion. Although much higher doses would be required than needed in Fig. 5, it appears feasible to ablate very thin layers of metals with very short pulselength optical lasers.

5. HYDRODYNAMIC CODE CALCULATIONS

Figures 6 and 7 present the results of hydrodynamic calculations, using the Chart D code[10] with the laser and material properties shown in those figures, to illustrate how the stress behavior changes when front surface spall occurs. The equation of state used was approximately that for aluminum, but it was artificially changed for illustrative purposes. In Fig. 6, a large spall strength was used so that spall would not occur; the stress is plotted in dynes/cm^2 (10^6 dynes/cm^2 = 1 bar) as a function of position in the material at times of 1.59 ns, 1.90 ns, 10 ns and 20 ns. In these plots the front surface was at x = 0.2 cm and x decreases as the distance into the material increases. In Fig. 7, where a spall strength of 3 kbar was used, the stress is plotted at essentially these same times plus at two additional times. The dashed vertical lines in Fig. 7 indicate the location at which spall planes develop. Spall first occurs at a time of 1.54 ns when the tensile stress reaches 3 kbar. The plot at 1.90 ns shows that the tensile stress has again built up to nearly 3 kbar. The plot at 1.92 ns shows that a second spall plane has developed. At 10 ns, six spall planes have developed and another is about to occur. At 10.9 ns, the seventh and last spall plane has occurred. After spall, the residual stress pulse in each spall layer reflects repeatedly within that layer. The momentum of each spall layer is about the same, but the thickness increases exponentially with successive layers so that the velocity

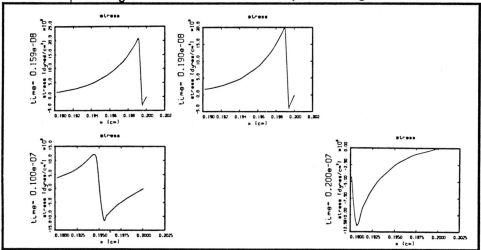

Figure 6. Hydrodynamic code calculations for no front surface spallation.

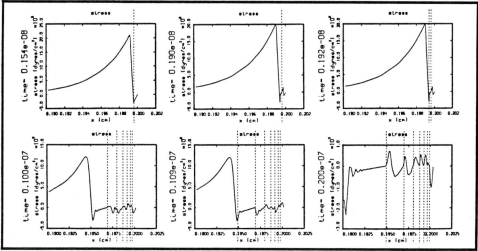

Figure 7. Hydrodynamic code calculations with front surface spallation.

decreases exponentially with successive layers. This signature might be something to look for experimentally. The position, x, in Figs. 6 & 7 is a Lagrangian coordinate (which is fixed to the mass elements rather than representing location in space) so that the gaps between spall layers do not appear in the plots.

6. SUMMARY

Front surface spallation is a potential mechanism for laser ablation of materials that may have a number of significant applications, such as medical surgery[5]. Perhaps it could be used also for desorption or to launch material (for example, large biological molecules) from a surface into a region where other techniques might be used to study the launched material. For example, the launched material, whose stoichiometry should be well determined, might be vaporized by exposure to another laser beam for production of superconducting thin films. It might also have value in cutting or shaping materials, including precision etching, such as for lithography.

7. REFERENCES

1. A. N. Pirri, *Theory for momentum transfer to a surface with a high-power laser*, Physics of Fluids, Vol. 16, no. 9, pp. 1435-1440, 1973.

2. R. S. Dingus and S. R. Goldman, *Plasma energy balance model for optical-laser induced impulse in vacuo*, Proceedings of the International Conference on Lasers '86, pp. 111-122, Orlando, Florida, 1986.

3. R. S. Dingus et al, *Pulsed Laser Effects Phenomenology*, in M. J. Berry and G. M. Harpole, editors, SPIE Proceedings of Thermal and Optical Interactions with Biological and Related Composite Materials, Vol. 1064, 1989.

4. R. S. Dingus and B. P. Shafer, *Laser-induced shock wave effects in materials*, SPIE Proceedings of Laser-Tissue Interaction, Vol. 1202, 1990.

5. R. S. Dingus and R. J. Scammon, *Grüneisen-stress induced ablation of biological tissue*, to be published in SPIE Proceedings of Laser-Tissue Interaction II, Vol. 1427, 1991.

6. P. E. Dyer and R. K. Al-Dhahir, *Transient photoacoustic studies of laser tissue ablation*, Proceedings of Laser-Tissue Interaction, SPIE Vol. 1202, 1990.

7. F. R. Tuler and B. M. Butcher, *A criterion for the time dependence of dynamic fracture*, J. of Fracture Mech. 4-4, 431-437, Dec, 1968.

8. J. C. Bushnell and D. J. McCloskey, *Thermoelastic stress production in solids*, J. Appl. Phys. 39, 5541-5546, 1968.

9. S. L. Thompson and H. S. Lauson, *Improvements in the CHART D Radiation-Hydrodynamic Code II: A Revised Program*, SC-DR-710715, Sandia Laboratories, Albuquerque, NM, February 1972.

10. D. E. Grady, *The Spall Strength of Condensed Matter*, J. Mech. Phys. Solids 36:3, 353-384 (1988).

Part V

Fundamental Physics

LASER ABLATION AND OPTICAL SURFACE DAMAGE

L. L. Chase, A. V. Hamza, and H. W. H. Lee
University of California
Lawrence Livermore National Laboratory
Livermore, CA 94550

ABSTRACT

Laser ablation usually accompanies optical surface damage to bare surfaces and coatings. Investigations of optical damage mechanisms by observation of ablation processes at laser fluences very close to the optical damage threshold are described. Several promising surface characterization methods for investigating damage mechanisms are also described. The possible role of laser ablation in initiating or promoting optical surface damage is discussed.

I. Introduction

Optical damage to nominally transparent bare surfaces and optical coatings is a form of laser ablation that is a major technological problem limiting the cost and performance of high power lasers. Despite the fact that optical materials often have bandgap energies that exceed the photon energies by severalfold, there are several possible sources of laser-surface interaction, and several of these may be present simultaneously on a particular surface. Generally speaking, any surface or near-surface feature that causes or enhances the deposition of laser energy can initiate surface damage or affect the damage threshold fluence. Examples of features that cause absorption are bulk and surface electronic states, electronic defects, impurities, and absorbing particulate inclusions. Defects that enhance absorption include cracks, cleavages, voids, grain boundaries, steps, and other types of surface roughness or irregularity. Although it has sometimes been possible to correlate optical damage thresholds with some of these causes, the physical processes that occur in the damage events are not understood. Without such an understanding, only very empirical approaches are possible to develop improved surface processing and coating methods to obtain higher damage thresholds. In this article several methods for investigating optical surface damage mechanisms and examples where laser ablation may play a role in the optical damage process are discussed.

II. Characteristics of optical surface damage

Optical damage is usually defined as a surface modification that is observable in an optical microscope. The rationale for this definition, which is not very precise, is that such modifications will result in observable changes in the optical transmission or scattering from the surface and impair its performance. Such visible surface modifications usually occur abruptly above a threshold optical fluence or intensity. Investigations with pulses of varying widths reveal that the threshold fluence varies with pulse width approximately as

$F \sim t_p^\alpha$ with $\alpha = 0.2 - 0.7$

and the threshold peak intensity, $I \sim F/t_p$, therefore varies as $t_p^{\alpha-1}$. This behavior shows that neither the fluence nor the intensity is the significant variable determining optical damage, and the ratio of intensity to fluence is an important variable. There are several physical mechanisms that could account for this behavior, including thermal diffusion and nonlinear absorption.

The threshold nature of optical damage creates a problem in searching for the surface characteristics and physical mechanisms that cause it. Several laser-surface interactions may be occurring below the damage threshold. This is illustrated in Fig. 1, which shows the types of interactions of the laser flux with the surface, assuming that the energy deposition rate in the surface region from the laser flux I(t) is a general power law, $dE/dt = c(t) I^n(t)$. If sufficient linear absorption is present, due to defects or

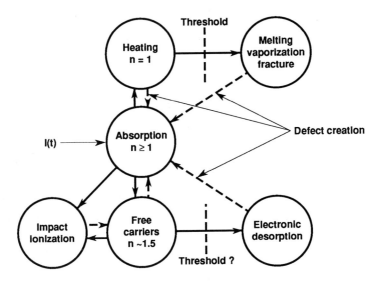

Figure 1: Various mechanisms by which the laser intensity I(t) is absorbed at a surface to cause optical damage or ablation.

nonstoichiometry, direct heating of the surface can lead to melting, vaporization, and fracture each of which can occur above a more or less sharp threshold fluence or intensity. Free carriers produced by linear or multiphoton absorption can contribute to optical damage initiation in several ways. The free carriers absorb energy with a power law exponent of 3/2 in the simplest model. This absorbed energy can cause direct heating. Also, at sufficiently high laser intensities, the carriers may gain enough kinetic energy to generate additional carriers via avalanche ionization and catastrophically increase the rate of energy deposition. The carriers can also cause desorption of surface atoms via "electronic" processes. Heating, melting, vaporization, fracture, and electronic desorption can create additional absorbing defects that modify the surface properties and lead to multiple pulse damage effects. Surface modification during the laser pulse may create the time dependence of the parameter c(t).

Observation of physical events occurring above the damage threshold or of the morphology of the damaged regions has not been very useful for investigating fundamental causes of surface damage. The problem is that these methods of investigation do not provide clues to the physical processes going on to the left of the "threshold lines" in Fig. 1. For this reason it is preferable to employ experimental techniques that probe events occurring at fluences below the damage threshold and to characterize the surface so that the initial surface properties can be correlated with optical damage sites and morphology. In the remainder of this article we give several examples of such investigations. These examples fall into three categories: 1) observation of surface ablation products, such as neutrals, ions and electrons; 2) in situ surface characterization; and 3) investigations with pairs of laser pulses that function somewhat like a "flight recorder", in analogy with airplane crashes.

III. Laser ablation as a probe of optical damage processes

Many efforts have been made to investigate optical surface damage processes by observing the characteristics of ablated particles, (electrons, ions, and neutral atoms and molecules) that are produced by laser excitation. Much of the recent work has been reviewed recently [1]. The identity, charge states, electronic excitation, yields, velocity distributions, and other significant properties of the ablated particles furnish a wealth of detailed information about ablation events, but detailed information on the effects of this ablation on the surface being investigated has rarely been obtained. Although the results have been useful, they have often been rather fragmentary and difficult to relate to the initial optical damage mechanism since the ablation itself very often occurs above a threshold fluence, which is very often the threshold for multiple-pulse damage.

A typical example of surface ablation occurring at fluences below the single pulse damage threshold is shown in Fig. 2. Here, emission of sodium neutrals produced by

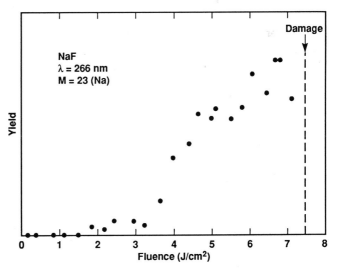

Figure 2: Yield of ablated neutral sodium atoms from a site on a polished NaF crystal surface irradiated by increasing laser fluence at 266 nm.

excitation with the fourth harmonic of a Nd:YAG laser at 266 nm was detected with a quadrupole mass spectrometer (QMS). The ablation begins at a threshold fluence of about 1.8 J/cm^2, and the damage threshold is at 7.5 J/cm^2. From time-of-flight measurements it is found that there are "slow" components (mean particle energies << 1 eV) and "fast" components (energies >> 1 eV). The slow component tends to bleach on successive laser shots at each fluence, and it is likely caused by desorption of weakly bound atoms on the surface. The fast component is very interesting; there is evidence that this is caused by "fractoemission" due to cracking or cleaving [2]. This high energy particle emission has been observed for several types of optical materials, and it usually appears just below the single-pulse damage threshold fluence.

Charged particles are also emitted in response to laser excitation. Very often, the total yield of this charged emission increases as a power law with the laser intensity, and it is tempting to associate this charged emission with electronic processes, such as photoemission [3] occurring on or near the surface. In several cases, however, it has been noted that the charged emission consists of roughly equal numbers of electrons and positive ions. The magnitudes of these charge components is, in fact, often so large that extremely high potentials would be produced by surface charging if this charge neutrality did not exist. These properties show that, in many cases, the charged emission results from photoionization of neutrals ablated from the surface [4]. In such instances, the neutral desorption is really the primary ablation process. This illustrates that caution must be exercised in interpreting such experiments.

Much has been learned from studies of laser ablation at fluences below the single-pulse damage threshold fluence. It is often found, however, that there is a threshold for this ablation and this is also the threshold for multiple-pulse optical damage. Although experiments with higher sensitivity may detect ablation at still lower fluences, it has not been established that these processes are connected with optical damage.

IV. In situ surface analysis

Fairly recently, sensitive methods of surface analysis have been employed to investigate the effects of laser irradiation on surface properties. Examples of these experiments include Auger spectroscopy, low energy electron diffraction (LEED), electron energy loss spectroscopy (EELS), laser ionization mass analysis (LIMA), and scanning force microscopy (SFM). The first three of these techniques are electron spectroscopies, which have somewhat limited effectiveness because of surface charging of many insulating optical materials. These have been used mostly for investigating basic laser-surface interaction mechanisms and surface and coating properties on relatively large surface areas. On some insulators, it is possible, although possibly very tedious, to use them for "before and after" investigations of the usually very small and randomly located optical damage sites.

LEED has been used to investigate the laser-induced modification of reconstructed surfaces on GaP [5] and Al_2O_3 [6] surfaces. In the case of the 12 X 4 reconstruction of a $(11\bar{2}0)$ surface of Al_2O_3, the modification was accompanied by emission of Al$^+$ ions.

The time-of-flight characteristics of these ions were measured with a quadrupole mass spectrometer; the data are shown in Fig. 3. The Al ions have a mean kinetic energy of about 8 eV, which is very reproducible at various laser fluences and for excitation by

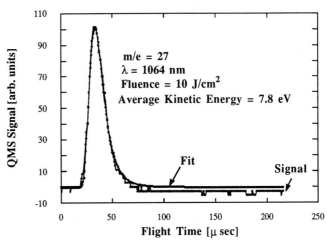

Figure 3: Time-of-flight of desorbing Al ions from Al$_2$O$_3$ after laser irradiation at 1064 nm.

the fundamental and third harmonic wavelengths of the Nd:YAG laser. This suggests that the desorption is caused by a well-defined electronic detachment process. Schildbach and Hamza [6] have suggested that most of the energy of an exciton localized on the Al ions is imparted as kinetic energy to a desorbing Al ion. This poses the interesting question of how a 1.17 eV photon from the YAG fundamental can impart 8 eV of energy to form such an exciton. Multiphoton absorption is one possibility. Another mechanism is suggested by the EELS spectrum obtained for this surface [6], which is shown in Fig. 4. In addition to the interband transitions beginning at about 9 eV

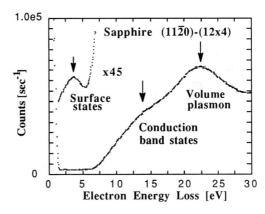

Figure 4: EELS spectrum of an Al$_2$O$_3$ surface illustrating plasmon peak, conduction band states, and surface states in the energy gap.

and the bulk plasmon at 22 eV, losses caused by surface states appear in the lower energy portion of the EELS spectrum recorded at higher sensitivity. If these surface states correspond to filled states below the conduction band, conduction electrons can be produced by direct one-photon excitation at low photon energies, and these electrons can be trapped to form the exciton. It is also interesting that Auger spectroscopy measurements show that electron bombardment of this reconstructed surface causes desorption of oxygen, whereas Al is desorbed by the much lower quantum energy imparted by the laser photons [6]. This illustrates the selectivity of the laser desorption mechanisms at photon energies much lower than the optical bandgap.

Another surface analytical technique that has recently been employed is atomic force microscopy. The most useful information can be obtained from this method by scanning a small region before and after laser excitation of a part of the region so that the influence of surface defects on laser damage mechanisms can be investigated. Recently, such measurements have been performed on coatings and bare surfaces [7].

V. Picosecond laser pulse-pair excitation

The basic mechanism of absorption of light in the surface region is of fundamental importance for understanding the causes of optical surface damage. Pump-and-probe measurements of laser ablation thresholds using picosecond laser pulses have recently been used on insulating optical materials to investigate basic absorption and damage processes [8]. These experiments can reveal whether the laser-surface interaction is linear or nonlinear in the laser intensity and the lifetimes of excitations (thermal, electronic, acoustic, etc.) produced by laser irradiation. The basic idea for this elementary pump-probe experiment is demonstrated in Fig. 5. It is assumed that the interaction of a single laser pulse with the surface produces some excitation that is the primary cause of the ablation or surface damage, and the magnitude of this excitation

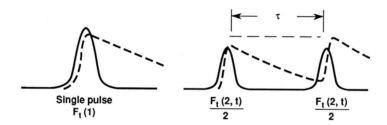

Figure 5: Single and double pulse formats and created excitation (dashed line).

(dashed line) decays in time after the laser pulse. It is further assumed that the excitation must exceed some threshold value in order to cause damage above the threshold fluence. When two pulses are applied, the excitation from the second pulse adds to the remaining contribution from the first pulse. Thus, although neither pulse exceeds the single pulse damage or ablation threshold, the threshold will nevertheless be exceeded by the pulse pair.

A simple model, assuming that the absorbed energy is proportional to a power law in the laser intensity, gives the following ratio of the damage threshold of the double pulse sequence, $F_t(2,\tau)$ to the single pulse threshold, $F_t(1)$

$$R(\tau) = \frac{F_t(2,\tau)}{F_t(1)} = \frac{2}{(1 + e^{-\tau/\tau_e})^{1/n}} \qquad (1)$$

where n is the order of the nonlinear absorption process and τ_e is the lifetime of the relevant excitation. For simplicity, we assume that the excitation decays exponentially in time; in general, much more complicated behavior is possible, including cooling of a heated region due to thermal diffusion and distributions of relaxation times for different surface regions and defects. The details of the model are not as important as the basic and simple idea behind the experiment itself: assuming that the pulses are identical in intensity and fluence, the ratio of fluence to intensity is varied by a factor of two, with the added feature of the time delay. At small time delays $\tau \ll \tau_e$, the ratio R becomes $2^{(1-1/n)}$, from which n can be determined. At sufficiently long time delays $\tau \gg \tau_e$, each pulse must independently exceed the threshold, and R = 2.

This experiment has been performed on several surfaces and coatings. The samples are mounted in a UHV chamber, and a quadrupole mass spectrometer provides a sensitive indication of when the damage or ablation threshold has been exceeded. Two examples are illustrated in Figs. 6 and 7. For the single crystal ZnS surface illuminated by picosecond dye laser pulses at 590 nm, the ratio R is observed to be close to unity over at least two orders of magnitude in time delay, and appears to

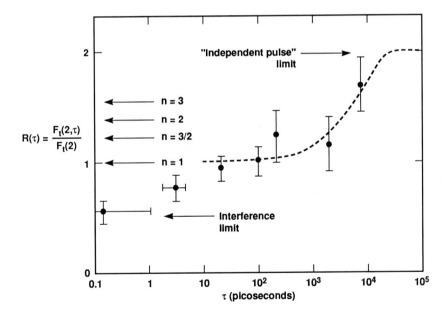

<u>Figure 6</u>: R(τ) as a function of time delay for a ZnS single crystal surface. Limits for various power law processes are indicated by arrows. A fit to Eq. 1 with n = 1 and τ_e = 5 ns is the dashed curve.

increase toward 2 on a nanosecond time scale. The linear nature of the laser-surface interaction in this case is clearly differentiated from nonlinear processes, for which R would have the values indicated by the arrows. At very short time delays ($\tau < 5$ ps), the temporal overlap of the dye laser pulses produces fringes on the surface, which double the laser intensity in the fringe maxima, and the ratio R actually decreases to 0.5, as expected [8]. This problem can be avoided by employing orthogonally polarized laser pulses or two different wavelengths for the pulses. In this case, the fact that the ratio R = 1 at longer time delays means that nothing further will be gained by doing this. These data indicate that linear absorption causes the optical damage, despite the fact that the peak intensity of the laser pulses was about 5×10^{11} W/cm^2 and the wavelength was in the range of very strong two-photon absorption. We believe that heating caused by the absorption is the cause of optical damage in this case. The relaxation time is estimated from the data to be about 5 ns, a typical thermal relaxation time for uniformly heated defects of submicron dimensions, assuming the bulk thermal diffusivity of ZnS.

Figure 7 shows the dependence of R on time delay for the $(11\bar{2}0)$ surface of an Al$_2$O$_3$ single crystal excited by 80 ps laser pulses of wavelength 1064 nm. R = 2 at pulse delays longer than a few ns, and it decreases at shorter delays. Temporal overlap of the pulses prevents measurements at shorter delays which would allow an unambiguous determination of the exponent n. The possible fits to the data shown in the figure

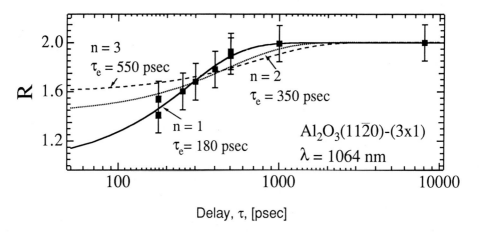

Figure 7: R(τ) for a single crystal Al$_2$O$_3$ surface irradiated by an 80 ps laser pulse at 1064 nm. Fits to Eq. 1 using various lifetimes and power laws are shown.

indicates that the most likely absorption process is linear, or at most a low-order nonlinear process. The short relaxation time of about 200 ps is most likely due to an electronic relaxation process. These observations are consistent with the explanation for Al ion desorption mentioned above in section IV. Other samples which have been investigated using this experiment include a borosilicate glass, which exhibited strongly nonlinear behavior [8] and single layer coatings of SiO_2 and HfO_2, for which linear absorption with relaxation times of ~35 ns (SiO_2) and ~1 ns (HfO_2) were inferred from the data [9].

We believe that this experiment, and possibly dual-wavelength variants of it, will be very useful for obtaining crucial information concerning optical damage and ablation processes. Caution must be exercised, however, in interpreting the data since the assumptions implicit in the model used to obtain n and τ_e may not be valid in all cases. Data taken with a range of pulse lengths may be necessary to build a complete picture of laser-surface interaction mechanisms in some cases. Although most of our measurements so far can be simply interpreted in terms of linear absorption, many complicated and interrelated events may be occurring on surfaces illuminated by intense laser beams, and a great variety of behavior might be anticipated.

VI. Conclusions

Considerable progress is being made in understanding the basic laser-surface interactions that initiate ablation and optical damage. The most difficult aspect of attacking the problem is the threshold behavior of these processes, which can be circumvented by a combination of in situ surface analysis, measurements of ablated particles and the effects of varying laser wavelengths, pulse formats, and pump-probe techniques.

Acknowledgments

This research was supported by the Office of Materials Sciences, Office of Basic Energy Sciences, U. S. Department of Energy and by Lawrence Livermore Laboratory under contract No. W-7405-ENG-48.

References
[1] E. Matthias and R. W. Dreyfus, Topics in Current Physics, Vol. 47, P. Hess, ed., Springer-Verlag (Berlin, 1989), p. 89.
[2] L. L. Chase and L. K. Smith, Laser-Induced Damage in Optical Materials: 1987, NIST Special Publication No. 756 (National Institute of Standards and Technology, Washington, 1988), p. 165.
[3] W. J. Siekhaus, J. H. Kinney, D. Milam, and L. L. Chase, Appl. Phys. A 39, 163 (1986).
[4] L. L. Chase, Nucl. Inst. Meth. B 46, 252 (1990).
[5] Y. Kumazaki, Y. Nakai, and N. Itoh, Phys. Rev. Lett. 59, 2883 (1987).
[6] M. A. Schildbach and A. V. Hamza, submitted to Physical Review B.
[7] W. J. Siekhaus, private communication.
[8] L. L. Chase, H. W. H. Lee, and Robert S. Hughes, Appl. Phys. Lett. 57, 443 (1990).
[9] L. L. Chase, H. W. H. Lee, and A. V. Hamza, to be published.

DISCLAIMER

This document was prepared as an account of work sponsored by an agency of the United States Government. Neither the United States Government nor the University of California nor any of their employees, makes any warranty, express or implied, or assumes any legal liability or responsibility for the accuracy, completeness, or usefulness of any information, apparatus, product, or process disclosed, or represents that its use would not infringe privately owned rights. Reference herein to any specific commercial products, process, or service by trade name, trademark, manufacturer, or otherwise, does not necessarily constitute or imply its endorsement, recommendation, or favoring by the United States Government or the University of California. The views and opinions of authors expressed herein do not necessarily state or reflect those of the United States Government or the University of California, and shall not be used for advertising or product endorsement purposes.

LASER INDUCED PHOTODISSOCIATION, DESORPTION AND SURFACE REACTION DYNAMICS

T. J. Chuang

IBM Almaden Research Center
650 Harry Road
San Jose, CA 95120-6099

The basic processes of surface photochemistry induced by ultraviolet light involve photodissociation, trapping of photofragments, reaction of surface species among the adsorbates or between the adsorbate and the adsorbent, as well as the desorption of reaction products and/or the excited parent molecules. The gross features of these processes have been investigated to some extent in the last decade by many conventional surface analytical techniques. For instance, we have used electron spectroscopies, thermal desorption and time-of-flight mass spectrometry to study the UV radiation effect on a model system: CH_2I_2 molecules adsorbed on Al_2O_3, Al and Ag surfaces from submonolayer to multilayer (thin film) coverages [1-3]. The photofragmentation patterns, the mass and the translational energy distributions, the surface reaction pathways and the collision effects are deduced from such studies. The characteristic electronic, thermal and explosive desorption behaviors have also been identified.

In the continuing effort to gain deeper understanding of the basic energy transformation process originated by the absorption of a UV photon and the subsequent surface reaction dynamics, we have recently employed a state-selective and time-resolved optical spectroscopic technique to determine the energy as well as the angular distribution of desorbed species. The experimental approach is based on a two dimensional (2D) imaging method to measure the kinetic energy, the internal state and the angular distributions of species photogenerated and/or photodesorbed from surface using two pulsed lasers. The photochemical event is initiated by an excimer laser pump pulse irradiated on the surface at a given coverage of adsorbed molecules. The desorbed photofragments, reaction products or parent molecules are selectively ionized via resonance-enhanced multiphoton ionization (REMPI) in the gas phase by an instantaneous sheet of probe light which is tunable in wavelength and is formed in front of the surface. By varying the delay time of the probe pulse with respect to the pump pulse, the translational energy distribution can be deduced. The internal (electronic, vibrational and rotational) state distribution is provided by the REMPI spectroscopy by scanning the wavelength of the probe laser. The ions generated in the field-free region in front of the substrate by the sheet of probe pulse contains two-dimentional spatial information directly correlated to the angular distribution of the desorbed species. After passing through a ground-grid, the ions are accelerated toward a microchannel plate (MCP). The 2D image of the phosphor behind the MCP is then captured by a CCD-camera, digitized and processed by a computer. Some recent results of the NO/Pt(111) and the $NO_2/Al_2O_3(11\bar{2}0)$ systems [4,5] are presented here.

The experiments are carried out in a UHV chamber equipped with x-ray photoemission (XPS), Auger electron spectroscopy (AES), thermal desorption spectroscopy (TDS) with a rf-induction heater, an ion gun for surface cleaning, a quadrupole mass spectrometer (QMS) for both TDS and time-of-flight mass spectrometry (TOF-MS) and a port specially designed for the laser excitation and REMPI imaging study. The cleaning and annealing procedure for the Pt(111) surface is well established. The UV-grade $Al_2O_3(11\bar{2}0)$ crystal is cleaned by heating to 750K. The substrate is cooled by liquid nitrogen (LN_2) to about 90K and exposed to the gas through a dosing tube facing the sample. At the saturation coverage for NO, about 0.5 of a monolayer is adsorbed on the Pt(111). NO molecules do not adsorb on Al_2O_3 at 90K as determined by XPS and TDS. In contrast, NO_2 can be easily adsorbed or condensed on the substrate.

For the REMPI-2D imaging experiments, an excimer laser operated at 193nm is used as the pump laser. The wavelength tunable UV probe laser is generated by a XeCl excimer laser pumped dye laser with frequency doubling by a BBO crystal. The beam is expanded and passes through a rectangular slit before being focused with a quartz cylindrical lens. It is reflected by a dielectric mirror in the UHV chamber and travels vertically in front of the sample surface forming a strip of light with a width of 0.1-0.2mm and a length of 10.0mm. It is then reflected by another mirror out of the chamber. A distance of 4.5mm is set between the probe light and the sample resulted in the REMPI probing angles of $0°\pm 65°$ in the vertical direction and $0°\pm 48°$ in the horizontal direction.

The laser desorbed species in a given electronic, vibrational and rotational state can be ionized by multiphoton absorption of the probe beam. For NO, the (1+1)-REMPI spectroscopy has been well studied and used in the present experiment. The ions generated by the probe laser continue to fly in the field-free region for 5.0mm before entering a stainless steel TOF tube applied with a negative voltage. The ions are accelerated toward a microchannel plate (MCP, Galileo, Chevron-type) of 40mm in diameter producing luminescence signal on a phosphor screen behind the MCP. The two dimensional (2D) pattern on the phosphor is detected by a cooled CCD-camera (Hamamatsu) coupled to an image digitizer and processor controlled by an IBM PC-AT computer which is connected to a host computer. The high voltages for both the MCP and the phosphor are gated with respect to the firing of laser pulses. This can eliminate the spurious background signals.

A typical REMPI-2D image of NO photodesorbed from Pt(111) is displayed in Fig. 1. The probe laser is set at R_{11} ($J''=7.5$, $V''=0$) rotational line and the delay time t_d at 3.3μs following the pump laser which is fixed at a fluence of $F=2.9$mJ/cm^2. The shapes of the images from $J''=0.5$ to 20.5 remains practically the same. The signal clusters around the center and is quite symmetrical with respect to the surface normal. To deduce the angular distribution from a 2D image, we take a narrow vertical strip passing through the center as depicted by the parallel lines in Fig. 1. These pair of lines cover $\theta=0°\pm8°$ in the horizontal direction and $\theta=0°\pm65°$ in the vertical direction. The profile of the strip is shown in Fig. 2a. Clearly the desorption flux peaks toward the surface normal. The best fit of the profile as a function of desorption angle is $\cos^4\theta$ (represented by the dashed curve in Fig. 2a). Figure 2b shows a similar profile obtained at $F=6.5$mJ/cm^2. In the fluence range used in the present study, i.e., from 0.5 mJ/cm^2 to 10 mJ/cm^2, the REMPI signal strength increases linearly with F. Again, most image profiles can be best fitted into a $\cos^4\theta$ distribution.

The results obtained so far by the REMPI method on NO/Pt(111) can be summarized as follows: (1) The NO desorption signal from Pt(111) induced by 193nm laser pulses has a quantum yield in the order of 10^{-4} - 10^{-5}. The signal strength increases linearly with the laser intensity. (2) A single TOF component is observed with a non-Maxwell velocity distribution. The translational energy is independent of laser intensity and much higher than the substrate temperature. (3) The rotational state distribution appears non-Boltzmann in general with the high J" levels much more populated than those in equilibrium with the substrate temperature. (4) The desorption flux peaks toward the surface normal showing a $\cos^4\theta$ distribution. The desorption yield is so small (less than 3×10^{-3} monolayers per pulse) that collisions above the surface can be ruled out.

Clearly the photodesorption signal observed by REMPI is induced by electronic excitation effect. From the present results and the existing data in the literature, it is still difficult to deduce the precise mechanism responsible for NO photodesorption. At present, both valence excitation of the NO-surface complex and temporary ion resonance involving "hot" electrons from the metal substrate have been proposed to be responsible for the behavior. Obviously more quantitative results, in particular, the quantum yields as well as the energy distributions as a function of the laser excitation wavelength, the incidence angle and the polarization need to be obtained before a definitive conclusion can be drawn. Likewise, the results on the angular distribution of

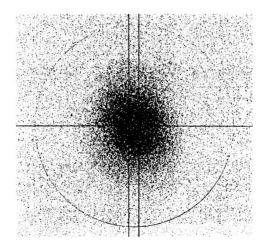

Figure 1. A typical REMPI-2D image of NO desorbed from Pt (111) at 90K by 193nm laser pulses with F= 2.9 mJ/cm². The probe laser is set at the $R_{11}(7.5)$ rotational line with a 3.3μs delay time following the pump pulse. The circle represents the boundary of the MCP-phosphor detection area ranging from 0° at the center to 65° at the boundary. (Data according to Ref. 5)

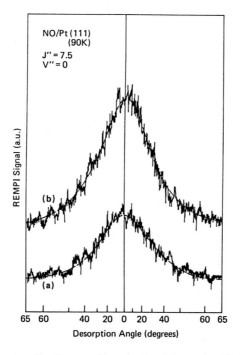

Figure 2. The angular distribution profiles obtained from the 2D images as shown in Fig. 1. The profiles are the signals as a function of desorption angle integrated between the two vertical lines also displayed in Fig. 1: (a) pump laser F= 2.9 mJ/ cm² and (b) F= 6.5mJ/cm². The dashed curves represent $\cos^4\theta$ distribution. (Data according to Ref. 5)

desorbed NO are not easy to understand. It is known that the molecule sits along the surface normal with N atom bonded to the metal substrate. Thus it is reasonable to expect NO to desorb along the normal direction. But why $\cos^4\theta$ distribution? Again, only thorough measurements, particularly the angular distribution as a function of the surface coverage and the laser excitation parameters, can provide the answer as to the correlation between the observed distribution and the surface bonding geometry as well as the desorption dynamics. Our results clearly demonstrate that the 2D-imaging technique can yield detailed information about the basic surface process.

For the $NO_2/Al_2O_3(11\bar{2}0)$ system, we have investigated the UV radiation effect with the TOF-MS apparatus and the 193, 248 and 308nm lasers. The sapphire crystal is transparent at these wavelengths. It is clear that the UV photons can effectively photodissociate the adsorbed NO_2 molecules into NO molecules and O atoms which subsequently desorb into the gas phase. At a monolayer coverage or less, NO and O are the only species detected by the QMS. The NO signal strength increases linearly with the laser intensity, but the TOF peak position is independent of F. At a given F, the 193nm photons generate a much higher NO yield than the photons at the other two wavelengths. Furthermore, the TOF peak position shifts toward a shorter time as the excitation wavelength is decreased. The typical translational energies at the peaks (E_m) for NO and O at 193nm are about 0.07eV (corresponding to about 428K) and 0.082eV (474K), respectively, with the uncertainty of ±0.005eV. The E_m value for NO decreases to 0.057eV at 248nm and 0.035eV at 308nm. The desorption yields of NO and O per pulse at a multilayer coverage are substantially higher than those at a monolayer or less. At a very high surface coverage, NO_2 molecular desorption can also be detected by the QMS. From the decrease of the TOF signal as a function of laser pulses and the corresponding reduction of NO_2 surface coverage measured by XPS, we can determine the NO production-desorption cross sections. At an initial coverage of 3-5 monolayers, the observed cross sections are about $1\times10^{-17}cm^2$ at 193nm, $2\times10^{-19}cm^2$ at 248nm and $1\times10^{-19}cm^2$ at 308nm.

It is evident that the observed NO production and desorption is induced by direct photofragmentation of adsorbed NO_2 molecules. The mass and the translational energy distributions of the adsorbate-mediated process are apparently quite different from those of the substrate-mediated process. The present results further suggest that a very large fraction of the excitation energy from the absorbed photon has been transformed into the internal excitation of the NO fragment. It should be very interesting to determine how much of the excess energy has been transferred separately into the electronic, vibrational and rotational states of the product molecule. The system is being further investigated by the REMPI-2D imaging method and the results will be reported later.

Acknowledgments:

The author wishes to thank R. Schwarzwald and A. Mödl for their contribution to the work presented here, and J. Goitia for the technical assistance with the experiment.

[1] K. Domen and T. J. Chuang, J. Chem. Phys. 90 (1989) 3318; and 90 (1989) 3332.
[2] A. Mödl, K. Domen and T. J. Chuang, Chem. Phys. Lett. 154 (1989) 187.
[3] T. J. Chuang and K. Domen, J. Vac. Sci. Technol. B7 (1989) 1200.
[4] R. Schwarzwald, A. Mödl and T.J. Chuang, Surf. Sci. 242 (1991) 437.
[5] T. J. Chuang, R. Schwarzwald and A. Mödl, J. Vac. Sci. Technol. A (1991, in press).

MECHANISMS OF LASER ABLATION OF MONOLAYERS AS DETERMINED BY LASER-INDUCED FLUORESCENCE MEASUREMENTS

Russell W. Dreyfus
IBM Research Division, Thomas J. Watson Research Center
P.O. Box 218, Yorktown Heights, NY 10598

INTRODUCTION

Identifying pathways for UV laser energy to remove monolayers, $\mathcal{ML}$, from a specific surface is a continuing quest. Most results fall in the category of thermal vaporization or photochemistry, see Fig. 1. Plasma effects are designated as a third pathway because of the strong coupling between photons and electrons in the plume, which only later produces thermal effects in the surface. Since we are presently considering removal of only $\sim\mathcal{ML}$, we shall not include macroscopic processes e.g. exfoliation, splashing or subsurface arcing [1].

The primary variables which determine the pathway are the nature of the surface (inc. defects, impurities and morphology), and the laser's power, wavelength and pulse length in decreasing order of importance. Even then, as implied in Fig. 2, transferring the concepts to other experiments or to similar materials may not always be practical because of the many orders of magnitude of power densities and etch depths. Also, interpreting results in terms of only the above four variables may be an oversimplification as both the optical absorption and stoichiometry of the surface may be time dependent. Thus, the surface is actually a time dependent variable, an important feature we shall return to later.

THE EXPERIMENT AND THE RESULTS

Laser-induced fluorescence, LIF, provides a wealth of information about the plume and hence about the ablation process. Details about the relative densities of species, electronic states, rotational and vibrational temperatures and kinetic energies are readily forthcoming, **see the table.** These quantities are determined for atoms, ions and diatomics some 1 to 2 cm above the surface and 0.5 to 20 μs after the ablation. Recording such specific information to identify ablation pathways is highly useful as all three pathways in Fig. 1 are observable with 4 to 6.4 eV photons. In simplest terms, high rotational/vibrational energies

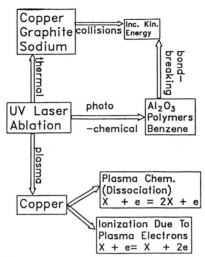

Fig. 1. Schematic diagram listing various materials investigated as to the details of processes leading to laser ablation. The dominate processes are tabulated along the pathways (arrows). The primary condition controlling the pathway is the nature of the surface and the laser's wavelength, power and pulse length in decreasing order of importance. In the few J/cm^2 laser fluence range (i.e. 100 to 500MW/cm^2 range for $\gtrsim$10ns pulse lengths) thermal processes are favored by long wavelengths and long pulse lengths. Photochemical processes are promoted by short (UV) wavelengths and high peak powers. Plasma processes are favored by high peak powers and intermediate wavelengths. At very high powers ($>>$ 500MW/cm^2 plasma processes usually dominate.

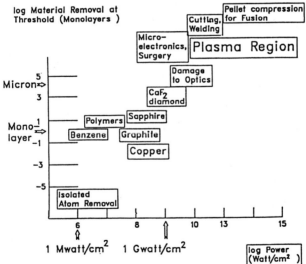

Fig. 2. Schematic illustration of the fact that laser ablation covers many decades of amount of material removed due to the extended range of power levels utilized for ablation. Failure to appreciate the extent of these ranges and how rapidly (threshhold-like) a transition from one ablation pathway to another can occur underlies many of the ongoing disagreements about one mechanism vs. another, even for a specific surface. Furthermore identifying a specific mechanism may be complicated by the fact that the surface may go through various morphological states and at various power densities in a matter of 10 to 30ns.

(high being taken as energies equivalent to temperatures which would remove significant material during the duration of the pulse; i.e. for 10ns and removal of a ML the temperature corresponds to the atmospheric pressure boiling point). Also, moderate kinetic energies are taken as evidence of a purely thermal process, whereas low internal energies but high (> 1 eV) kinetic energies are taken to indicate photochemistry. While LIF provides a framework of results, the complementary experiments are equally important to extend and to confirm these results.

Up to this point we have summarized the pathways between laser energy absorption and the production and energies of plume species. This is not a complete story yet as the question of the actual absorption process is often not an obvious one. Changes in absorption have been noted to occur over many laser pulses in polymeric materials, or alternately as a transient of only nanoseconds duration in sapphire crystals. In the case of CaF_2, many laser pulses (up to 3000) may be required before efficient generation of Ca and CaF are observed [9]. One possible explanation is that color centers (related to anion vacancies) accumulate in the surfaces and enhance the photon absorption. LIF measurements reinforce this view as large densities of anion deficient radicals, e.g. Ca, CaF, are observed in the plume. Both sapphire and polymers [10] (polyethylene terephthalate and polyimide) show transient darkening of nanosecond to microsecond duration. The former appears to represent "pumping-up" of an absorbing electronic state as the absorption disappears in < 20 ns. In the case of the polymers, it appears that the effect is due to elevated temperatures as the absorption spreads out significant distances from the irradiated area into the surrounding material and bleaches out only slowly, ~μs.

In contrast to questions about transient absorption in the surface layer, absorption of energy by electrons in the plume appears to follow well known and predictable concepts [3]. At low powers, < 100 MW/cm^2, virtually no ions are present even though ablation of neutrals may be extensive, therefore direct ejection of ions from the surface usually does not occur. At moderate power densities (~200 to 500 MW/cm^2) multiphoton ionization (within the plume) sets in, at higher powers ($\gtrsim$500 MW/cm^2) inelastic photon-electron losses dominate to the point that at powers > > 500 MW/cm^2 the plume becomes an opaque[11] plasma. The progression through these three regions has a great similarity to breakdown in gases, only with excimer lasers the power required is lower, presumably due to the random mode pattern of the laser and ease of ionizing metal vapors as compared to inert gases.

SUMMARY

Examples of photochemistry, thermal effects and plasma processes have all be identified in UV laser ablation by the application of LIF measurements. On the other hand, both transient and long lasting chemical and physical changes in the energy absorbing surface are only now coming under examination.

Table. 193 nm Ablation Processes as Determined by LIF

Material [Species]	Measured Energy Internal (K)	Kinetic (eV)	Ablation Process	Complementary Experiments
Graphite[2] [C_2,C_3]	~3800	≳0.4	Thermal	SEM of Surface
Copper[3] [$Cu°,Cu^+,Cu_2$]	(1800-2600)	≥0.2	Thermal + plasma	Langmuir Probe[4] Doppler linewidth
Benzene[5] [C_2]	3800	0.5	Photochemistry	Frozen ML and gas phase photochemistry
Sapphire[6] [AlO,Al]	≳500	1-2	Photochemistry	Oxidized Al Photomechanical[7]
Polyimide[8] [CN,C_2]	~1000 to 1400	2.8	Photochemical Thermal	SEM
CaF_2[9] [$Ca°,Ca^+,CaF$]	~1600	2	Photochemical + Thermal (?)	Incubation then Ablation

REFERENCES

1. Roger Kelly, J. J. Cuomo, P. A. Leary, J. Rothenberg, B. E. Braren and C. F. Aliotta, Nucl. Instr. and Meth., **B9**, 329 (1985).
2. R. W. Dreyfus, Roger Kelly and R. E. Walkup, *ibid* **B23**, 557 (1987) and R. W. Dreyfus, Roger Kelly, R. E. Walkup and R. Srinivasan, Proc. of Society Photo Optical Instrumentation Engineers **710**, 46 (1987).
3. R. W. Dreyfus, J. Appl. Phys. **69**, 1721 (1991).
4. R. J. von Gutfeld and R. W. Dreyfus, Appl. Phys. Letts. **54**, 1212 (1989).
5. R. W. Dreyfus and R. Srinivasan, Conf. on Lasers and Electro Optics, Opt. Soc. Amer., Anaheim, CA, 25 April 1988.
6. R. W. Dreyfus, Roger Kelly and R. E. Walkup, Appl. Phys. Lett. **49**, 1478 (1986).
7. R. W. Dreyfus, F. A. McDonald and R. J. von Gutfeld, Appl. Phys. Lett. **50**, 1491 (1987).
8. R. Srinivasan, B. Braren and R. W. Dreyfus, J. Appl. Phys. **61**, 372 (1987).
9. S. Petzoldt, E. Matthias and R. W. Dreyfus, Conf. on Lasers and Electro-Optics, Opt. Soc. Amer., Anaheim, CA, 21 May 1990,
10. R. Srinivasan, K. Casey, B. Braren and M. Yeh, J. Appl. Phys. **67**, 1604 (1990).
11. C. R. Phipps, T. P. Turner, R. F. Harrison, G. W. York, W. Z. Ozborne, G. K. Anderson, X. F. Corlis, L. C. Haynes, H. S. Steele and K. C. Spicochi, J. Appl. Phys. **64**, 1083 (1988).

Laser-Induced Particle Emission from Surfaces of Non-Metallic Solids: A Search for Primary Processes of Laser Ablation

Noriaki Itoh, Ken Hattori, Yasuo Nakai, Jyun'ichi Kanasaki and Akiko Okano
Department of Physics, Faculty of Science, Nagoya University
Furo-cho, Chikusa-ku, Nagoya 464-01, Japan

Richard F. Haglund, Jr.
Department of Physics and Astronomy
Vanderbilt University, Nashville, TN 37235 USA

Abstract. We review the phenomenology of laser-induced particle emission at sub-bandgap photon energies from the surfaces of halides and compound semiconductors, which are representative of solids with strong and weak electron-lattice coupling, respectively. In the former a multi-hole localized state generated by cascade excitation of self-trapped excitons can cause nonlinear dependence of the desorption or ablation yield on the laser fluence. In weak-coupling systems, multi-hole localization on surface defect sites by the negative-U interaction can lead to ejection of atoms. The contribution of these electronic processes to laser ablation is discussed.

Introduction

There is growing interest in laser ablation of solids, but significant questions about the underlying mechanisms remain unanswered, particularly in the case of the nanosecond laser pulses generally used in such experiments [1]. Laser ablation is initiated by the interactions of laser photons with electrons in the solid, and is characterized by a superlinear dependence of the cumulative effects of electronic excitation and relaxation processes on the laser intensity. These cumulative effects may be either thermal or electronic.

In laser ablation, the primary interaction between the laser light and the solid surface induces electronic transitions. The laser-surface interaction depends strongly on the relative magnitudes of the laser photon energy $E_{h\nu}$ and the solid band-gap energy E_G: If $E_{h\nu} < E_G$ the laser causes one-photon transitions involving surface states or defect sites, or multi-photon band-to-band transitions; when $E_{h\nu} > E_G$, the laser produces single-photon band-to-band transitions [2]. The details of laser-induced particle emission also depend strongly on the material response to optical excitation. We shall find it convenient to classify solids as type I or type II, depending on whether or not excitons and/or holes *are not* or *are* self-trapped, respectively. This division is consistent with known relationships between the transfer energy J and the electron-lattice relaxation energy E_{LR} of an exciton or hole [3,4,5]: In type I solids, $J > E_{LR}$, while in type II solids, $J < E_{LR}$.

In our discussion of electronic processes we are particularly interested in laser-induced particle emission for $E_{h\nu} < E_G$, for which it is clear that thermal effects cannot be the principal issue [6]. We first review the general features of ablation for this case, and then discuss the possible contribution of electronic effects to laser ablation in type I and type II solids. We shall confine our discussion to compound semiconductors and halides, which belong to type I and II respectively.

General Remarks about Laser Ablation

Laser ablation may be defined as light-induced particle emission from surfaces where there is significant residual surface damage. It is important to note that the term as usually employed connotes both a process of particle emission and a residual damaged surface, frequently with an accompanying laser-generated plasma. We prefer to use the terms laser-induced particle emission or laser sputtering to describe the removal process, understanding that the condition of the surface is a separate question and may not always involve detectable damage. We note that the surface morphology changes continuously during a laser pulse; the laser ablation yield thus involves an integral of the surface damage up to each stage of damage evolution, while the particle emission is characteristic of the surface at each stage of damage evolution. In this paper, we shall concentrate on the initial stage of laser ablation: particle emission from atomistically clean surfaces and the resulting evolution of surface damage, including changes in the surface structure and composition.

The possibility of electronic processes under dense electronic excitation, such as defect migration and particle emission, was first suggested by Van Vechten [7] as the mechanism for laser annealing of semiconductors. It was assumed that if n electron-hole pairs per unit volume are produced in the conduction band of a solid with N molecules per unit volume, the strength of each bond will be weakened on average depending on n/N, reducing the activation barrier for particle hopping and thus increasing the frequency of thermal migration. The effects cannot be very significant in this case, since n/N can hardly be greater than 0.1. In fact, we now understand that the van Vechten mechanism can be excluded as the cause of the laser annealing, since the time for annealing to take place is generally shorter than the lifetime of the excited state [8]. The approach taken in the present paper [9,10] is different from that proposed by van Vechten and emphasizes bond breaking due to localization of holes by an electron-lattice interaction.

The rubric "thermal ablation" refers to a mechanism in which phonons generated as a result of non-radiative transitions or of electron-lattice interactions are accumulated leading to particle emission by vibrational motion. In electronically induced laser ablation, on the other hand, an accumulation of electronic excitations leads directly to bond-breaking. Electronic mechanisms for laser ablation have not yet been explored to a significant degree; it remains the conventional wisdom that heating effects are the dominant factor in laser ablation even for $E_{h\nu} < E_G$.

However, a point of view based on statistical distribution of thermal energy to all the atoms within the absorption zone is not consistent with experimental evidence that selective electronic processes are the rule rather than the exception in laser ablation [11]. It is now well accepted that coupling between electronic excitation and the vibrational degrees of freedom of the solid causes substantial lattice distortion, leading to atomic motion. A well-known example is the phenomenon of self-trapping of excitons in type-II solids. The localization of an exciton implies that the hole is localized in a specific lattice point; it corresponds to an electron-hole plasma with $n/N = 1$, even though the bond weakening is local. In type I solids, in which self-trapping of excitons does not occur, defect or impurity sites are known to act as hole-localization sites and consequently produce Jahn-Teller distortions [12,13]. As it already known in the Knotek-Feibelman mechanism [14], the two-hole localized state leads to atomic ejection via a Coulomb explosion. Thus one expects that two-hole localized states, if generated by the negative-U interaction of two holes in the valence bands, can cause particle emission.

Laser-induced Particle Emission: Solids with Strong Electron-Lattice Coupling

In type-II solids, such as alkali halides, alkaline earth fluorides, and silicon dioxide, band-to-band transitions forming STEs near the surface produce particle emission in proportion to the density of excitation, that is, proportional to the intensity for one-photon excitation [15] and to the nth power of the intensity for n-photon excitation [16], as shown in Fig. 1. Because of this characteristic dependence on density of excitation, we refer to this as the "linear regime." The effective depth from which atoms are ejected by electronic excitation in the linear regime has been evaluated to be about 3-5 atomic layers; most of the STEs generated in this depth result in particle emission [17]. In alkali halides, because the decay of the STEs leads to preferential ejection of the non-metallic atom, the surface becomes enriched in the alkali metal [18]. Ejection of metal atoms occurs at high temperatures by evaporation [19], but the rate-limiting process is in fact the ejection of halogen atoms or ions due to defects in the halogen sublattice created by the decay of STEs.

Intense laser irradiation at sub-bandgap energies also induces nonlinear processes. A typical result for NaF is shown in Fig. 2, in which the linear regime is apparently suppressed [20]. In this nonlinear regime, ejection of neutral alkali atoms in both the ground and excited states as well as metal ions has been observed [21,22,23,24]. In all of these works, one of the important results is that the yield vs. laser-intensity curves show a superlinear relationship: the yield is proportional to a high power of the density of excitation. Measurements of the laser wavelength dependence of electron emission from BaF_2 have shown that the yield of the laser-induced electron and ion emission has a resonant behavior which may be related to the formation of surface F-centers [25]. The self-trapped excitons, which are the nearest-neighbor F-H pairs, are known to be converted to distant F-H pairs as the lattice relaxes to a local equilibrium.

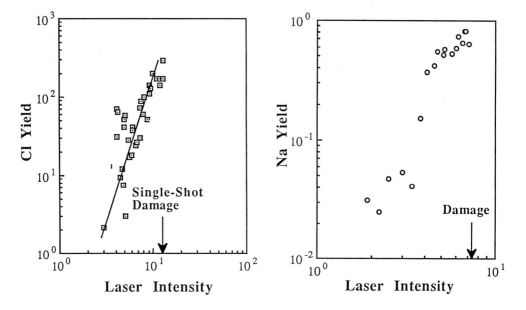

Figure 1. Yield of Cl neutrals from KCl as a function of laser intensity (or fluence) for a laser wavelength of 696 nm (ruby). Adapted from Ref. [16]; used by permission.

Figure 2. Yield of Na neutrals from NaF as a function of laser intensity at $\lambda = 266$ nm (4th harmonic of Nd:YAG). Adapted from Ref. [20]; used by permission.

In solids of type II, self-trapping of excitons induces strong lattice rearrangement as demonstrated in Fig. 3, which illustrates the geometric structure of the self-trapped exciton in CaF_2 [26]. The STE in CaF_2 is essentially the nearest vacancy-interstitial pair (F-H center pair; an F-center is an electron trapped by an anion vacancy, and an H-center is an X_2^- molecular ion where X denotes a halogen atom). The structure is well established from experimental observations using electron spin resonance. The STEs are generated with high efficiency throughout type-II solids by band-to band excitation. We assume that the STEs near the top layer are spontaneously decomposed [27,28,29] and are the origin of the linear behavior as a function of excitation density. Since the decomposition of an STE in the top layers (region 1) produces an F-center (neutral halogen vacancy) and an emitted halogen atom, its decomposition automatically results in the generation of alkali-rich surface layers.

Figure 3. Schematic diagram of the self-trapped exciton (STE) in calcium fluoride. The STE is the nearest-neighbor vacancy-interstitial pair (*F-H* center pair). See reference [26].

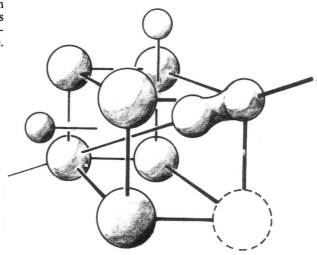

The STEs *underneath* region 1 may remain stable or be converted to distant *F-H* center pairs. We suggest that the self-trapped excitons and *F-H* pairs in the underlying region (region 2) can act as a source of additional one-photon or small-size-multi-photon transitions. According to a semi-empirical molecular orbital calculation of the energetics of the ejection of a halogen atom from a cluster of alkali halide crystal including an STE [30], the energy needed to remove a halogen atom from the surface is nearly the same as the energy of a self-trapped exciton at the lowest state in the bulk. Thus it is likely that the STEs in region 2, when excited, result in emission of a halogen atom. As one might guess from the atomic structure of the STEs, they usually exhibit three types of optical absorption bands: visible and infrared absorption due to the electron transitions (e^*h); ultraviolet absorption due to hole transitions (eh^*); and absorption near the band edge due to STE or defect-perturbed exciton transitions [31]. Electronic excitation of the latter two types can create localized hole pairs, which may weaken the bond further and cause additional ejection of atoms. It has been shown that these excitations of the STEs in the bulk of RbI and CaF_2 create *F-H* pairs in the halogen sublattice [32]. In view of this evidence, we suggest that the STEs and distant *F-H* pairs generated in region 2 can be the source of multi-hole localized states and hence of particle emission.

We now discuss the progressive change of the alkali halide surface layers which may occur during irradiation with an intense laser pulse of subgap energies. At the beginning of the laser pulse, the optical absorption coefficient is small and most photons penetrate through the bulk, except those absorbed in multi-photon band-to-band transitions. In the linear regime of excitation density, region 1 (near the surface) is enriched with metal and the STEs and *F-H* pairs accumulate in the underlying region 2. In the nonlinear regime, the deviation from non-stoichiometry extends from the surface into region 2, and consequently the bulk optical absorption coefficient increases. This leads to increasingly effective laser heating of the solid at the later stages of the nanosecond laser pulse.

Laser-induced Particle Emission: Solids with Weak Electron-Lattice Coupling

In solids with weak electron-lattice coupling, such as semiconductors, laser-induced particle emission is not observed at low fluences. A typical example of the dependence of the yield of Ga^0 from GaP irradiated by subgap photon-energy laser pulses is shown in Fig. 4 [33]. Although a two-photon band-to-band transition is energetically allowed, the yield data do not exhibit the characteristic quadratic dependence of yield on laser intensity. Thus clearly no linear regime exists for type-I solids. Fluence dependence of this type has usually been taken as a clear indication of laser-induced heating. However, since the optical absorption coefficient drops by a factor of 10^{-3} below the bandgap for GaP [34], laser-induced ablation by photons of subgap energies cannot be thermal in origin and hence must be electronic.

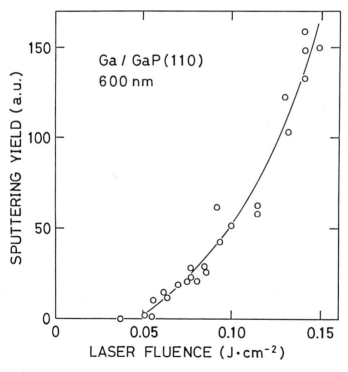

Figure 4. Yield of Ga^0 emitted at a laser wavelength of 600 nm ($h\nu < E_G$) from the (110) surface of GaP, plotted *versus* laser fluence. Note the absence of second-order dependence on laser intensity which would be characteristic of two-photon processes.

Recently, high-sensitivity measurements have been carried out of Ga⁰ emission induced by nanosecond laser pulses of subgap photon energies from clean GaP(110) surfaces exhibiting a (1x1) LEED pattern. In these experiments, particle emission is observed *below* the damage or ablation threshold; the sensitivity of detection is so high that the same spot of the surface can be irradiated repeatedly without destroying the surface. Below threshold, the LEED patterns continue to show a (1x1) pattern, signifying that the surface remains intact - say, to within 0.1 monolayer - for up to 10^4 laser shots, while the yield Y decreases with the number of laser pulses n_p. A distinct threshold laser fluence for evolution of damage has been found in these experiments [35], above which Y increases rapidly with n_p. (Fig. 5.)

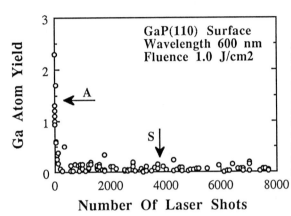

Figure 5. (Upper graph) Yield of Ga⁰ from GaP(110) surface for a laser wavelength of 600 nm (less than bandgap energy) at a fluence of 1.0 J·cm⁻², plotted as a function of laser shot number. This fluence is below the macroscopic damage threshold. The designations "A" and "S" refer respectively to the rapidly decreasing and the slowly decreasing components of the yield.

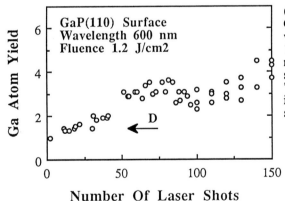

(Lower graph) Yield of Ga⁰ from the GaP(110) surface for a laser wavelength of 600 nm at a fluence of 1.2 J·cm⁻², above the threshold for macroscopic damage to the surface, such as gross metallization. The label "D" suggests that the source of the Ga⁰ in this case is a defect site, possibly a surface vacancy or vacancy cluster.

It has been proposed that particle emission below the damage threshold originates at defect sites at which a Ga atom is relatively loosely bound and which are eliminated once Ga⁰ emission takes place. A typical example of such a defect is an adatom. Ga⁰ emission from step sites has been proposed as the source of the emission which is almost constant $Y(n_p)$. Particle emission above the damage threshold, on the other hand, is believed to be initiated at vacancy sites on the surface. Ejection of a Ga atom from a vacancy creates a di-vacancy. Successive emission from a vacancy will enlarge the lateral size of the vacancy cluster; furthermore, as the vacancy cluster in the surface deepens, vacancies in the second layer begin to be exposed as well. Thus the damage cluster can evolve perpendicular to the surface as well as laterally.

The lack of the linear regime in solids of Type I is correlated with the absence of the self-trapped exciton. It follows that one-hole localization at defect sites does not lead to particle emission in these solids. Thus we must ask whether multi-hole localized states can be created in solids of type II under laser irradiation. As discussed already, electron-lattice coupling at some defects in Type I solids is so strong that defect structure is heavily altered by localization of a hole and/or electron [36]. Two possible ways of generating multi-hole localization in type-II solids are plausible: (1) successive photo-induced hole excitation of a defect with a trapped hole; and (2) negative-U-type two-hole localization followed by photo-induced hole excitation. Negative-U hole localization can occur when the lattice relaxation energy of the two-hole localized state exceeds the on-site Coulomb repulsion energy [37,38], and when the rate of localization depends strongly on the density of electron-hole pairs because of the charge screening in the electron-hole plasma [39].

Thus it has been suggested that the two-hole localized state formed by a negative-U interaction eventually leads to ejection of atoms [40], as has been demonstrated by bond-orbital model calculations [41]. Photoexcitation of one- or two-hole localized states will be significant if the lifetime of the localized state is longer than 0.1 ns, as described above. Since the lifetime of the hole-localized states is determined primarily by Auger processes, we expect longer lifetimes for those states with larger lattice relaxation which are subsequently decoupled from surface states, although no quantitative analysis is yet available. Summarizing, we suggest that the negative-U-type hole localization is the cause of the superlinear particle emission in type I solids, although the possible enhancement of the yield by successive photo-excitation events cannot be excluded.

Recently it has been shown that the damage threshold laser fluence does not change appreciably when the photon energy is increased above the band-gap energy in a type-I solid [42]. Since lattice relaxation after self-trapping occurs very rapidly, generally within 1 ps, at the initial stage of an intense laser pulse with photon energies greater than the band-gap energy, the electronic processes described above break the bond and produce substantial damage on the surface. This in turn will increase the *one*-photon optical absorption constant and making the heating more effective. At a later stage, electronic excitation is induced in a heated surface layer. The contribution of electronic excitations at this stage is less obvious, but should still be seriously considered.

Summary

We have reviewed experimental evidence for the existence of laser-induced, non-thermal, electronic particle emission exhibiting a superlinear fluence dependence in solids of both types I and II. The immediate precursor to desorption is a multi-hole localized state in each case, but the mechanism of particle emission is slightly different in the two types of solids: In type I solids, the multihole state is caused by a negative-U effect accompanied by significant lattice relaxation; in type II solids, the multi-hole state is the result of successive, multiphoton excitation of excitons trapped by Coulomb distortions of the lattice. These features are summarized in the following Table. Even though clearest indications of the electronic processes can be obtained using lasers with subgap photon energies, we point out that electronic mechanisms can also contribute to laser ablation even for intense laser pulses at above-bandgap photon energies.

Comparison of Laser-Induced Particle Emission from Type-I and Type-II Solids when $h\nu < E_G$

	Linear Regime	Nonlinear Regime
Type I	none	Below damage threshold: Decreasing yield with repeated laser irradiation. Emission initiated from adatom and step (kink) sites.
		Above damage threshold: Increasing yield with repeated laser irradiation. Emission possibly initiated at surface vacancy sites.
		Plausible mechanism: Negative-U-type two-hole localization of multi-hole complexes generated by photo-excitation of quasi-localized state.
Type II	Below damage threshold: Preferential emission of anions, thermal emission of metal atoms.	Above damage threshold: Observed above a threshold below the bulk damage threshold
	Plausible mechanism: Spontaneous disintegration of self-trapped excitons at surfaces, generated by multiphoton transitions.	Plausible mechanism: Spontaneous disintegration due to photo-induced multi-hole generation at self-trapped excitons and defect sites produced by multiphoton band-to-band transitions.

Acknowledgement

This work was supported in part by the National Science Foundation Division of International Programs, Grant Number 89-16097, and by the Japan Society for the Promotion of Science.

References

1. M. von Allmen, *Laser-Beam Interactions with Materials* (Springer-Verlag, Berlin, 1987).

2. N. Itoh and K. Tanimura, Rad. Eff. **98** (1986) 269 and N. Itoh and T. Nakayama, Nucl. Instrum. Meth. in Phys. Research **B13** (1986) 550..

3. Y. Toyozawa, *Vacuum Ultraviolet Radiation Physics*, eds. E. E. Koch, R. Haensel and C. Kunz (Pergamon, Braunschweig, 1974), p. 317.

4. E. I. Rashba, *Excitons*, eds. E. I. Rashba and M. D. Sturge (North Holland, Amsterdam, 1982), p. 543.

5. W. Hayes and A. M. Stoneham, *Defects and Defect Processes in Nonmetallic Solids* (John Wiley, New York, 1984).

6. Y. Nakai, K. Hattori and N. Itoh, *Desorption Induced by Electronic Transitions: DIET IV*, eds. G. Betz and P. Varga (Springer-Verlag, Berlin, 1990).

7. A. Van Vechten, R. Tsu and F. W. Saris, Phys. Lett. **74A** (1979) 422.

8. A. M. Malvezzi, H. Kurz and N. Bloemberge, Appl. Phys. **A36** (1985) 143.

9. N. Itoh, Nucl. Instrum. Methods **B27** (1987) 155.

10. Y. Nakai, K. Hattori, A. Okano, N. Itoh and R. F. Haglund, Jr., Nucl. Instrum. Methods in Phys. Research B, in press.

11. R. Dreyfus, in *Desorption Induced by Electronic Transitions: DIET IV*, eds. G. Betz and P. Varga (Springer-Verlag, Berlin, 1990).

12. G. D. Watkins, Mat. Sci. Forum **38-41** (1989) 39.

13. G. D. Watkins, Rev. Solid State Sci. **4** (1990) 279.

14. M. L. Knotek and P. J. Feibelman, Phys. Rev. Lett. (1978); P. J. Feibelman and M. L. Knotek, Phys. Rev. B **18** (1978) 6531.

15. H. Kanzaki and T. Mori, Phys. Rev. B **29** (1984) 3573.

16. A. Schmid, P. Bräunlich and P. K. Rol, Phys. Rev. Lett. **35** (1975) 1382.

17. M. Szymonski, in *Desorption Induced by Electronic Transitions, DIET IV*, eds. G. Betz and P. Varga (Springer-Verlag, Berlin, 1990).

18. P. H. Bunton, R. F. Haglund, Jr., D. Liu and N. H. Tolk, Surf. Sci. **243**, 227 (1991). N. G. Stoffel, R. Riedel, E. Colavita, G. Margaritondo, R. F. Haglund, Jr., E. Taglauer and N. H. Tolk, Phys. Rev. B **32**, 6805 (1985).

19. D. J. Elliott and P. D. Townsend, Phil. Mag. **23** (1971) 249.

20. L. L. Chase and L. I. Smith, *NIST Spec. Publ.* **756** (1986) 165.

21. J. Reif, Opt. Eng. **28** (1989) 1122.

22. R. F. Haglund, Jr., K. Tang, P. H. Bunton and L.-J. Wang, Proc. SPIE **1441** (1991) 732.

23. E. Matthias and T. A. Green, in *Desorption Induced by Electronic Transitions: DIET IV*, eds. G. Betz and P. Varga (Springer-Verlag, Berlin, 1990), p. 112.

24. E. Matthias, H. B. Nielsen, J. Reif, A. Rosen and E. Westin, J. Vac. Sci. Technol. **B5** (1987) 1415.

25. J. Reif, H. Fallgren, W. E. Cooke and E. Matthias, Appl. Phys. Lett. **49** (1986) 770.

26. R. T. Williams. M. N. Kabler, W. Hayes and J. P. Stott, Phys. Rev. B **15** (1977) 725.

27. P. D. Townsend and F. Lama, in *Desorption Induced by Electronic Transitions, DIET I*, eds. N. H. Tolk, M. M. Traum, J. C. Tully and T. E. Madey (Springer-Verlag, Berlin, 1983), p. 220.

28. R. T. Williams and K. S. Song, J. Phys. Chem. Solids **51** (1990) 679.

29. N. Itoh and K. Tanimura, J. Phys. Chem. Solids **51** (1990) 717.

30. N. Itoh, A. M. Stoneham and A. H. Harker, Surf. Sci. **217** (1989) 573.

31. R. T. Williams and M. N. Kabler, Phys. Rev. B **9** (1974) 1897.

32. K. Tanimura and N. Itoh, Phys. Rev. B **40** (1989) 1282. K. Tanimura and N. Itoh, Phys. Rev. Lett. **60** (1988) 2753. K. Tanimura and N. Itoh, Nucl. Instrum. Meth. in Phys. Res. **B32** (1988) 211.

33. N. Itoh, *Interfaces under Laser Irradiation*, eds. L. D. Laude, D. Bäuerle and M. Wautelet (Martinus Nijhoff, Dordrecht, 1987), p. 215.

34. J. E. Davey and T. Pankey, J. Appl. Phys. **40** (1969) 212.

35. K. Hattori, A. Okano, Y. Nakai, N. Itoh and R. F. Haglund, Jr., J. Phys. C Cond. Matter, to be published.

36. J. C. Bourgoin, Rad. Eff. Defects Solids **111/112** (1989) 20.

37. P. W. Anderson, Phys. Rev. Lett. **34** (1975) 953.

38. Y. Toyozawa, J. Phys. Soc. Japan **50** (1981) 1861.

39. N. Itoh and T. Nakayama, Phys. Lett. **92A** (1982) 471.

40. N. Itoh, K. Nakayama and T. Tombrello, Phys. Lett. **108A** (1985) 480.

41. R. F. Haglund, Jr., K. Hattori, N. Itoh and Y. Nakai, J. Vac. Sci. Technol. B, to be published.

42. A. Okano, K. Hattori, Y. Nakai and N. Itoh, Surf. Sci., to be published.

CHARGED PARTICLE EMISSION BY LASER IRRADIATED SURFACES

Guillaume Petite, Philippe Martin, Rusty Trainham, Pierre Agostini,
Stephane Guizard, François Jollet et Jean Paul Duraud
Service de Recherche sur les Surfaces et l'Irradiation de la Matière
DRECAM – CEN Saclay – 91191 Gif sur Yvette, France
Klaus Böhmer and Jurgen Reif
Fachbereich Physik, Frei Universität
Arnimallee 14, D–3000 Berlin, DFG

In laser–surface interactions at optical frequencies, the fundamental coupling is that of the electromagnetic field with the electrons in the solid. Energy is absorbed by means of transitions usually involving multiphoton absorption. It is the subsequent relaxation of the energy absorbed, eventually to the lattice, which determines the final result of the interaction. Some electrons however may have gained enough energy to overcome the surface barrier and escape from the solid: this is the multiphoton counterpart of the well documented photoelectric effect[1-3]. As usual in this type of emission processes, the photoelectron energy spectrum will carry informations on the different states (initial, intermediate or final) involved in the process. This paper reports a study of multiphoton photoemission (MPE) on metals, and preliminary results concerning insulators. Ion emission, another outcome of the laser surface interaction, will also be studied, since it also carries essential informations on the interaction process.

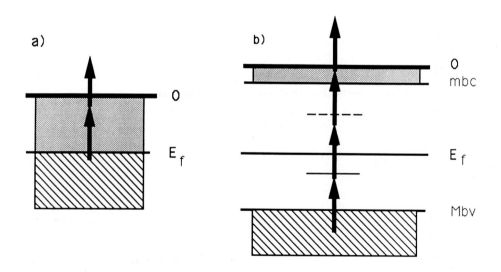

figure 1: Multiphoton transitions in (a): the conduction band of a metal, (b) an insulator's bandgap

Two different cases of MPE are depicted on figure 1. Case a) is that of a metal. Transitions here take place in the conduction band. The metal work function is usually of the order of 5 eV, so that for visible or near U.V. light, MPE processes are of a low order (typically two or three photons are needed to overcome the work function). The electrons concerned by the MPE process are essentially those whose energy lies within one photon of the Fermi level. Indeed, Pauli exclusion principle forbids absorption of one photon for those with a lower energy. They could still be excited via a two photon non resonant process, but with a much lower probability and this can be for the moment neglected. Case b) is that of a wide bandgap insulator (such as Al_2O_3 or SiO_2). Here, the multiphoton transition takes place in a forbidden band, and is thus expected to be much less probable since it is essentially non–resonant. One should not forget localized defect states in the band gap which can serve as initial or relay states. The same restriction as in the case of metals apply to the electrons in the valence band, but it is of less importance because even transitions starting from the top of the valence band are non resonant, so that the only difference is in the order of the transition for electrons located deeper in the band.

Experimental setup

The laser used in this experiment is a commercial Nd:YAG system starting with a passively/actively mode–locked oscillator which delivers pulses whose duration can be selected between 50 and 150 ps. After selection, one of these pulses is amplified to an energy of 100 mJ. The amplified output can be frequency doubled or tripled. A set of Glan prisms and a rotating half–wave plate is used to adjust the intensity, and the measured average intensity is fed back to the half–wave plate to compensate for any long–term intensity drift. Pulse to pulse fluctuations are of the order of ± 5% at the fundamental frequency. Since much excess energy is available, the laser output is severely spatially filtered and mildly focused on the sample. This secures a smooth intensity distribution, essentially determined by diffraction laws.

Two types of samples have been used in this experiment. An aluminum single–crystal, cut along its (100) face, and prepared using the standard surface physics procedures. It is kept in a background pressure of less than 10^{-10} Torr. Insulators were high purity (>99.999%) SiO_2 or Al_2O_3 single crystals, cut perpendiculary to the C axis, and prepared by heating at 900°C, at the athmospheric pressure (to avoid reconstruction of the surface). LEED patterns of these samples were taken before introduction in the irradiation chambre, and they were checked after irradiation to detect macroscopic damages (using SEM) or changes in the composition (using XPS). Samples were found unaltered, except in one occasion by pollution originating from sputtering of the stainless–steel sample holder.

MPE was studied by measuring the dependence of the total electron photocurrent on the laser intensity, the photoelectron spectrum, and the ion mass spectrum of the ablated species. The total photocurrent was measured by plugging the oscilloscope directly in the sample polarization circuit, when possible. Otherwise, the electrons were counted using the electron multiplier used in the photoelectron spectrometer described below.

Photoelectron spectroscopy and ion–mass spectrometry were performed using time–of–flight spectrometry. A straight time–of–flight tube was used for the electrons, and a reflectron spectrometer for ion–mass spectrometry. One important feature of the electron energy spectrometer is the input optics used to fight space–charge problems due to the high photocurrent in the case of Aluminum. It is made of an extraction grid which is set at a high positive potential, followed by an intermediate grid to prevent leakage of the high field in the flight tube. For electron energy spectroscopy, the electrons are brought back to a moderate potential in the tube, while a high voltage is applied to the flight tube when the apparatus is used for total photocurrent measurements. The spectrometer is represented on figure 2.

The photocurrent was measured in the Al case by plugging the oscilloscope in the sample polarization circuit. Its dependence on the extraction field resembles that of a planar diode: first an increase, then a plateau region in which all the electrons are extracted from the metal. The photocurrent measurement were made in the plateau region, known as the "space charge saturation" region. Space charge also was found to affect the electron energy spectrum and , with a slight anticipation, we just note that the electron energy was found to increase with the extraction field and then to stabilise for fields of the order of 3×10^5 V/m. Therefore all electron energy spectra were measured at an extraction field of 4×10^5 V/m.

Experimental results: aluminum

We first studied the case of a metal (aluminum), and measured the dependence of the photocurrent as a function of the laser intensity. Figure 3 shows this dependence at a 2.34 eV photon energy, for a pulse duration of 35 ps. In this case, the dependence is that expected for a two–photon process (a I^2 power law). This is not the case at all photon energies, since for instance the order of non–linearity measured at 1.17 eV is much higher than 4, the expected result. This

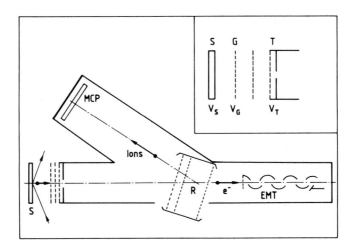

figure 2: Reflectron time–of–flight spectrometer

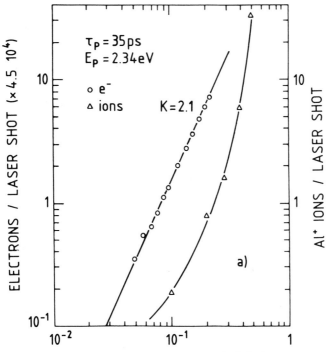

figure 3: Dependence of the MPE photocurrent and of the ion current on the laser intensity (in GW.cm^{-2})

can however be understood as the result of a thermally assisted multiphoton process[4]: the modifications of the Fermi distribution induced by laser heating of the sample make lower order (more probable) processes possible, and these act on a population which grows exponentially, hence a steep increase of the photocurrent.

Figure 3 also shows the variations of the number of Al$^+$ ions desorbed with the laser intensity. In the case shown here (photon energy $E_P = 2.34$ eV, pulse duration $\tau_P = 35$ ps) electron emission strongly dominates ion emission. This is a rather general rule, and the following remark can be made concerning the compared magnitudes of electron and ion emission: longer pulses and longer wavelengths favor ion emission. Neutrals are not detected in this experiment, and there is no way of deciding whether the ions are desorbed as such from the sample or if they are desorbed as neutrals and subsequently ionised either by the laser or in a collision with an electron with enough kinetic energy. Let us just mention that the trends noted above, together with the observation of thermally assisted effect at $E_P = 1.17$ eV (at which the ion current is the most important) are in agreement with a thermal effect for the ion emission. Energy is absorbed from the laser field by the electrons and electron–phonon relaxation (a fast process on the laser pulse time–scale) leads to a local and transient heating of the lattice. Let us note here that even if the photocurrent is non linear in essence, most of the electrons absorb only one photon. These electrons therefore cannot escape the metal and necessarily relax their energy to the lattice. Therefore the laser induced heating of the sample is fundamentally a linear process. The "threshold–like" behavior observed on figure 3 illustrates the fact that dependence of the number of ions ejected on the energy absorbed by the sample is complicated (at threshold, at least).

The ion mass spectrum observed in the intensity range of figure 3, such as shown on figure 4 ($I_1 = 3\times10^8$ W.cm^{-2}), contains some important informations. First about the state of the sample surface, which seems free of pollution, if one neglects the small K$^+$ component. It should be said that such spectra can only be obtained after a certain time of continuous irradiation (typically 15 min., that is a few 10^4 laser shots). Laser irradiation seems to be an efficient mean of cleaning the surface. The spectrum contains essentially Al$^+$ ions, and a few Al$_2^+$ and Al$_3^+$ clusters. It is important to note that no multiply charged Al ions are observed (they appear at much higher intensities, for which a plasma plume can be observed). If one assumes that the laser pulse is long enough for Local Thermal Equilibrium to be reached, this sets an upper limit to the electron energy at such laser intensities (which should be smaller than the Al$^+$ ionization threshold: 18.8 eV).

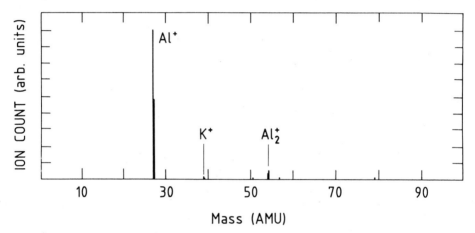

figure 4: ion mass spectrum (aluminum) at a laser intensity of 0.35 GW.cm^{-2}

Electron Energy Spectra:

Figure 5 shows three electron energy spectra taken for a photon energy of 2.34 eV, a pulse duration of 35 ps, and intensities from 10^6 to 1.2×10^8 W.cm^{-2}. They show a strong increase of the maximum electron energy from about 1 eV (compatible with a two–photon absorption) to about 70 eV. This behavior has been observed by different authors, in the same intensity range, for both picosecond and nanosecond pulses[5-6]. In our case, electron acceleration by stimulated inverse breemsstrahlung is ruled out, since we know that at 10^8 W.cm^{-2}, no plasma is formed in front of the surface. It was proposed that absorption of a high number of photons, a process known in gas phase interactions as "Above Threshold Ionization (ATI)"[7], but at much higher intensities (around 10^{13} W.cm^{-2}), would occur in laser surface interactions at lower intensities. In the case of Aluminum, since we are dealing with quasi–free electrons, an estimation based on a stimulated

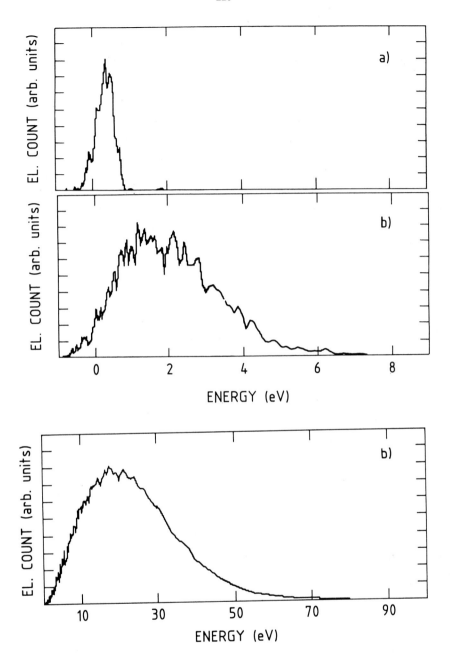

figure 5: three electron energy spectra at three different laser intensities: (a) 5x10⁶ W.cm⁻², (b) 3.5x10⁷ W.cm⁻², (c) 10⁸ W.cm⁻²

inverse breemstrahlung process inside the metal should give a good order of magnitude of the cross–sections for absorption of many photons. We performed this calculation, using the well known Kroll & Watson formula[8], which gives the result shown on table I. The outcome is that the cross–section decreases very rapidly when the order of the absorption process increases, which is nothing that what was expected at our intensities. The intensity dependence of the cross section, which though this calculation is non pertrbative in the intensity, shows the behavior expected from perturbation theory. Also note that the scaling of the cross–section with the order does not depend of the potential used in the calculation (impurity, surface potential, or Al muffin–tin potential), which is a known result of the Kroll & Watson formalism, and results from the first Born approximation. So, absorption of many photons does not seem to be able to explain the high electron energies we measure. This is indeed in agreement with the observation of Al$^+$ only, and with the square dependence of the photocurrent with the laser intensity.

The high electron energies measured here (and in the other experiments as well) are most probably a consequence of the transport of the electrons and of the extraction field applied in the collection optics. Numerical model calculations[9] are necessary to obtain the orders of the energies that can be extracted this way, but the essential physics can be sketched as follows: the elementary increase of the electron energy can be obtained from the first principle of mechanics as the work of the external forces

$$dE = \mathbf{F}.d\mathbf{r} \qquad (1)$$

where $\mathbf{r}$ is the electron position in the laboratory frame. Forces that act on the electron are due to the electric fields, external ($\mathbf{E}_{ext}$) and due to the other electrons ($\mathbf{E}_c$). So that we can write (1) under the form

$$dE = -e\,(\mathbf{E}_{ext}+\mathbf{E}_c).(\mathbf{V}_{com}+\mathbf{V}_c)\,dt \qquad (2)$$

where the electron velocity has been decomposed in the center–of–mass velocity and the

Pot.	Range	N = 1	N = 2	N = 5	N = 10
(1)	5 Å	.45x10^{-4}	.85x10^{-9}	.94x10^{-22}	.34x10^{-45}
(2)	5 Å	.72x10^{-7}	.85x10^{-12}	.13x10^{-26}	.30x10^{-52}
(3)	2.8 Å	.40x10^{0}	.49x10^{-4}	.75x10^{-17}	.30x10^{-40}

Table I: Cross section for inelastic scattering of electrons with absorption of N laser photons on (1) a Yukawa potential (impurity), (2) a step potential (surface), (3) the aluminum muffin–tin potential (an aluminum ion)

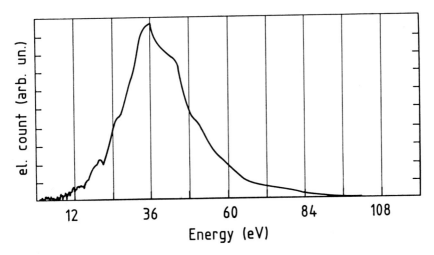

figure 6: electron energy spectrum. The flight tube is set at a 35 V potential, showing the symmetrical broadening of the spectrum.

"Coulomb explosion" velocity. Two terms can be identified. One due to the external field which is just the transformation of potential energy in kinetic energy, taken into account in the calibration of our spectrometer. What is not taken into account is the effect of the Coulomb field (the space charge). If no external field is applied, it results in the transformation of the "Coulomb" potential energy into kinetic energy (due to the term $\mathbf{E_c}.\mathbf{V_c}$). Numerical calculations give 4 eV for this quantity, which is not enough to explain our data, and moreover does not depend on the extraction field (we have to assume that the electron energy spectrum dependence on the extraction field also comes from transport effects). The contribution of the cross term $\mathbf{E_c}.\mathbf{V_{com}}$, on the contrary, does depend on the extraction field, and is larger that the pure Coulomb energy (since the extraction field is larger – 10^5 V/m – than the Coulomb field – 8×10^3 V/m – and so are the corresponding velocities). When this term is taken into account, the calculation gives a maximum electron energy which in our experimental conditions can reach 50 eV, and whose behavior as a function of the different experimental parameters fits that we observ. As a consequence, we attribute the high energies we measure as due to the work of the space charge induced Coulomb field on the center–of–mass trajectory inposed by the external field. All experiments measuring high electron energies are in the same situation of using extraction fields to collect the many electrons. If they do not, the electron energy measured is smaller, and probably simply equal to the initial coulomb energy of the particle. A definite proof of this origin of the electron energy is found in the spectrum of figure 6. It was obtained by setting the flight tube at a 35 V potential. Indeed, what is expected if the above ideas are right is a symmetrical spectrum. The coulomb field for a particle in the leading part of the cloud is directed towards the detector, and accelerates the electron, whereas in the trailing part of the cloud it points in the opposite direction and slows the electron down (this in fact secures the overall energy conservation: the enegy gained by an electron is loosed by another). In the case of figure 5, the electrons having been decelerated simply do not enter the flight tube. When we make sure that all electrons are collected, we observe, as shown on fig. 6, a symmetrical spectrum, centered around an energy equal to the tube potential corresponding to electrons with basically no initial kinetic energy. Moreover, the width of the electron spectrum in this case increases with both the extraction field and the laser intensity.

The final conclusion of these studies is that that none of the observations obtained so far really question the fact that, in the moderate intensity range explored here which is the one concerned in most laser ablation experiments, electron emission, and more generally electronic excitations (even if of the multiphoton type), are well described in the framework of perturbation theory. This conclusion certainly applies to all interband or intraband transitions in solids.

Wide bandgap Oxides

We have studied electron and ion emission by wide bandgap Oxides, namely sapphire (Al_2O_3) and α–quartz (SiO_2). We give here an account of the preliminary results obtained in this case. For both of these materials the binding energy for valence electrons is high so that a five (resp. 4) photon process is expected for a photon energy of 2.34 eV (resp. 3.51 eV). Table II gives a comparison of the orders of non–linearity observed for different photon energies with their expected values. In all cases the observed value is lower than the expected one, indicating that the electrons do not originate from the valence band but rather from defect states in the bandgap. The nature of this states is questionable. They can be due to both defects initially present in the sample (impurities, vacancies,...) or to laser induced defects. Such a behavior has been observed on amorphous silica, and was attributed to oxygen vacancy states[10].

E_p (eV)	3.51	2.34	1.98
K(theo)	4	5	6
K(expt)	2 (±.2)	2.8(±.3)	3.8(±.3)

Table II: comparison of the orders of non linearity
(theo): for electrons of the top of SiO_2 valence band
(exp): as measured in the experiment.

The role of defects in the ion emission is evidenced by the following observation: the ion emission mass spetra obtained on an Al_2O_3 and a 99.99% pure SiO_2 sample look exactly alike in the case of mildly focused laser beams (intensity of one GW.cm^{-2} or less). No Si ions can be observed with the quartz sample, but only Al^+ and alkali ions (essentially K^+). Aluminum is known to substitute for some Si ions in quartz, and require an alkali as charge compensator, so that it is not a surprise to find these ions associated in the ion emission process. In the case of tightly focused beams, Si and O ions were indeed observed, but the ions current was high so that we do not know if these ions originate from collisional ionisation of neutrals in a surface plasma (in which case, the neutrals could be present in the lower intensity interactions too) or if high intensities are necessary to extract any Silicon from a quartz surface. In the last case, the difference would then come from the somewhat different natures of the Al–O and Si–O bonding.

Another striking feature of the Oxide emission is the strong temperature dependence of all the observations. The electron photocurrent increases by two orders of magnitudes when the sample temperature is raised from room temperature to 200°C. The electron energy spectrum shows the behavior already observed in earlier studies on amorphous silica[11], for low temperatures (<300°C), one observes a low energy shift of the spectrum of up to a few eV, which can be attributed to a progressive buildup of a positive charge on the sample. This disappears at high temperatures (>300°C), which can be interpreted as the result of a temperature induced conductivity of the sample. Is this due to an ionic conductivity or to the impossibility to localize the positive charges left behind by the electron emission at high temperature is still undecidable.

Conclusions

Electron and ion emission by laser irradiated surfaces have been studied in the case of both metal surfaces and wide bandgap oxides. In the case of metals, we observe high kinetic energies at moderate intensities, but we have shown that they are due to transport phenomenons, so that perturbation theory is certainly valid here, as it is for gas phase interactions at these intensities, and can be used to obtain the cross–section for multiphoton electronic excitations in solids. Preliminary results obtained on insulators seem to emphasize the role of defect states in the initial laser solid interaction.

References

1 – E. M. Logothetis and P. L. Hartman, Phys. Rev., **187**, 460 (1969)
2 – Gy. Farkas, in Multiphoton Processes, Eds.: J. H. Eberly and P. Lambropoulos, (Wiley, New York, 1978) pp. 81 –100
3 – Ph.Martin, R. Trainham, P. Agostini and G. Petite, submitted to Phys. Rev. B.;
4 – R. Yen, J. Liu and N. Bloembergen, Opt. Comm., **35**, 277 (1980)
5 – S. Luan, R. Hippler, H. Schwier and H. O. Lutz, Europhys. Lett., **9**, 489 (1989)
6 – Gy. Farkas and Cs. Toth, Phys. Rev., **A41**, 4123 (1990)
7 – P. Agostini and G. Petite, Contemp. Phys., **29**, 57 (1988)
8 – N.K. Kroll and K.M. Watson, Phys. Rev. A, **8**, 804 (1973)
9 – H.J. Drouin and Ph. Brechet, Appl. Phys. Lett., **56**, 2152 (1990), see also J.P. Girardeau–Montaut and C. Girardeau–Montaut, J. Appl. Phys., **65**, 2889 (1989) and *to be published in* Phys. Rev. A..
10 – K. Arai, H. Imai, H. Hosono, Y. Abe and H. Imagawa, Appl. Phys. lett., **53**, 1891 (1988)
11 – G. Petite, P. Agostini, C. Boiziau, J. P. Vigouroux, C. Le Gressus and J. P. Duraud, Opt. Comm., **53**, 189 (1985)

LASER EJECTION OF Ag$^+$ IONS FROM A ROUGHENED SILVER SURFACE: ROLE OF THE SURFACE PLASMON

M. J. Shea*
Vanderbilt University
Nashville, Tennessee 37235

R. N. Compton
Oak Ridge National Laboratory
Oak Ridge, Tennessee 37831-6125

Introduction

Laser irradiation of metal surfaces with sufficiently intense light gives rise to laser-produced plasmas. Previously, we have studied plasmas from the metals Mg, Bi, Pd, Al, Cu, and Au. The laser plasma occurred with a sudden threshold in laser power. Laser irradiation yielded thermal ions at laser powers less than the plasma threshold, and full plasmas occurred upon reaching the plasma threshold. A detailed study of laser plasma production for palladium loaded with hydrogen (deuterium) has been published [1]. In the present study, the mass-resolved ion kinetic energy distributions are determined in order to study the physics of laser ablation. The kinetic energy distributions (KEDs) of silver ions exhibit a marked departure from the KEDs of other laser-produced plasmas. In addition to the thermal ions, a second peak at 3.6 eV (FWHM = 2 eV) appeared for laser energies below the onset of plasma formation.

Experimental

The laser employed in these studies was a Quantel YG571C Nd:YAG laser capable of 30 psec or 10 nsec light pulses. The second, third, and fourth harmonics were used with wavelengths of 532 nm, 355 nm, and 266 nm, respectively. Ions were detected through a 160° spherical-sector electrostatic energy analyzer, which also acted as a time-of-flight mass spectrometer to identify different species of ions.

The procedure was to first roughen the sample by using the laser at high power density ($> 5 \times 10^9$ W/cm^2), which was sufficient to form a visible plasma. The plasma appeared as a flash of light that extended many millimeters away from the surface. One feature of laser-produced plasmas is that about 15% of the ablated material is redeposited back onto the surface. When the plasma is formed, it is jettisoned away from the solid with a center-of-mass (COM) velocity, while the target recoils to conserve momentum. A Maxwell-Boltzmann distribution asserts itself around this COM velocity, and some species have a velocity in the backward direction greater than the COM velocity in the forward direction. They strike the target, and many of them stick.

---

*Now at GTE, Electrical Products Group, Danvers, Massachusetts 01923

"The submitted manuscript has been authored by a contractor of the U.S. Government under contract No. DE-AC05-84OR21400. Accordingly, the U.S. Government retains a nonexclusive, royalty-free license to publish or reproduce the published form of this contribution, or allow others to do so, for U.S. Government purposes."

Material from the plasma was redeposited on the silver surface as a variety of features, ranging from a "coral reef" structure to individual, well defined spheroids residing on the surface. The surface was examined with a scanning electron microscope (SEM). The important feature is that spheroids were formed on the surface with diameters comparable to and smaller than the wavelength of laser light. Such structures allow direct coupling of laser light to surface plasmons. After forming the roughened surface, the laser power density was reduced to the point where no ions were detected and gradually increased until the thermal peak was observed. A further increase in laser power showed the emergence of a peak at 3.6 eV (later to be called the plasmon peak). Continued increase of the laser power eventually resulted in the abrupt formation of a hot plasma and much higher Ag^+ ion kinetic energies.

Results

Ion kinetic energy distributions (KEDs) for Ag^+ ions for a variety of laser power densities, wavelengths, and pulse durations are shown in Figures 1, 2, and 3. The peak at low energy (< 1 eV) is called the "thermal peak" and has been described [1] as thermionic emission from a hot surface. The second peak centered around 3.6 eV is called the "plasmon peak."

If the surface was not previously roughened, no plasmon peak was observed at any laser power. Instead, the thermal peak grew in height until a plasma was ignited. When this occurred, the KED would switch abruptly from showing only a thermal peak to a distribution that extended to many tens of eV, just as with the other metals mentioned above. The structures in the plasmon peak are irreproducible. They are random fluctuations in the signal. The KED shown in Figure 3 exhibits

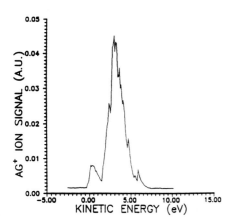
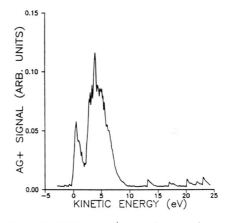

Figure 1. KED of Ag^+ ions showing thermal ions and ions with the plasmon energy of 3.6 eV. The laser wavelength was 532 nm, pulse duration 10 ns, and power density 1.1×10^6 W/cm^2.

Figure 2. KED of Ag^+ ions showing thermal ions and ions with the plasmon energy of 3.6 eV. The laser wavelength was 355 nm, pulse duration 10 ns, and power density 3.2×10^6 W/cm^2.

a slightly higher energy tail than the lower power KEDs. It is possible that a second peak is appearing at ~7 eV. Much more experimental data would be required to attribute this feature to two-plasmon decay.

The plasmon peak was also observed for both 10 nanosecond and 30 picosecond pulse durations and wavelengths of 532 nm and 355 nm. For a wavelength of 532 nm and a pulse duration of 10 nanoseconds, a KED similar to that obtained with a laser power density of 1.1×10^6 W/cm^2 (Figure 1) was observed. With a wavelength of 355 nm and a pulse duration of 10 nanoseconds, the laser power densities were 3.2×10^6 W/cm^2 (Figure 2), and a wavelength of 532 nm and a pulse duration of 30 ps required 7.5×10^7 W/cm^2 (Figure 3). Thus the trend is established that the longer pulse duration requires a lower laser power density, and shorter wavelengths also require a lower laser power density to form the plasmon peak. The thermal peak grew linearly with laser power. The plasmon peak, on the other hand, grew roughly as the cube of the laser power and eventually dwarfed the thermal peak. With 266 nm radiation, the thermal peak was evident, but no plasmon peak was observed. The coupling of light to surface plasmons thus exhibits a marked wavelength dependence. The enhancement as a function of wavelength was calculated from $P = \frac{3}{4\pi}\left(\frac{\epsilon-1}{\epsilon+2}\right)E_0$ using the field enhancement for a sphere together with the experimentally determined dielectric function for silver [2]. Figure 4 shows a maximum at 350 nm. In our experiments we used three wavelengths, one of which was 355 nm (3.49 eV) which lies very close to the calculated maximum. But we also created the plasmon peak (albeit weaker than for 355 nm) with a wavelength of 532 nm (2.33 eV), which is only on the fringe of the strong enhancement wavelength region. The calculation shown in Figure 4 assumes perfect spheres. However, the enhancement does not require a perfect sphere; ellipsoids work as well and also give a calculable

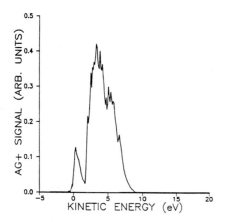 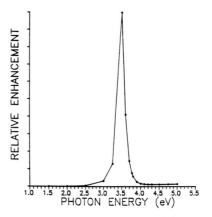

Figure 3. KED of Ag$^+$ ions showing thermal ions and ions with the plasmon energy of 3.6 eV. The laser wavelength was 532 nm, pulse duration 30 ps, and power density 7.5×10^7 W/cm^2.

Figure 4. Electric field enhancement for a dielectric sphere in a harmonic electric field as a function of wavelength. Then enhancement is a maximum for a wavelength of 355 nm.

enhancement. For silver ellipsoids, the range for the dielectric function to form a resonance is $\epsilon_1 = -2$ to $\epsilon_1 = -20$ [3]. Over this range the imaginary part ϵ_2 remains small. Some typical calculations treat the roughened surface as a distribution of spheres or spheroids of various sizes [4] and each spheroid as a point dipole with a depolarization factor that accounts for the deviation from a perfectly spherical shape [2]. The result is that longer wavelengths (up to 700 nm) also show a strong enhancement. We also used 266 nm (4.66 eV) light, with no resultant plasmon peak, which is reasonable since this shift is to shorter wavelengths, and 266 nm does not fall into the enhanced region.

The polarization dependence favors p-polarized light for the production of surface plasmons [5]. p- and s-polarization are undefined for a single sphere suspended in space. But when spheres or spheroids are laid down on a plane, symmetry is reduced, and the definitions of polarization regain meaning. We found that p-polarization was most effective in generating the plasmon peak. The ratio of p- to s-polarization was ~ 10. The thermal peak also showed the same polarization dependence as the plasmon peak. This is expected since more energy is deposited to the surface for p-polarized light. Thus, the fact that the plasmon peak is favored for p-polarized light is a necessary, but not sufficient, condition for being attributed to surface plasmon decay. Our results differ from a previous report of plasmon-induced ejection of Ag^+ and Ag_2^+ ions [5]. In this study, much lower light intensity at 351 and 248 nm was directed upon a single crystal (unroughened) silver surface. Ag^+ ions of 9 eV was detected. Others have reported the ejection of neutrals due to plasmon decay [6,7]. In these two cases, the kinetic energy of the neutrals ejected was considerably less than the surface plasmon energy.

The occurrence of the Ag^+ peak at 3.6 Ev has a simple physical interpretation. Ions which are in the process of being thermally liberated from the roughened surface subsequently absorb a quantum of energy from the surface plasmon. The presence of the thermal ions combined with the fact that the plasmon peak does not extend down to zero suggests that free ions residing near the surface receive a liberating impulse from the plasmon decay. Ritchie, Manson, and Echenique (to be published) have considered such an ion ejection as an inverse bremsstrahlung-type absorption of the surface plasmon quantum by an Ag^+ ion which undergoes a small impact collision with another ion or atom in the vicinity of the surface. Such a model predicts that the Ag^+ ion signal should be proportional to the cube of the laser power, in agreement with our observations. It is expected that neutral silver atoms are also ejected from the surface with an energy equal to or less than (plasmon energy minus the Ag^0 surface binding energy) the surface plasmon energy. Future studies are designed to examine neutral ejection under low laser power conditions.

Acknowledgments

This publication is based on work performed in the Laboratory Graduate Participation Program under contract number DE-AC05-76OR00033 between the U.S. Department of Energy and Oak Ridge Associated Universities and was also sponsored by the Office of Health and Environmental Research, U.S. Department of Energy under contract number DE-AC05-84OR21400 with Martin Marietta Energy Systems, Inc. We would also like to thank

Henry Helvajian for the loan of his silver single crystal. We acknowledge important discussions with R. H. Ritchie, J. C. Ashley, T. L. Ferrell, and J. R. Manson during the course of this study.

References

1. M. J. Shea, R. N. Compton, and R. L. Hettich, *Phys. Rev. A* **42**, 3579 (1990).

2. H.-J. Hagemann, W. Gudat, and C. Kung, *DESY-report SR-7417*, Deutsches Elektronen-Synchrotron, May 1974.

3. P. F. Liao, J. G. Bergman, D. S. Chemla, A. Wokaun, J. Melngailis, A. M. Hawryluk, and N. P. Economon, *Chem. Phys. Lett.* **82**, 355 (1981).

4. S. L. McCall, P. M. Platzman, and P. A. Wolf, *Physical Letters* **77**, 381 (1980).

5. S. W. Kennerly, J. W. Little, R. J. Warmack, and T. L. Ferrell, *Phys. Rev. B* **29**, 2926 (1984).

6. I. Lee, J. E. Parks, II, T. A. Callcott, and E. T. Arakawa, *Phys. Rev. B* **39**, 8012 (1989).

7. W. Hoheisel, K. Jungmann, M. Vollmer, R. Weidenauer, and F. Träger, *Phys. Rev. Lett.* **60**, 1649 (1988).

A SURFACE PLASMON MODEL FOR LASER ABLATION OF Ag[+] IONS FROM A ROUGHENED Ag SURFACE

R. H. Ritchie[*], J. R. Manson[+] and P. M. Echenique[++]

*Oak Ridge National Laboratory, P. O. Box 2008
Oak Ridge, TN 37831-6123
and
Department of Physics, University of Tennessee
Knoxville, TN 37966

+Department of Physics, Clemson University
Clemson SC, 29631

++Faculdad de Quimica, Universidad Pais Vasco
Apto. 1072, San Sebastian, Spain

ABSTRACT- Experimental work by Shea and Compton[1] suggests that Ag[+] ions emitted from a roughened Ag surface irradiated by a nanosecond or picosecond laser beam may absorb the full energy of the Ag surface plasmon (SP). We have modeled this process under the assumption that it proceeds through an inverse bremsstrahlung-type absorption of the SP quantum by an Ag[+] ion which also undergoes a small-impact parameter collision with another ion or atom in the vicinity of the surface. We give a quantitative estimate of the absorption probability and find reasonable agreement with the Shea-Compton results.

I. Introduction

In a very interesting recent set of experiments, Shea and Compton[1] have studied the distribution in energy of Ag[+] ions emitted from a roughened Ag surface irradiated by photons. They find copious emission of Ag[+] ions when they use a picosecond laser with a photon energy of 3.49 eV at a power level of 3×10^7 W/cm^2 with a pulse duration of 30 ps. The energy distribution consists of a peak at an ion energy of $\approx$ 0.5 eV, which they term the thermal peak, and a broader peak with a mean energy of $\approx$ 3.5 eV. The yield of ions in the higher energy peak is found to scale approximately as the third power of the laser power level, while the emission of ions in the thermal peak is linear in the laser power. When the photon energy is increased to 4.66 eV the thermal peak is observed but the peak at 3.5 eV (the 'SP' peak) is not seen. At a photon energy of 2.33 eV both the thermal and the SP peaks are observed. Shea and Compton hypothesize that the higher-energy peak in the ion distribution is due to the decay of the well-known Ag surface plasmon. This is made plausible by the observed polarization dependence of the yield.[1] To account for the dependence of the emission probability on photon energy, they argue that the SP is excited with only small probability at the higher energy and that at the lower energy, a complex, two-photon absorption may be occurring. Their data at the lower energy are not sufficiently detailed to show whether or not the yield of ions in the higher peak scales as a higher power of the laser power level. Observation of such a dependence would tend to support the mechanism that they propose.

We have made a quantitative estimate of the probability that the peak in the ionic distribution at ≈ 3.5 eV is due to the annihilation of surface plasmons through the decay channel that yields ions possessing the full SP energy.

Direct conversion of the SP energy into the kinetic energy of an Ag^+ ion is quite unlikely due to the strong mismatch in momentum between the SP and an ion. The SP carries momentum ranging from zero to less than ≈ 1 Å^{-1}, while that of an Ag^+ ion with the SP energy is several hundred times larger than this maximum value. To conserve momentum, collision with a third body is necessary. We assume that a second Ag^+ ion or an Ag atom is present to participate in a three-body collision with the SP and the ion. We evaluate the decay rate of the SP through this channel using quantal perturbation-theoretic methods and compare the computed probability of ion emission with that measured by Shea and Compton.

Theory

A photon, incident from vacuum on the plane surface of a medium capable of supporting a surface plasmon, cannot create a quantum of the SP field by annihilation.[2] This is not allowed because the phase velocity of the SP at a vacuum-bounded surface is always less than the velocity of light in vacuo. Speaking quantally, there is a mismatch in momentum; the photon momentum is always less than that of the SP. Such an interaction is not forbidden: (a) if the photon is incident on the surface from a medium in which its speed is less than c, (b) if there are irregularities in structure (roughness) on a planar surface, (c) or if the surface plasmon exists on a finite body, such as a spheroid.[3]

Here we assume that in the Shea-Compton work the Ag surface has been sufficiently roughened by prior laser irradiation that the conversion of a photon to a surface plasmon is readily accomplished. We are not concerned here with the details of this process, but will focus on the SP-ion transformation, deriving an analytical expression for the damping rate of the SP to a final state consisting of at least one Ag^+ ion.

Figure 1 shows schematically: (a) a portion of a metal surface, (b) an evanescent SP field associated with the metal, and (c) Ag^+ ions and atoms emerging from the surface. The latter are assumed to be emitted from the surface independently of the SP field and are due to ordinary evaporation processes occurring in the transiently heated surface region. The electric field associated with the SP decreases in strength with increasing distance from the surface. At a plane surface, or at the surface of a cluster of atoms with radius large compared with the wavelength of light at the SP eigenfrequency, the field strength is expected to decrease exponentially with distance from the surface.

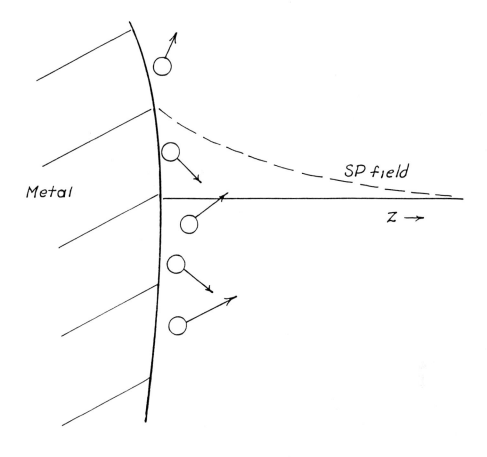

Figure 1. A schematic picture of a section of a metal surface, subjected to laser irradiation, that supports a surface plasma oscillation. The z axis is taken normal to the surface. The surface plasmon field decreases in strength with increasing distance from the surface and is indicated by a dashed line. Atoms and ions evaporating from the surface are shown as circles with quasi-randomly directed velocities. An ion may gain the energy of the surface plasmon under the influence of its electric field if a momentum-conserving collision with another ion or atom takes place.

The Feynman diagrams of Fig. 2 illustrate two different channels by which a photon can annihilate through creation of a surface plasmon, followed by decay of the latter. In Fig. 2a the final product is an electron-hole pair, the electron of which goes to a final state of the solid or can be emitted into the vacuum. The diagonal dashed line represents the photon propagator, the wavy line that of the SP, and the solid line that of an electron. The double horizontal line depicts the interaction between the SP and the photon that can be mediated, e. g., by surface irregularities. The momentum necessary for this process is amply available through an interband transition of the electron or through thermal diffuse scattering on the phonon field of the solid. In Figs. 2b and 2c the final state is one in which an ion and an atom or another ion are emitted into the vacuum. The double lines represent ions or atoms that participate in the interaction and the single dashed horizontal line the interaction between them.

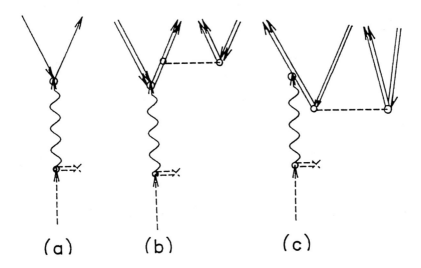

Figure 2. Open Feynman diagrams representing various processes by which a photon, shown as a diagonal dashed line, creates a surface plasmon, indicated by a wavy line, that decays into various final states. Conversion of the photon into a plasmon can be mediated by irregularities in the surface, indicated by the crosses. Figure 2a depicts decay of the surface plasmon into a single electron-hole pair. The solid line represents an electron if it is directed upward, denoting propagation forward in time. If directed downward, a solid line denotes a hole in the solid. Momentum is conserved between the plasmon and the electron-hole pair through collision of the electron with the lattice. Figure 2b shows plasmon decay through generation of an ion that takes the full plasmon energy by experiencing a momentum-conserving collision with another ion or an atom outside of the solid. The horizontal dashed line represents the instantaneous ion-atom or ion-ion interaction. Figure 2c shows the same final state but with time-reversed ordering of the plasmon-ion and the ion-atom interactions.

We use a schematic model to estimate the probability of the processes shown in Figs. 2b and 2c. The coupling between the SP field and the ion is taken to be linear in the field coordinates and in the charge on the ion.

The second-order quantal decay rate of the SP due to ion emission is found to be

$$\gamma_{ion} = \frac{4}{3} \frac{e^2 p_f V_{p_f}^2}{\hbar^3 \omega_s^2} n^2. \qquad (1)$$

In obtaining this result we have made a number of approximations. A full account of this theory will appear elsewhere.[4] In this equation: (a) ω_s is the eigenfrequency of a surface plasmon, here taken to be dispersionless, (b) V_{p_f} is the momentum representation of the interaction potential for ion-atom scattering evaluated at the momentum $\hbar \vec{p}_f$ of an ion with the full SP energy, and (c) n is the density of ions or atoms in the neighborhood of the surface.

The probability that a surface plasmon will decay by giving its energy to an ion through the process under consideration, is given by the ratio of the rate calculated above to the total decay rate γ_T of the SP due to all processes, viz., $P = \dfrac{\gamma_{ion}}{\gamma_T}$.

To make an estimate of n appropriate to the Shea-Compton experiment, we use the standard evaporation model.[5] In this, the number of atoms ϕ_e evaporated from a solid surface per unit area per unit time is given in terms of p, the vapor pressure of the solid, M, the mass of the atoms evaporated, and their speed v. The relation is $\phi_e = \dfrac{fp}{Mv}$, where f is the fraction sticking to the surface upon striking it. The usual expression for p is $p = Ce^{-W/kT}$, where W is the binding energy of the atom in the solid, and C is a constant. Fitting experimental data[6] we find $p = e^{20.4 - 2.86/kT}$, where p is in mm Hg and kT is in eV. To find the density of atoms in the neighborhood of the surface we divide ϕ_e by v

and convert to cgs units to obtain $n=\dfrac{1327.5f}{Mv^2}e^{20.4-2.86/kT}$. Taking the effective temperature of the surface to be 0.5 eV during the laser pulse and setting f=1, one finds n=2.3x10^{21} atoms/cm^3.

Substituting into Eq. 1, using atomic units and employing a screened Coulombic form for the interaction potential, we find γ_{ion}=1.9x10^{-7}a.u. If one takes γ_T=1.5x10^{-2}, then P=4.3x10^{-2}. This result depends sensitively on the value assumed for the effective temperature of the surface. Using kT=0.4 eV one finds P=3.8x10^{-3}.

Shea and Compton estimate that a typical yield in their experiments is ≈10^{-8} ions/incident photon.[1] Dividing this value by our computed ratio of ion to total decay rate of the SP, we find that ≈10^{-5} surface plasmons are created per incident photon in the Shea-Compton experiments. This value is perhaps not unreasonable considering that scanning electron microscope pictures show that the structures on the Ag surfaces used by Shea and Compton[1] that are assumed to be responsible for absorption of photons into surface plasmon oscillations are, on the average, quite large compared with the photon wavelength and that they cover only a few percent of the surface. These surfaces are apparently quite different from those used in the experiments of Hoheisel, et al[7], according to the analysis by Monreal and Appel.[8]

SUMMARY

Using quantal perturbation theory we have made an estimate of the probability of decay of a surface plasmon at an Ag surface through an inverse bremsstrahlung-type process. The decay is taken to be to a final state in which an Ag$^+$ ion gains the energy of the surface plasmon and collides with an atom or another ion to conserve momentum. We argue that these results are consistent with observations that the emission of Ag$^+$ ions in the laser irradiation of a roughened Ag surface is proportional to the third power of the intensity of the photon field and with the absolute emission probability measured by Shea and Compton.[1]

The approximations used here limit our theory to prediction of the total number of ions in the SP peak. We plan to extend this work to account for the finite width of the Ag$^+$ energy distribution; this will involve including damping of the SP and collisional and Doppler broadening of the ion peak, as well as representing the SP field more realistically.

ACKNOWLEDGEMENTS

Thanks are due to Bob Compton and Mike Shea for many helpful conversations and suggestions during the course of this work. This research was sponsored jointly by the Office of Health and Environmental Research, U. S. Department of Energy, under contract number DE-AC05-84OR21400 with Martin Marietta Energy Systems, Inc.

REFERENCES

1) M. J. Shea and R. N. Compton, "Surface Plasmon Ejection of Ag^+ Ions from Laser Irradiation of a Roughened Silver Surface"-these Proceedings; M. J. Shea, Ph.D. Thesis, Vanderbilt University

2) See, e. g., R. H. Ritchie, Surf. Sci. 34, 1 (1973)

3) R. H. Ritchie, J. C. Ashley and T. L. Ferrell, "The Interaction of Photons with Surface Plasmons", in Electromagnetic Surface Modes, Ed., A. D. Boardman, John Wiley and Sons, (1982)

4) R. H. Ritchie, J. R. Manson and P. M. Echenique, to be published.

5) N. F. Mott and H. S. R. Massey, The Theory of Atomic Collisions, third ed., Oxford.

6) Handbook of Chemistry and Physics, Ed. R. C. Weast, Chemical Rubber Co., 51st ed., p. D145

7) W. Hoheisel, K. Jungmann, M. Vollmer, R. Weidenauer, and F. Trager, Phys. Rev. Lett. 60, 1649 (1988)

8) R. Monreal and S. P. Apell, Phys. Rev. 41, 7852 (1990)

UV LASER ABLATION FROM IONIC SOLIDS

Richard F. Haglund, Jr., Mario Affatigato, James Arps and Kai Tang
Department of Physics and Astronomy
Vanderbilt University, Nashville, TN 37235

Abstract. We compare ultraviolet laser ablation at a wavelength of 308 nm from ionic solids where the photon energy is much less than (KCl), or approximately equal to (LiNbO$_3$), the bulk bandgap energy. The results are interpreted in the framework of simple charge-transfer models.

Studies of ablation products from KCl and LiNbO$_3$ surfaces subjected to ultraviolet laser irradiation indicate that relative yields of different species depend both on laser intensity and on the history of the irradiated spot. This suggests that the ablation process involves laser-induced defects as well as the electronic structure of the perfect lattice sites.

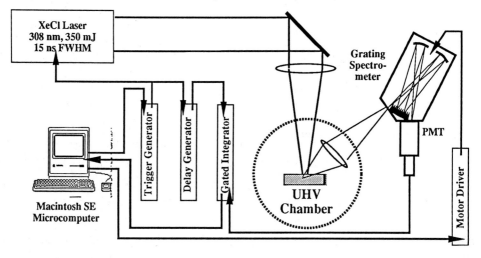

Figure 1. Apparatus for measurements of laser-ablated excited atoms from KCl.

In our experiments, single-crystal targets of either single-crystal KCl cleaved from bulk material or of optically polished, Y-cut LiNbO$_3$ were mounted on a manipulator in high or ultrahigh vacuum. Visible fluorescence from ablated K* atoms was detected by an 0.3 meter monochromator and a photomultiplier operated in pulse-height analysis mode; spectra were stored in a Macintosh microcomputer. In the LiNbO$_3$ experiments, neutrals and ions were detected by a quadrupole mass spectrometer (QMS) with a channeltron detector; peak height data for each identified mass were stored on-line in an IBM microcomputer for subsequent analysis.

Laser Ablation of KCl

In KCl, with a bandgap of approximately 8 eV, two photons are required to generate electron-hole pairs by valence-band excitation. Desorption occurs *via* the decay of the self-trapped exciton and the subsequent relaxation of *F*-center-*H*-center pairs near the surface. For example, diffusion of *F*-centers to the surface from the near-surface bulk neutralizes alkali ions on the surface which then desorb thermally, with a significant fraction in excited states [1]. Figure 2 shows the fluorescence spectrum of excited potassium atoms near the first resonance lines at 7668 Å, at laser intensities on the order of 1 GW·cm^{-2}. Each point on the spectrum is from a single laser shot; there are two hundred points in each spectral scan from 7600 to 7800 Å. Thus some 140 shots intervene between the upper spectrum and the lower. The decrease in K* yield with increasing photon dose

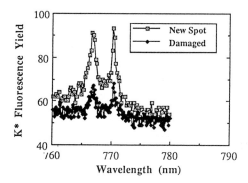
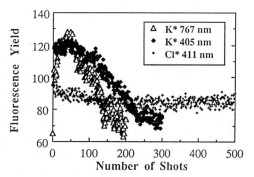

Figure 2. Spectrum of K* atoms desorbed from KCl at an intensity of 1 GW·cm^{-2} on successive scans, starting from a pristine spot. The atomic transition is 4p → 4s.

Figure 3. Spectrum of K* (405 and 767 nm) and Cl* (411 nm) atoms ablated from KCl at an intensity near 1 GW·cm^{-2}, as a function of shot number.

is well known from PSD studies of alkali halides where valence-band ultraviolet irradiation leads to a steady decrease of excited-atom fluorescence due to metallization of the surface [2]. This excess metal allows resonant ionization of desorbing K* atoms, thus reducing the total excited-state yield. Figure 3 shows a comparison of relative yields of K* (from the 5p → 4s and 4p → 4s transitions) and of Cl* emission originating from transitions in the ($3p^45P → 3p^44S$) manifold. Again, the K* emission vanishes after a few hundred shots, while the Cl* yield is virtually constant as a function of total dose. Note that the 767 nm K* transition vanishes before the 405 nm transition, consistent with the picture of the Fermi level rising into the insulator bandgap to ionize the excited atoms. The parent states of the Cl emission line are 12 eV above the ground state, suggesting that the Cl* arises from multiphoton excitation of ablated *ground-state* Cl in the laser-produced plasma. This is consistent with the generally accepted picture that ground-state Cl emission is the primary ablative process, and results from the decay of the *H*-centers [3].

Laser Ablation of LiNbO$_3$

Laser ablation of lithium niobate is currently of interest partly because LiNbO$_3$ is not easily processed by chemical or mechanical means, so that lasers are being used for micromachining [4] and direct writing in optoelectronic circuits [5]. Laser-ablation-assisted deposition is also being used to fabricate ferroelectric thin films for optoelectronic applications.

Samples used in our experiments were polished Y-cut LiNbO$_3$ crystal wafers from Crystal Technologies, Inc., with a nominal bandgap energy of 4.0 eV. A Lumonics XeCl laser with 15 mJ maximum energy in a 15 ns pulse was focused on the targets in a vacuum chamber with a base pressure of $2 \cdot 10^{-6}$ torr. The beam was focused on the target at angles of incidence between 45° and 30° with a 175-mm focal length lens and a focal spot area of 0.2 to 7.5 mm^2. These parameters yield a range of laser intensities spanning the plasma formation threshold, varying from $1.9 \cdot 10^7$ to as much as $6.7 \cdot 10^8$ W·cm^{-2}.

We observed a number of ablated atomic, ionic and molecular species by means of quadrupole mass spectroscopy, including Li$^+$, O$^+$, Nb and Nb$^+$, NbO and NbO$^+$, NbO$_2$, and LiNbO$_3$. The threshold for the LiNbO$_3$ molecular desorption was below the threshold for all other species observed, as discussed elsewhere [6]. The intensity dependence of the yields of the NbO and NbO$_2$ neutral molecules are shown in Figure 4. We suggest that the observed saturation of the NbO$_2$ yield arises as this species reaches an equilibrium population in the ablation plume due to impact dissociation by hot electrons. As shown in Figure 5, a comparison of Li$^+$ and Nb$^+$ yields for undamaged and damaged spots shows a moderate lowering of the ablation intensity threshold for the damaged surface in both cases and evidence for saturation at nearly the same intensity level.

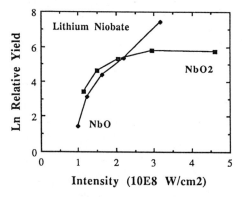
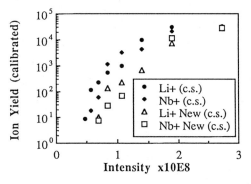
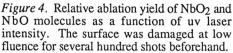

Figure 4. Relative ablation yield of NbO$_2$ and NbO molecules as a function of uv laser intensity. The surface was damaged at low fluence for several hundred shots beforehand.

Figure 5. Comparison of the intensity dependence of Li$^+$ and Nb$^+$ ion yields as a function of laser intensity for damaged and undamaged spots.

Discussion

In both the alkali halides and the ferroelectrics, the initial response of the material to photons is the creation of an electron-hole pair by a band-to-band transition: by two-photon absorption in KCl and by one-photon absorption in $LiNbO_3$. Also, in both materials, the valence band is formed primarily from the non-metallic ion: the $3p$ Cl orbital in the case of KCl, and the O $2p$ orbital in $LiNbO_3$. The optical response in these ionic insulators is primarily an inter-atomic charge transfer arising from ineffective screening of the valence-band hole. The inter-atomic charge transfer upsets the balance of Coulomb forces in the ionic crystal, initiating ion motion and desorption as in the Knotek-Feibelman mechanism. Further irradiation produces mobile holes in a dense electron-hole plasma which can be localized at the defects created during the latent phase of the damage.

However, whereas in alkali halides the decay of permanent electronic defects formed from the decay of the self-trapped exciton produces the characteristic features of the desorption yield, such as surface metallization. In the ferroelectric oxides, however, it appears that excitation of pre-existing defects and molecular emission may be involved significantly in the desorption process. The super-linear intensity dependence of the ablation yield in the oxides may reflect the exponential variation of the screening potential for charge trapping at defect sites as a function of the electron-hole plasma density.

Acknowledgements

We thank Prof. W. Heiland of the University of Osnabrück for supplying the $LiNbO_3$ samples, and Dr. C. H. Chen of Oak Ridge National Laboratory for the use of his laboratory for portions of this work. Our research was partially supported by the Office of Naval Research through the Vanderbilt University Free-Electron Laser Center, and by the SURA/ORNL/ORAU Summer Cooperative Research Program. One of us (MA) gratefully acknowledges a scholarship from the Society for Photo-Optical Instrumentation Engineering.

References

1. P. H. Bunton, R. F. Haglund, Jr., D. Liu and N. H. Tolk, submitted to Phys. Rev. B.
2. P. H. Bunton, R. F. Haglund, Jr., D. Liu and N. H. Tolk, Surf. Sci. **243**, 227 (1991).
3. R. F. Haglund, Jr., K. Tang, L.-J. Wang and P. H. Bunton, SPIE Proc. **1441** (1991).
4. M. Eyett and D. Bäuerle, Appl. Phys Lett. **51**, 2054 (1987).
5. T. Krauss, A. Speth, M. M. Opryski, B. Fan and K. Grebe, Appl. Phys. Lett. **53**, 947 (1988).
6. K. Tang, M. Affatigato, R. F. Haglund, Jr. and C. H. Chen, submitted to Appl. Phys. Lett.

PHYSICS OF PULSED LASER ABLATION AT 248 NM: PLASMA ENERGETICS AND LORENTZ INTERACTIONS

L. Lynds and B. R. Weinberger
United Technologies Research Center, E. Hartford, CT 06108

ABSTRACT

Pulsed KrF (248 nm) excimer laser ablation of targets can generate quasi-neutral ground state atomic particle plasmas with energies in the range of 200 -1100 eV. Onset of plasma formation occurs when optical intensities exceed 10^8 W/cm^2. The expanding ensemble has a pronounced angular energy dependence, the highest energies peaked in the direction normal to the target. Correlation of the temporal nature of the laser pulse and the ensuing ion mass pulses suggest that transfer of energy is, indeed, a very rapid process -probably less than 5 ns. A wide range of elemental and several compound targets were studied to determine how ion energy distributions were influenced by atomic mass. The energetic plasma beams interact with magnetic fields as predicted from the Lorentz force. Results are consistent with an inverse bremsstrahlung and Lorentz force mechanism for particle acceleration.

INTRODUCTION

The laser induced extraction of material from a target is now commonly used as a source for depositing a wide variety of thin films extending from superconductors to dielectric materials [1]. With the advent of high temperature superconductivity (HTS), much attention has been drawn toward laser ablation as a general technique to create quality HTS thin films [2,3]. Following the discovery of giant pulse lasers, there was a considerable effort to understand their interactions with condensed matter [4-18]. The nature of radiation from excimer lasers is quite different than its longer wavelength predecessors. It is important to explore the underlying physics of the target interactions, the ensuing energetic plasmas and, eventually, how particles condense on a surface acting as a third body.

EXPERIMENTAL

Pulsed laser ablation of targets was performed in a cylindrical vacuum chamber [3] with background pressures in the range of 10^{-6} to 10^{-7} torr. Fixed angle laser ablation was performed with a 30° angle of incidence. Optical emission from the target surface and surrounding plume was selectively viewed with a fast Si diode detector. Radiation at 248 nm was generated with a Lambda Physik LPX 200 excimer laser. Mass and optical data were acquired with a Tektronix DSA 602 digitizing signal analyzer which provides a temporal resolution of 1 ns. Ion kinetic energy distributions were determined by time of flight (TOF) measurements. A drift tube terminated with a mass quadrupole filter (MQF) was affixed to a chamber port providing direct line of sight to the target centers. Inserts were used to adjust the working distance to the detector to be either 115 cm or 238 cm. The pulse temporal dynamics were controlled by using either a He or Ne diluent together with the usual excimer precursor (Kr/F_2 in this case). Stable (SR) or unstable resonator (UR) configurations were used to change the temporal laser light characteristics. The (UR) radiation is less divergent and can be focussed into a smaller area giving higher target fluences. Pulses generated with the He diluent consistently have a 14 ns rise time, a series of 4 or 5 diminishing peaks with 8 ns spacing between maxima and an overall FWHM of about 34 ns. In contrast, pulses generated with Ne diluent have similar rise times but have smoothly decaying contours and the pulse FWHM is effected by the discharge energy.

RESULTS AND DISCUSSION

Pulse Dynamics

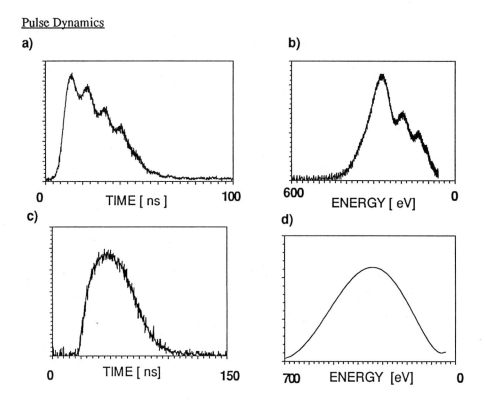

Fig. 1 a) UR/Ne laser pulse; b) Y^+ mass pulse at 238 cm. c) SR/He laser pulse; d) Y^+ mass pulse at 238 cm.

Experiments discussed, herein, were constrained to fluences normally used to prepare thin film devices (above 1 J/cm^2). The kinetic energy distributions of Y^+ as derived from the 238 cm drift tube experiments are compared in Fig. 1 to the temporal nature of the laser pulse which impinges on Y targets. A remarkable similarity is noted between the temporal evolution of the corresponding pulses (i.e. Fig. 1a-b and Fig. 1c-d). The particle energy distribution in Fig. 1b consists of a descending series of peaks separated by approximately 20 µs and is reminiscent of the (UR/Ne) laser pulse depicted in Fig. 1a. Separate experiments utilizing a single maxima laser pulse (SR/He) for target ablation creates a similar mass pulse as indicated in Fig. 1c-d. We note that mass density fluctuations are occurring on a time scale of approximately 10^3 greater than the laser relaxation fluctuations. Typical TOF pulse widths (FWHM) acquired at two distances, 115 and 238 cm, from Ta and Au targets are presented in Table I.

TABLE I: PLASMA DYNAMICS

ION	DIST [cm]	$\Delta\tau_1$ [µs]	DIST [cm]	$\Delta\tau_2$ [µs]	$\Delta\tau_2/\Delta\tau_1$
Ta^+	115	13	238	65	5.0
Ta^{+2}	115	12	238	32	2.6
Ta^{+3}	115	10	238	17	1.7
Au^+	115	13	238	70	5.3
Au^{+2}	115	12	238	35	2.9
Au^{+3}	115	10	238	20	2.0
Au^{+4}	115	11	238	35	3.2

A spreading of the plasma envelope in time is observed over these distances and is somewhat non-linear suggesting that dynamical processes are still occurring beyond 115 cm as the plasma propagates away from the target. At 115 cm the pulse widths [FWHM] were 10 - 13µs whereas at the longer distance, 238 cm, the widths were somewhat wider than expected for a fixed velocity distribution. Moreover, a time regression of the [FWHM] linewidths suggests that the effluents probably originate on about the same time scale as the laser pulse. Implied from these observations is that the ions are launched at different velocities due to local phase space temporal variations of the optical field. At longer distances (238 cm, in this case) it becomes easier to resolve the density variations. The optical field - target interaction is certainly three dimensional and, consequently, it is probably more meaningful to connect the dynamical properties of the effluents to an instantaneous flux density, F_v, rather than the usual time-averaged fluence. Implications are that photon energy transfer to kinetic energy of accelerating particles is occurring on a time scale of less than 5 nsec.

Angular Distribution of Mass and Energy

Relative flux of the condensed phase (Au) collected on pyrex plates were spatially determined by performing optical transmission measurements. Vertical and horizontal cross

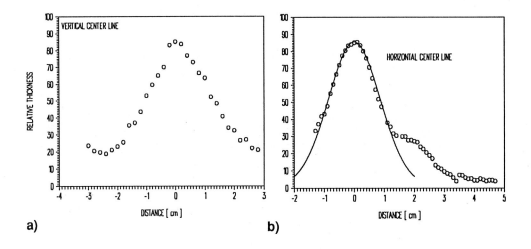

Fig. 2 Spatial distribution and relative thickness of Au condensed on pyrex plate located 3.3 cm from target. Theoretical curve (solid line) is included in 2b.

sections through the center line normal to the target are shown in Fig. 2a and 2b, respectively. The distribution is sharply directed along the target normal and is common for this kind of laser source for an angle of incidence of 45° or less. It is interesting to note that the deposition density shows an "elliptical" asymmetry which suggests that the plume expands more rapidly from the narrow dimension (vertical, in this case) of the rectangular far-field pattern on the target. The shoulder on the horizontal distribution is caused by small rectangular "hot-spots" which reside on each end of the pattern. Half of the ablated mass is confined to ± 20° from the target normal. Mass flux distributions may be accounted for by a simple statistical gas phase collisional model described in earlier work [19,20]. A radial expansion of small neutral particle domains can be written in terms of a vector momentum field, $\mathbf{P} = \mathbf{P}_0 \exp[-\alpha z/V\cos\theta]$ where α is the collision rate below the "surface," z is the depth of the interaction volume below this "surface," V is the particle velocity and θ is the trajectory angle relative to the target normal. Assuming $V = 2 \times 10^6$ cm/s (measured by TOF), $z = 3 \times 10^{-5}$ cm (from mass loss and spatial measurements) and $\alpha = 10^{12}$ (collisions per sec), Fig. 2b includes a theoretical distribution that fits the data quite well. In the presence of a constant magnetic field, a significant $\mathbf{V} \times \mathbf{B}$ interaction on the beam trajectories was observed as derived from the physical displacement of the deposit topography. A comparison of deposit cross sections from a 0.15 T and a weak 0.005 T interaction are shown in Fig. 3a and 3b, respectively. The displacement is about 0.3 cm as predicted from a $\mathbf{V} \times \mathbf{B}$

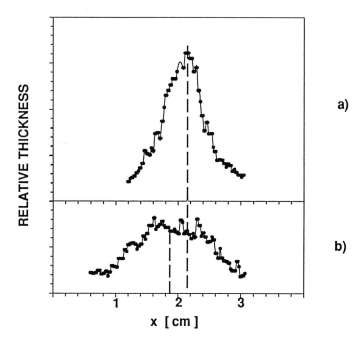

Fig. 3 Effect of magnetic field on trajectories. Spatial distribution and relative thickness of Au condensed on pyrex plate located 5.0 cm from target. a) B = 0.15 T; b) B = 0.005 T.

Lorentz force acting on a heavy mass (AMU = 197) with the given experimental parameters and geometry.

Particle Energetics

The general relationship between kinetic energy, mass and ion charge of ablated atoms was investigated for a wide mass range. Excimer ablation in this range is dominated by single and multiply charged ground state ionic species in the form of a neutral plasma. Effects of ion charge was mainly limited to single and multiply charged ions of Au and Ta. Maximum m/q ion signal was realized when the target normal at the point of laser contact was aligned concentrically with the center of the drift tube. Targets included Si, Cu, Y, Y_2O_3, $YBa_2Cu_3O_{7-x}$, Ba, Ta, W, Au and Bi. Integrated mass flux, determined from TOF measurements, increases linearly with fluence. Singly and multiply charged ions were observed and a clear spatial separation of ions occurred in the expanding plasma. The higher charged species appeared in the leading edge of

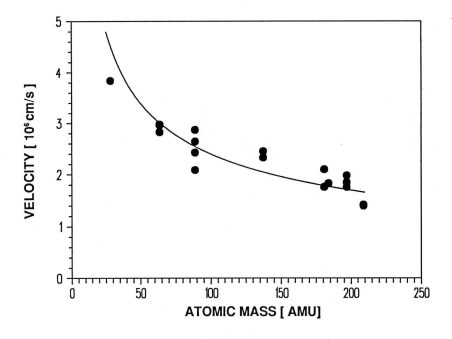

Fig. 4 Ion (M^+) velocities vs atomic mass

the expanding plasma; the singly charged species in the trailing edge. Energies calculated from TOF measurements varied from 200 to 1200 eV. In most cases, mass flux of $q = 2$ ions was a little larger than the $q = 1$ states followed, then, by the $q = 3$ and 4 states. In fixed fluence experiments, ion energies ($q = 1$) were almost constant (slight increase) with mass and their velocities, as shown in Fig. 4, decreased inversely as the square root of the mass. Included in Fig. 4 are energies of Y^+ from metallic Y (280 eV), $YBa_2Cu_3O_{7-x}$ (270 eV) and Y_2O_3 (370 eV) targets. Our experience has been that the physical nature of the target has little bearing on the plasma energetics. A comparison of ion energy centroids and ion charge for Au^{+n} and Ta^{+n} generated from metallic targets are illustrated in Fig. 5. In both cases, the relationship is clearly linear but not quite 1:1. Energy balance experiments were performed on Au targets with 200 mJ per pulse laser radiation focussed to rectangular areas of about 0.01 cm^2. Targets were weighed using an analytical balance before and after exposure to multiple shots (1000, 2000, etc.). Average mass loss was 1.0 μg per laser shot or 3×10^{15} atoms of Au. Energy required for the phase transition, $Au(s) \longrightarrow Au^+(g)$, is 6.1 mJ per mass pulse. Together with the average kinetic energy based on a measured centroid velocity of 2×10^6 cm/sec, the total ensemble

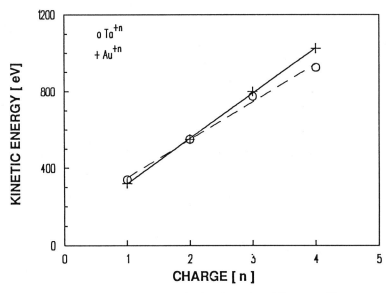

Fig. 5 Kinetic Energy Distribution of Au^{+n} and Ta^{+n}

energy for the propagating plasma is approximately 130 mJ. The calculated per shot "crater" depth on the Au surface is thus 3200 Å and there are about 55 photons per atom in the volume of interaction. These observations are consistent with a multiphoton process leading to the acceleration of particles.

Theory

A simple Lorentz force model can be constructed for charge separated species occurring within the Debye length during the initial stages of energy transfer. Thus Newtonian physics connects mass, acceleration and the position/time dependent E field. If E(r,t) is derived from a potential $(-\nabla\Phi)$, integration and rearrangement leads to a useful relationships between experimental and atomic parameters. In the absence of magnetic fields (B = 0).

$$U_k = q<\Phi> \qquad (1)$$

$$<V> = [2|q|<\Phi>/m]^{1/2} \qquad (2)$$

where U_k is the kinetic energy, q is the charge and $<\Phi>$ is the average potential difference within the Debye length. Referring back to Fig 4 and 5, Eqns 1 and 2 are consistent with the experimental observations that particle velocity decreases approximately as $m^{-1/2}$ and kinetic energy increases linearly with ion charge.

SUMMARY

In summary, there are significant differences between the characteristics of ablation from excimer lasers, described herein, and from longer wavelength lasers (Ref. 19 as an example). Plasmas created by target ablation at 248 nm are dominated by single and multiply charged ground state ions in an energy range of 200 to 1200 eV. The self- similarty of the laser pulse and mass pulse dynamics suggests that the energy transfer process is occurring on a time scale of 5 nsec or less. Instantaneous ablation rates are about 10^{22} atoms/sec. Particle energy appears to be related to the rate of photon absorption per unit volume. Energy balance indicates a multiphoton process occurs involving approximately 50 photons per atom. The expanding plasma is structured with the highest charged particles in the leading edge and the lowest in the trailing edge. Velocity distributions are distinctly non-Maxwell-Boltzmann and are consistent with a non-equilibrium ablation mechanism. About 3×10^{15} atoms were ablated per laser pulse leading to gas phase densities in the range of 10^{15}-10^{16} /cm^3 within 10 cm from the target. Analysis of mass temporal behavior at two distances indicates that plasma interactive collisions continue far beyond the Debye shielding length. Particle trajectories are strongly directed along the target normal. In the presence of an applied magnetic field, they are significantly deflected by Lorentz forces consistent with heavy ion interactions. Linearity of energy and ion charge together with the mass dependent velocities argues toward a Lorentz type force that accelerates particles. Energetic "reagents" generated by laser ablation offer new and interesting dimensions for chemistry and the alteration of surfaces.

REFERENCES

1. J. T. Cheung and H. Sankur, CRC Critical Review (Solid State Material Science) **15**, 109 (1988).

2. A. Inam, M. S. Hegde, X. D. Wu, T. Venkatesan, P.England, P. F. Miceli, E. W. Chase, C. C. Chang, J. M.Tarascon and J. B. Wachtman Appl. Phys Lett. 53, 908(1988); references therein.

3. L. Lynds, B. R. Weinberger, G. G. Peterson and H. A. Krasinski, Appl. Phys. Lett. 52, 320 (1988).

4. N. R. Isenor, Appl. Phys. Lett., 4, 152(1964).

5. N. R. Isenor, Can. J. Phys., 42, 1413 (1964).

6. H. Puell, , Z. Naturforsh., 25A, 1807 (1970).

7. H. Puell, H. J. Neusser and W. Kaiser, Z. Naturforsch., 25A, 1815 (1970).

8. Yu. A. Bykovskii, V. I. Dorofeev, V. I. Dymovich, B. I. Nikolaev, S. Ryzhikh, S. M. Sil'nov, Zh. Tekh. Fiz. 38, 1194 (1968). [Transl: Sov. Phys. Tech. Phys.14, 1269 (1970)]

9. B. E. Paton and N. R. Isenor, E., Can. J. Phys. 46, 1237 (1968).

10. S. Namba, P. H. Kim and A. Mitsuyama, J. Appl. Phys., 37, 3330 (1966).

11. C. Faure, A. Perez, G. Tonon, B. Aveneau and D. Parisot, Phys. Lett., 34A, 313 (1971).

12. D. W. Gregg and S. J. Thomas, J. Appl. Phys., 37, 2787 (1966).

13. D. W. Gregg and S. J. Thomas, J. Appl. Phys., 37, 4313 (1966).

14.. D. W. Gregg, S. J. Thomas, J. Appl. Phys., 38, 1729 (1967).

15. J. F. Ready, *Effects of High-power Laser Radiation*. New York, Academic (1971).

16. G. J. Pert, J. Phys, A., 5, 506 -15 (1972)..

17. S. Rand, Phys. Rev., 136, 231 (1964).

18. G. J. Pert, J. Phys, A., 5, 506 (1972).

19. L. Lynds, B. R. Weinberger, D.M. Potrepka, G. G. Peterson and M. P. Lindsay, Physica C 159, 61 (1989).

20. F. Reif, *Fundamentals of Statistical and Thermal Physics* (McGraw-Hill, New York, 1965) pp.461.

EXCIMER LASER ABLATION OF FERRITE CERAMICS

A. C. Tam, W. P. Leung and D. Krajnovich
IBM Research Division
Almaden Research Center
650 Harry Road
San Jose, California 95120-6099

ABSTRACT

We study the ablation of Ni-Zn or Mn-Zn ferrites by 248-nm KrF excimer laser irradiation for high-resolution patterning. A transfer lens system is used to project the image of a mask irradiated by the pulsed KrF laser onto the ferrite sample. The threshold fluence for ablation of the ferrite surface is about 0.3 J/cm^2. A typical fluence of 1 J/cm^2 is used to produce good-quality patterning. Scanning electron microscopy of the ablated area shows a "glassy" skin with extensive microcracks and solidified droplets being ejected that is frozen in action. This skin can be removed by ultrasonic cleaning.

INTRODUCTION

Laser etching of ferrites was previously done[1-4] by scanning a focused continuous-wave laser beam on a ferrite sample in a chemical environment. We apply here a new technique using the efficient excimer laser dry etching[5-7] for micromachining ferrites in air, thus eliminating the need of chemical assistance. Furthermore, improved throughput with good-quality patterning is obtained here by "parallel processing", i.e., projectional imaging of the ablation pattern onto the sample, rather than by "serial processing", i.e., drawing the etched pattern by a focused laser spot. Features of 20 μm width and as deep as 70 μm are made here, but much finer features are possible.

EXPERIMENT

The experimental arrangement is shown in Fig. 1. The KrF laser (Lumonics HYPEREX 460) produces pulses of duration 16 ns and energy up to 400 mJ. A pulse repetition rate of 30 Hz is typically used. The beam is made spatially uniform by a fly's eye beam homogenizer, and then incident on a dielectric transmission mask, which is imaged onto the ferrite sample by a 4-element, −0.5× lens system made of UV grade fused silica with antireflection coating for 248 nm. The image resolution is about 5 μm, and the depth of focus is about 50 μm. A fast air jet is directed near grazing incidence at the ablation spot to reduce ablation debris deposited onto the sample. Also, a polyester film (Scotch tape) of 45 μm thickness is applied onto the sample before laser sputtering to protect the unexposed area from collecting ablation debris. The initial 300 shots are used to etch through this film, which can be peeled off as one piece after the ablation. The ablation pattern studied here is 55-μm-wide etched grooves separated by unexposed areas of the same width. Typical quality for the ablative patterning after stripping the protective coating and ultrasonic cleaning is shown in Fig. 2.

The ablated surface *as-is* after 900 pulses is shown in Fig. 3. Here, the etch depth is about 17 μm, as measured by a stylus profilometer. It is clear that the ferrite surface is covered with a skin with extensive microcracking. More interestingly, "frozen-in-action" droplets of sizes on the order of microns can be seen throughout this area. Some of these droplets are connected to the surface with a "necked-down" stem, indicating that such droplets were almost departing from the surface but barely did not make it. The data strongly suggest that one mechanism responsible for ablative etching of ferrites is the hydrodynamic ejection of droplets from the surface.

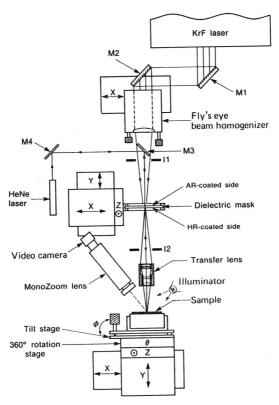

Figure 1. Experimental setup for projection ablation of ferrite ceramics using KrF excimer laser. M1, M2 are dielectric mirrors, M3 is a dichroic mirror, M4 is an aluminized mirror and I1, I2 are irises.

Figure 2. Groove patterns ablated in the Ni-Zn ferrite sample after ultrasonic cleaning. The etched depth is 56 μm.

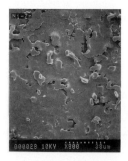 

Figure 3. SEM photos showing the microcracked skin of the *as-is* ablated Ni-Zn ferrite and emitted droplets that are "frozen in motion."

When the fluence is below the ablation threshold of 0.3 J/cm^2, the laser pulses do not produce any physical depression but instead only roughen the sample surface, preferentially at grain boundaries. For example, the initially invisible grain boundaries of the polycrystalline ferrite become clearly visible in a microscope after 1800 pulses at 0.265 J/cm^2. This effect only happens at a very narrow fluence range, between 0.2 − 0.3 J/cm^2 for our samples. These results indicate that pulsed laser irradiation of a polished polycrystalline surface with suitable fluence can produce the interesting and useful result of "staining" the grain boundaries, making them visible.

CONCLUSION

We describe the phenomenon of ultraviolet excimer laser photoablation of ferrites in air, and show that micropatterning of a large area of the ferrite ceramic surface can be done by projecting the image of a mask irradiated by the laser pulses onto the sample surface[8]. The ablation threshold is 0.3 J/cm^2, and a rather efficient ablation rate of 100 Å/pulse is produced at an incident fluence of 1 J/cm^2. Fluences slightly below the ablation threshold are found to produce roughening or darkening of the grain boundaries, revealing the grains of the polycrystalline sample. Scanning electron microscopy of the ablated surface shows the presence of frozen droplets attempting to leave the "glassy skin" of the ablated ferrite surface.

REFERENCES

1. F. A. Houle, SPIE 385, 127 (1983).
2. E. K. Yung, B. W. Hussey, A. Gupta and L. T. Romankiw, J. Electrochem. Soc. 136, 665 (1989).
3. Y. F. Lu, M. Takai, S. Nagatomo, T. Minamisono and S. Namba, Jpn. J. Appl. Phys. 28, 2151 (1989).
4. S. Nagatomo, M. Takai, H. Sandaiji and S. Ohara, "Method of Producing a Core for Magnetic Head," U. S. Patent 4751779, (June 21, 1988).
5. B. Braren, P. Leary-Renick and R. Srinivasan, IBM Tech. Discl. Bull. 29, 4258 (1987).
6. J. E. Rothenberg and R. Kelly, Nucl. Instrum. Methods B 1, 291 (1984).
7. R. W. Dreyfus, R. Kelly and R. E. Walkup, Appl. Phys. Lett. 49, 1478 (1986).
8. A. C. Tam, W. P. Leung, and D. Krajnovich, J. Appl. Phys. 69, Feb, 1991 (in press).

Part VI

Poster Presentations

CHARGE EMISSION FROM SILICON AND GERMANIUM SURFACES IRRADIATED WITH KRF EXCIMER LASER PULSES

M. M. Bialkowski, G. S. Hurst, and J. E. Parks
Institute of Resonance Ionization Spectroscopy
University of Tennessee, Knoxville, TN 37932

and

D. H. Lowndes and G. E. Jellison, Jr.
Solid State Division
Oak Ridge National Laboratory, Oak Ridge, TN 37831

ABSTRACT

Time-resolved measurements of the emission of positive and negative charges from Si and Ge surfaces irradiated with 248-nm KrF excimer laser pulses is reported.[1] With pulse energies both below and above the melting threshold, the time evolution of the emission currents is complex and strikingly different for Si and Ge. The positive ion emission signal from Ge persists only for the duration of the laser pulse (<60 ns), but in sharp contrast, the signal from Si continues for several microseconds. A tentative suggestion is made that the positive ions encounter a Knudsen layer created just above the surface of the Si target.

INTRODUCTION

The effects of laser irradiation on a material depend on properties of the target, such as its reflectivity, density, specific heat, latent heat, temperature of melting, etc., parameters of the laser beam, such as wavelength and energy density, and the time of interaction. Among the most prominent of these effects is the emission of electrons, positively charged ions and neutral atoms.

EXPERIMENTAL SETUP

The experimental setup is shown in Fig. 1. Thin wafers of Si and Ge were irradiated with pulses of 248 nm radiation from a KrF excimer laser. A lens and an aperture were used to produce a uniform intensity within a spot several mm in diameter. The laser energy density E_L at the target was adjusted by placing pairs of quartz plates in the laser beam.

The current flowing, due to the charge emission, and the laser pulses as a function of time, were measured and recorded for later analysis. The reflected intensity of a HeNe laser beam centered on the irradiated spot was also recorded and used to determine the surface melt duration.[2] The lowest E_L values were chosen for both Ge and Si to be just below the surface melting threshold. For E_L values above the melting threshold, the reflectivity signal provides an accurate in situ determination of E_L, since the E_L vs. melt duration relation is known for Si and Ge.[2]

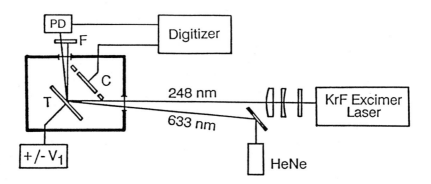

FIG. 1. Experimental arrangement.

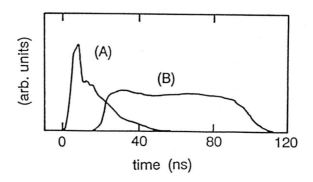

FIG. 2. KrF excimer laser signal (A) and reflectivity change (B) of the HeNe probing laser during the melting of Ge with the laser pulse $E_L = 0.60$ J/cm².

SUMMARY OF OBSERVATIONS

The emission of positive charge from a Ge surface just below the melting threshold[2] for crystalline Ge ($E_L \sim 0.3$ J/cm^2) was measured when the target was biased with a positive voltage. As shown in Fig. 3(a), the charge collection signals follow almost exactly the KrF excimer laser pulse; changing the bias voltage influences only the amplitude. In the case of Si just below the melting threshold[2] ($E_L \sim 0.7$ J/cm^2) and biased with a positive voltage, we observed a significant difference in comparison with Ge. As we can see in Fig. 3(b), the charge collection signals reach their maximum values much later than the laser pulse, and they extend into the microsecond range, well beyond their calculated 300 ns collection time.

For both Ge and Si targets biased with a negative voltage just below the melting threshold, the signals recorded are composed of two parts, as shown in Fig. 3(c) and Fig. 3(d). The initial part lasts as long as the KrF laser illuminates the surface, independent of the bias voltage on the target. The second part has a lower amplitude; its duration strongly depends on the value of applied voltage.

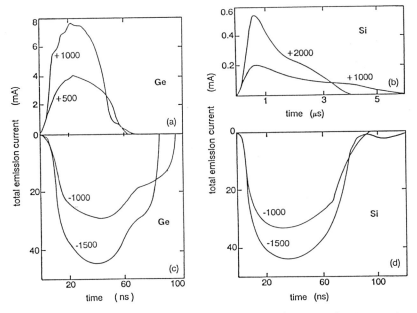

FIG. 3. Selected emission signals from unmelted Ge and Si targets (both very close to the melting point) biased with different voltages V_1 (shown in the figure in volts). (a) Positive charge emission from Ge; (b) Positive charge emission from Si; (c) Negative charge emission from Ge, (d) Negative charge emission from Si.

To confirm near-surface melting and to calculate the laser E_L, we measured the duration of the transient reflectivity change with a continuous HeNe laser beam.[2] For a positive bias voltage, charge emission signals from melted Ge, Fig. 4(a), and Si, Fig. 4(b), are very similar to the corresponding cases below the melting threshold, but they have higher amplitudes and obviously larger areas under the signal curves. A significant change comes when positively biased Si is melted with laser pulses of $E_L \sim 1.5$ J/cm^2, in which case the emission signals have strong oscillatory character at the early times.

For negative bias voltage, charge emission signals from melted Ge, Fig. 4(c), and Si, Fig. 4(d), also have two parts: an initial peak that appears to follow the laser pulse and a second component of longer duration that becomes more pronounced with E_L. For higher E_L the emission signals from Ge and Si strongly resemble each other. It is especially interesting that the number of negative charges was generally an order of magnitude larger than the number of positive charges.

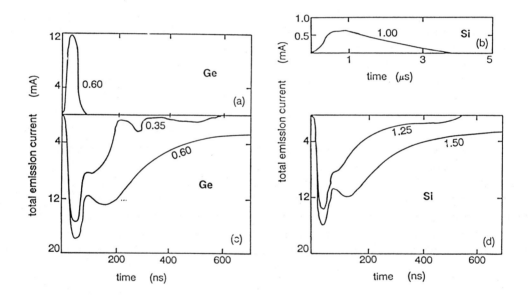

FIG. 4. Selected emission signals from melted Ge and Si targets biased with voltages $V_1 = -500$ V [(a) and (b)] and $V_1 = +1000$ V [(c) and (d)]. Energy densities in J/cm^2 given as numbers in the figure. The melting thresholds of Ge and Si are, respectively, 0.3 J/cm^2 and 0.7 J/cm^2.

DISCUSSION

Bloembergen's group has made use of picosecond lasers to obtain information on excitation and relaxation processes. They found that the number of negative and positive charges were about equal near or above the melting threshold of Si.[3] One of our results is that we always see, with the slower (60 ns) laser pulse, an order of magnitude more electrons than positive ions. Yet the most surprising feature of our results has to do with the time behavior of the positive ion signals. For instance, the current pulse for positively biased Ge has approximately the duration of the laser pulse, under conditions for which ions should be collected for about 500 ns.

We comment further on the sharply contrasting data for positive charge emission from Ge and Si. For Ge, the positive emission persists for the duration of the laser pulse, and the electron emission seems to reflect much of the same time behavior although the total number of electrons is about three times that for the positive emission. The most interesting feature of the Ge data is that while at 1000 volts it would take 300 ns to collect even the lightest ion, the time duration of the positive ion signal is just that of the 60 ns laser pulse. The most likely reason is the recombination of positive ions with electrons, which is consistent with the time behavior for both of the positive and negative signals. The alternative idea that for the Ge^+ ions to have enough kinetic energy for them to cross the detector in less than 60 ns would require that most of the ions have several keV of initial energy. The Ge data of Fig. 3 are consistent with a recombination mechanism.

The time-resolved emission of negative species from both Ge and Si becomes more complex above the melting threshold. In both cases there are multiple peaks whose positions and magnitude are quite sensitive to the values of the laser energy density. This may be connected with plasma processes which become important when the electron density increases such that the Debye length is reduced to a value comparable to the plume dimensions. It is difficult to estimate the Debye length because of large uncertainties in the concentrations of the charge species, and in the electron temperature. While it is well known that both Ge^- and Si^- are stable species with electron affinities equal to 1.23 and 1.39 eV respectively,[4] it is not likely that these play a dominant role in the time-resolved emission of the negative species.

The Si data are the real puzzle since the positive ion signal persists for about 3 μs, whereas the electron signal follows the laser pulse. Consider a simple mechanism that delays the collection of Si^+ ions, since in Fig. 2 the positive signal in Si doesn't even peak until about 700 ns. A clue was found simply by comparing the physical features of Si and Ge, and in particular it was noted that the vapor pressures at the respective melting points are radically different in the two cases; 10^{-3} torr for Si and 7×10^{-7} torr for Ge.

The model that we propose to explain these results is the following. Suppose that the density of neutral particles when Si is laser annealed is large enough to form a Knudsen layer where the fast particles are slowed down by the slow ones and the slow ones are speeded up by the collisions. (Such effects are known to be important in estimating the temperature of laser irradiated surfaces from time-of-flight measurements on emitted particles.[5]) Positive ions of Si are caught up into this Knudsen layer where the collisions are frequent enough to prevent acceleration by the external electric field. In other words, the positive ions in the Knudsen layer are behaving as if they were under the familiar swarm condition where mobilites are quite low.

To illustrate this model, suppose 1 monolayer of Si atoms is removed in a pulse. During the 60-ns laser pulse the Knudsen layer grows to about 10^{-2} cm, and thus represents a neutral particle density of about 10^{17} cm^{-3} (about 3 Torr of gas pressure). In this near-surface layer, a Si$^+$ ion would make 10-100 collisions and would therefore "drift" in the electric field. At the pressure-reduced electric field (E/P) value of 300 V cm^{-1} Torr^{-1}, a drift velocity of 10^4 cm s^{-1} would be reasonable. Thus, the "delay time" for the Si$^+$ to make its way through the Knudsen layer would be on the order of microseconds, and with this view, the observed time delays were found to be reasonable[1].

REFERENCES

1. M. M. Bialkowski, G. S. Hurst, J. E. Parks, D. H. Lowndes, and G. E. Jellison, Jr., J. Appl. Phys. 68, 4795 (1990).
2. G.E. Jellison, Jr., D.H. Lowndes, D.N. Mashburn, and R.F. Wood, Phys. Rev. B 34, 2407 (1986).
3. N. Bloembergen, Mat. Res. Soc. Symp. Proc. 51, 3 (1985).
4. S.G. Lias, J.E. Bartmess, J.F. Liebman, J.L. Holmes, R.D. Levin, and W.G. Mallard, J. Physical Chemistry Reference Data 17, Supp. 1 (1988).
5. R. Kelly and R. W. Dreyfuss, Surf. Sci. 198, 263 (1988).

PULSED LASER DEPOSITION OF TRIBOLOGICAL MATERIALS

M.S. Donley*, J.S. Zabinski*, V.J. Dyhouse**, P.J. John**, P.T. Murray** and N.T. McDevitt***
* WRDC/MLBT, Wright Laboratory, Wright-Patterson AFB, OH 45433-6533, ** University of Dayton Research Institute, Dayton, OH 45431, *** RAMSPEC Research, 4399 E. Mohave Dr., Dayton, OH 45431

ABSTRACT

Pulsed laser deposition (PLD) of tribological materials is an emerging technology that offers the possibility to tailor film properties for an application. Early research efforts to deposit MoS_2 utilized a frequency doubled Nd:YAG laser, while recent effort is focussed on laser processing of tribological materials using the UV wavelengths available from an excimer laser. PLD provides a mechanism to deposit highly adherent thin films of a variety of tribological materials. The materials of interest include metal dichalcogenide solid lubricants such as MoS_2 and wear resistant carbides such as TiC and Cr_3C_2. Applications of interest include solid lubricants for satellite precision direction mechanisms, wear resistant coatings for turbine engine components, and protective coatings for harsh environments. PLD films often exhibit superior performance, as compared to films deposited by more traditional methods. Improved film performance is due to increased adhesion, full density - low porosity, and optimized crystal structure and morphology. PLD is also being utilized to develop new materials formed by laser processing. Based on the results with lubricious films, PLD offers the possibility to tailor film properties by the appropriate choice of substrate materials, deposition parameters and post deposition treatments. The properties of films deposited by PLD are investigated as a function of: (1) substrate material, (2) laser deposition parameters (wavelength, pulse energy, fluence, and rep rate), (3) duration of post deposition laser annealing treatments, and (4) substrate temperature during deposition.

INTRODUCTION

Molybdenite (MoS_2) has been effectively applied to wear surfaces as a lubricant by mechanical burnishing, binding agents and lubricant compacts [1]. However, lubrication of spaceborn precision direction mechanisms requires a high degree of control over film thickness, excellent film adhesion, and minimal film porosity. In addition, component operating temperatures in other high precision applications may be limited by available binding agents used in the film growth. Some of these deficiences were addressed by the application of sputter deposition (SD) techniques to lubricant coatings. This technology permitted reliable control of film thickness and eliminates temperature limitations imposed by binding agents [2]. Although SD lubricant coatings are technologically important, they possess several characteristics that limit their effectiveness. SD films are frequently contaminated by background gases in the deposition chamber (i.e., H_2O, O_2, etc.) [3-5]. While oxygen impurities may not be detrimental to film integrity for room temperature applications, recent studies suggest that they may accelerate lubricant decomposition at elevated temperatures [6]. SD films typically have a platelike, porous morphology with the MoS_2 basal planes oriented perpendicular to the substrate surface. This orientation exposes reactive edge planes to the environment and increases their reactivity towards oxidants [7-10].

Pulsed laser deposition (PLD) is an attractive alternate technique for the deposition of lubricant materials that does not rely on physical sputtering processes. This technique permits the deposition of dense, near stoichiometric MoS_2 lubricant coatings that have the optimal basal orientation [11-14]. Early studies of the PLD technique utilized the 532 nm laser radiation from the frequency doubled output of a Nd:YAG laser to deposit MoS_2 onto 440C stainless steel (SS) substrates [11,12]. These films exhibited good adhesion, based on the wear life of the film, and a low coefficient of friction (0.15). Films deposited at 200 - 300°C had a S/Mo ratio of 1.8 and exhibited the best and most consistent friction and wear properties. However, the films contained large (1 to 3 um diameter), spherical surface features, which were due to a phenomenon called "splashing". Here, light energy impinging on the target produces subsurface heating of the target, and eventual melting. This molten material is ejected from the target and "splashes" onto the substrate. Later efforts eliminated the splashing problem by using the shorter, more energetic 193 and 248 nm radiation available from an excimer laser operating with ArF and KrF gas, respectively. Recent work has focussed on investigating the properties of MoS_2 films deposited by the excimer laser [13,14].

Laser-induced material removal has been found to occur by at least three mechanisms, depending mostly on photon energy. The first and lowest energy mechanism is laser desorption. Here, the photons provide only enough energy to desorb weakly bound species from the solid surface. The second mechanism is laser evaporation, where the photons have enough energy to remove chemically bonded species from the surface. In this mode, the laser acts as a thermal source and the irradiated area is in approximate thermodynamic equilibrium. The third and most energetic mechanism is laser ablation. In this mechanism, photons have enough energy to break chemical bonds. Garrison and Srinivasan have performed molecular dynamics calculations on the evaporation and the ablation mechanisms [15]. These calculations describe evaporation as local melting of the target, where material is ejected in a broad angular distribution. Ablation is described as a more energetic process where material is removed layer by layer, forming well-defined pits in the target. Material is ejected in a narrow angular distribution, similar to a supersonic expansion. The average perpendicular velocity of the ablated species was predicted to be on the order of 1-2 km/s.

Deposition of tribological materials by PLD offers several potential advantages: (1) films may be deposited in less than 10^{-5} Pa of background gases, producing high purity films and permitting precise control of dopant concentrations, (2) because photo ablation appears to be a congruent process, complex targets may be used to create complex films, (3) excellent film adhesion due to the energetics of the PLD process, (4) films can often be grown at lower substrate temperatures, as compared to more conventional methods, and (5) film properties may be controlled by proper selection of laser parameters, dopant gas, substrate temperature, and post laser anneals. The purpose of the work presented here is to describe in some detail the ongoing research in laser processing of tribological materials at Wright Laboratory.

EXPERIMENTAL

A schematic diagram of the Pulsed Laser Deposition System is presented in Figure 1. The system is comprised of four components: a laser, a series of optical components, a film deposition chamber, and an optional surface analysis system. At Wright Laboratory, early PLD work was accomplished with the output from a frequency doubled Nd:YAG laser, and current research utilizes either the 193 or the 248 nm output from an excimer laser. The beam is manipulated by a series of optical components. First, the beam is focussed to a 2.9 mm by 6.5 mm rectangle by a focussing lens (focal length of 1 m) and second, the beam is directed into the deposition chamber through a laser window by a 45° reflecting mirror. The 45° mirror also allows the beam to be rastered across the target.

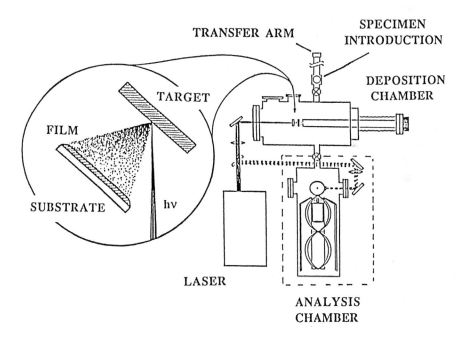

Figure 1. Schematic diagram of the Pulsed Laser Deposition System.

The Deposition System consists primarily of a stainless steel vacuum chamber and a 400 l/s turbomolecular pump. The system is capable of maintaining a base pressure of 8×10^{-7} Pa; the pressure during film deposition typically rises to 1×10^{-5} Pa. The deposition chamber includes a rotatable substrate stage with integral heater and a rotatable target mount. Resistive heating is employed to degas and/or anneal the specimens while their temperature is controlled using an infrared (IR) pyrometer. The laser beam is directed through the MgF_2 window and onto the target. For maximum transmission, the laser window construction requires the use of a window material such as MgF_2, suprasil, or sapphire, depending on the selected wavelength of light. The target and the substrate are rotated at about 10 rpm to ensure film uniformity. Film deposition rate and thickness are determined using a calibrated quartz crystal oscillator which is mounted in the deposition chamber. Using the ultraviolet (UV) output of the excimer with the beam being rastered, the material removal rate from the target decreases exponentially depending on the total number of pulses striking the target. This effect is due somewhat to the coating of the laser window, but mostly to target roughening. In the ablation mode, material removal occurs layer-by-layer, resulting in pitting of the target, and substantially increasing the target surface area. The material removal rate decreases because the net photons/cm^2 decreases with the total number of pulses. The target may be "preconditioned" to remove any surface contamination by ablating onto the back of the sample shutter prior to deposition. A Lambda Physik LPX 110i excimer laser is used to irradiate the target. The beam is focused to a 2.9 mm by 6.5 mm rectangle on the target face providing a total fluence of 530 mJ/cm^2. The unfocused laser beam described above provides a fluence of 38 mJ/cm^2 for post deposition anneals.

The surface analysis system is included for the chemical analysis of the PLD films. The analysis system is shown directly connected to the deposition chamber to allow film analysis without exposure to the ambient. This option is required in the deposition and analysis of many materials, especially nitrides and carbides, but is also essential for the initial film growth experiments of any new material. In PLD studies involving a material whose film deposition parameters are known, the surface analysis capability can provide quality assurance (QA) type analysis, but need not be vacuum connected. Surface analytical techniques are used extensively for film analysis in the PLD research ongoing at Wright Laboratory. Here, surface chemistry is studied with a Surface Science Instruments (SSI) M-probe X-ray photoelectron spectroscopy (XPS) instrument operated at 3×10^{-7} Pa providing an instrumental resolution of 0.87 eV. Binding energies are calibrated using the Cu 3s and Cu $2p_{3/2}$ peaks at 122.39 and 932.47 eV, respectively. In addition, small angle X-ray diffraction (XRD) and Raman spectroscopy are used to determine the crystal structure and chemistry of the PLD films. A Rigaku D/max -1B diffractometer equipped with a thin film attachment and a monochrometer is employed for XRD analysis. Raman spectroscopy is performed with a SPEX 1877 spectrometer which was calibrated with a CCl_4 standard. The tribological properties of the films are evaluated using a ball-on-flat tribometer to determine friction coefficients and wear lives as a function of film chemistry and crystal structure. During tribotesting, a 1/8" diameter 440C stainless steel ball is loaded with a 100 g weight, creating a Hertzian stress of approximately 1.00 Gpa. The substrate disks are rotated at 200 rpm in dry N_2 while the ball is positioned on a 3/8" diameter track. To investigate laser/target interactions, time-of-flight (TOF) analysis techniques are utilized to obtain the mass distribution and the velocity distribution of species removed from the target. The TOF apparatus is described elsewhere [12,16].

Hot-pressed MoS_2 powder from Climax Molybdenum Co. was used as the target material. Partially stabilized zirconia (PSZ) and 440C SS coupons were used as substrates and were polished to a final finish of 0.01 and 0.3 mm, respectively. Substrates were cleaned ultrasonically in hexane, acetone and methanol. Prior to deposition, the specimens were degased at ~ 573K for 1 h and allowed to cool. Substrates were coated with MoS_2 while the they were held at 300, 373, 473, and 573K. Films deposited at 300K were also annealed in 248 nm laser radiation for 1.5, 3.0, 6.0, and 12.0 m.

RESULTS AND DISCUSSION

To investigate the effects of photon energy, TOF analysis is compared for the PLD of MoS_2 using the frequency doubled Nd:YAG laser, $\lambda = 532$ nm [12], and the UV wavelengths of the excimer laser, $\lambda = 248$ and 193 nm [16]. Using photons of 532 nm wavelength to remove material, the velocity distribution of evaporated neutral S is well described by a Maxwell-Boltzman distribution, with an effective translational temperature of 1500K. These data indicate an evaporative mechanism. Using the 248 nm and the 193 nm photons of the excimer for material removal, the velocity distribution of S^+ and Mo^+ ions is well described by a drifted Maxwell-Boltzman, or a supersonic expansion. The effective translational temperature for the 248 and 193 nm wavelengths was calculated to be 63,000K and 111,000K, respectively. These data indicate an ablative mechanism. The TOF data are summarized in Table I. The evaporative process occurring at 532 nm wavelength and the ablative process occuring at 248 and 193 nm wavelengths are consistent with the proposed models of Garrison and Srinivasan [15].

Table I. Summary of TOF data.

Photon Energy (eV)	Wavelength (nm)	Fluence (mJ/cm2)	v_o (km/s)	T (K)
2.33	532	500	*	1,500
5.00	248	330	6.5	63,000
6.42	193	310	8.0	111,000

* Process is described by Maxwell-Boltxman distribution and has no v_o component.

To investigate the effects of laser fluence, laser wavelength, and substrate temperature, the Lambda Physik excimer laser was charged with KrF and with ArF for operations at 248 nm and 193 nm wavelength, respectively. The laser was operated at 10 Hz, at energy levels of 100 mJ and at 200 mJ, providing a total fluence of 530 mJ/cm^2 (@100 mJ) to irradiate the target. Based on XPS data, the films deposited at room temperature contained substantially more free S (less MoS$_2$-type sulfur) than films deposited at 573K, regardless of laser fluence. The films deposited at room temperature using 248 nm light were particularily sulfur-rich, with a calculated S/Mo ratio of 2.85 for the 100 mJ films and 2.24 for the 200 mJ films. This stoichiometry is not observed for films deposited at 573K. Here, the calculated S/Mo ratio was 2.2 +- 0.1 for all films, regardless of fluence and wavelength. Friction and wear testing did not reveal substantial differences for the films grown within the above deposition parameters. Raman spectra for films deposited by 248 nm wavelength light at 100 mJ are presented in Figure 2 for 20°C and 300°C substrate deposition temperature. As shown, no evidence of a crystalline structure is indicated in the 20°C spectrum. The spectrum of an MoS2 single crystal specimen is included for comparison. These results indicated that the substrate deposition temperature has a significant effect on the Mo-S chemistry of the films.

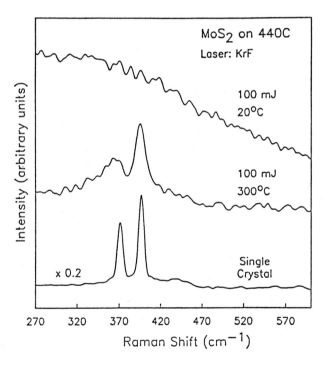

Figure 2. Raman spectra of MoS$_2$ films deposited by 248 nm wavelength light at 100 mJ for substrate temperatures of 20°C and 300°C, as indicated. Single crystal MoS$_2$ spectrum is included for comparison.

To investigate the effects of substrate type, substrate deposition temperature, and post-deposition laser annealing, the Lambda Physik excimer laser was charged with KrF and operated at 10 Hz and 100 mJ providing a total fluence of 530 mJ/cm^2 to irradiate the target. The deposition time was held constant at 45 min. to provide films that were about 350.0 nm thick. These films were thick enough to be of technological importance yet minimized variations in composition. The details of the XRD and XPS analysis are provided elsewhere [13]. The films deposited at room temperature onto SS and PSZ are amorphous, as demonstrated by featureless Raman and XRD spectra. The XPS data indicate that the film is primarily MoS_x where x is close to 2.0, but includes a multiplicity of chemical states. The S/Mo ratios of the film surfaces determined from XPS are 3.9 (SS) and 3.1 (PSZ). The S/Mo ratio of the film surface was found to increase somewhat as a function of deposition time. Some of the excess sulfur on the surface is in the form of metallic sulfur or a polysulfide species. The ratio of only MoS_x sulfur to Mo is 2.1, indicating that the films are "MoS_2-like" with some excess sulfur contained in the near surface region. Annealing the films to 373 and 473K does not produce crystalline features in the Raman or XRD spectra. The amount of metallic or polysulfide sulfur in the surface region decreases after thermal annealing, based on the XPS data. The loss of sulfur is likely caused by a reduction in the sulfur sticking coefficient.

Deposition of MoS_2 onto the SS substrate at 573K permitted Mo and S to nucleate into MoS_2 crystallites that were large enough (> 10.0 nm) to provide a broad (002) peak (basal plane) in the XRD spectrum. This demonstrates that the c-axis of the MoS_2 crystallites is predominantly aligned perpendicular to the substrate surface exposing the low friction, inert basal plane to the environment. Raman peaks at 383.9 and 407.7 cm^{-1} confirm the presence of MoS_2 crystals, as reported earlier [11,12]. Based on the XPS data, the S/Mo ratio of the film surface is 2.1; increasing the temperature caused a greater decrease of the sulfur sticking coefficient. The film deposited onto PSZ at 573K showed no Raman or XRD features. However, when a 75% thinner PSZ substrate was used for a 573K deposition, Raman peaks for MoS_2 were observed. It is concluded that the low thermal conductivity of PSZ prohibits the substrate surface from reaching the temperature measured for SS. Based on the observation above, the chemistry and crystal structure of films deposited onto SS and PSZ should be similar, provided that the annealing temperature is identical.

The chemistry and crystal structure of laser annealed MoS_2 films deposited at 300K are not significantly effected by substrate material. To compare the effects of film deposition temperature with laser annealing, Raman spectral peak intensities (arbitrary units) vs. laser anneal time (min.) and substrate deposition temperature (K) are presented in Figure 3. As indicated, Raman peak intensity increases with increased laser annealing time and with increased substrate deposition temperatures. As shown, films annealed for as little as 0.75 min. in 248 nm laser radiation provide Raman peaks at 384.7, and 408.5 cm^{-1}, representative of crystalline MoS_x (1.5 < x < 2.1). Laser annealing the films for increasing lengths of time increases the intensities of the peaks at 384.7 and 408.5 cm^{-1}. These data clearly demonstrate that laser annealing provides energy for the crystallization of amorphous MoS_x. However, XRD spectra remain featureless for annealing periods less than 12 min. indicating that the

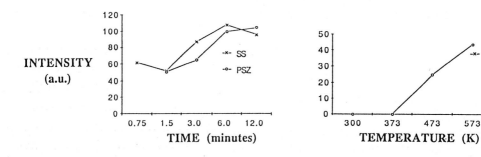

Figure 3. Raman spectral peak intensity (arbitrary units) vs. laser anneal time (min.) and substrate deposition temperature (K), as indicated.

MoS_x crystallites are significantly smaller than 10.0 nm. Laser annealing films for 12 min. and longer causes the chemical environment around the S and Mo atoms to become more uniform and more nearly like that of the hot pressed MoS_2 target. The S/Mo ratio of the film surface becomes nearly stoichiometric after 0.75 minutes of laser annealing and does not change significantly by annealing for longer times. In addition, the metallic or polysulfide species at the surface are removed by laser annealing, based on the XPS data. The laser energy may have permitted surface sulfur to diffuse into the film. In tribological testing, MoS_2 films deposited by PLD using 248 nm radiation are lubricious (coefficient of friction typically less than 0.1) and long lived when tested in dry N_2.

CONCLUSION

PLD is an attractive alternate technique for the deposition of lubricant materials such as MoS_2. This technique permits the deposition of dense, highly adherent, near stoichiometric MoS_2 lubricant coatings that have the optimal basal plane orientation. The films have low levels of O and C contamination and are resistant to oxidation in laboratory air. MoS_2 films deposited by PLD using 248 nm radiation are lubricious (coefficient of friction typically less than 0.1) and long lived. The chemistry, stoichiometry, and crystal structure of PLD films may be controlled by appropriate selection of thermal annealing treatments and post deposition laser anneals. The high degree of control of: (1) the concentration of contaminants/dopants in PLD films and (2) the chemistry and crystal structure of PLD films offers the potential to tailor lubricants to a specific application or to optimize their properties for exposure to multiple environments. Research is currently underway to utilize PLD for the thin film growth of new materials for a variety of tribological applications.

REFERENCES

1. F.J. Clauss, *Solid Lubricants and Self-Lubricating Solids*, Academic Press, New York, (1972).
2. T. Spalvins, ASLE Trans., **12**, 36 (1969).
3. V. Buck, Thin Solid Films, **139**, 157 (1986).
4. V. Buck, Wear, **114**, 263 (1987).
5. Jeffrey R. Lince, J. Mater. Res., **5(1)**, 218 (1990).
6. D.E. Pierce, R.P. Burns, H.M. Dauplaise, and L.J. Mizerka, U.S. Army Materials Technology Laboratory, Report # MTL TR 90-39 (1990).
7. J.R. Lince and P.D. Fleischauer, J. Mater. Res., **2**, 827 (1987).
8. P.D. Fleischauer, ASLE Trans., **27**, 82 (1984).
9. M. Matsunaga, T. Homma, and A. Tanaka, ASLE Trans., **25**, 323 (1982).
10. T. Spalvins, J. Vac. Sci. Technol. A, **5(2)**, 212 (1987).
11. M.S. Donley, P.T. Murray, S.A. Barber and T.W. Haas, Surf. and Coatings Technol., **36**, 329 (1988).
12. M.S. Donley, P.T. Murray and N.T. McDevitt in *New Materials Approaches to Tribology*, edited by L.E. Pope, L.L. Fehrenbacher and W.O. Winer, Mater. Res. Soc. Proc. 40, Boston, MA, (1989) pp. 277-282.
13. J.S. Zabinski, M.S. Donley, V.J. Dyhouse, and N.T. McDevitt, "Crystallization of Molybdenum Disulfide Films Deposited by Pulsed Laser Ablation", Mat. Res. Soc. Proc., Fall 1990.
14. P.J. John, V.J. Dyhouse, N.T. McDevitt, A. Safriet, J.S. Zabinski, and M.S. Donley, "Pulsed Laser Deposition of Metal Dichalcogenides on Stainless Steel", Mat. Res. Soc. Proc., Fall 1990.
15. B.J. Garrison and R. Srinivasan, J. Appl. Phys. **57**, 2909 (1985).
16. P.T. Murray, V.J. Dyhouse,, L. Grazulis, and D.R. Thomas, "Dynamics of MoS_2 Photoablation", Mat. Res. Soc. Proc., Fall 1990.

ION-MOLECULE REACTIONS OF CARBON CLUSTER ANIONS[*]

R. L. Hettich

Analytical Chemistry Division, Oak Ridge National Laboratory

Oak Ridge TN 37831-6120

INTRODUCTION

In an effort to use graphite as a substrate material for the laser desorption of biological compounds, we have noted the ease of generating carbon cluster ions by laser ablation of the graphite surface. Both positive and negative ion clusters can be generated by this process. The structures and energetics of neutral and ionic carbon clusters have been the source of many theoretical and experimental investigations. The importance of these clusters is manifest in many diverse areas, such as probing the nature of catalysis and combustion processes. While much research has been devoted to the examination of the structures of these clusters, less information is known about their reactivities. The focus of this report is to illustrate how laser ablation Fourier transform mass spectrometry can be used to generate carbon cluster anions and study their ion-molecule reactions.

EXPERIMENTAL

All experiments were performed with an Extrel FTMS-2000 Fourier transform mass spectrometer equipped with a Quanta Ray DCR-11 pulsed Nd:YAG laser. Carbon cluster ions were generated by 1064 nm laser ablation (10^8 W/cm^2) of a flat graphite foil. The resulting positive or negative ions were then trapped in the FTMS cell where they could be manipulated and examined in detail. No bath gas was used to collisionally cool

---

[*] Sponsored by the Office of Health and Environmental Research, U.S. Department of Energy under contract DE-AC05-84OR21400 with Martin Marietta Energy Systems, Inc.

the clusters during formation. The positive ion spectrum consisted primarily of carbon clusters C_n^+ where n = 7-20. Lowering the laser energy resulted in the formation of larger carbon cluster cations C_n^+ where n = 50-130, with C_{60}^+ (Buckminsterfullerene) observed as the most intense cluster ion. The negative ion spectrum revealed small carbon clusters C_n^- extending to n = 13, as shown in Figure 1. Collisional dissociation experiments were performed by accelerating a given ion into argon as a collision gas. Ion-molecule reactions were examined by trapping the carbon cluster anions in a static pressure of the reactant gas.

RESULTS

In order to probe ion-molecule reactions, these carbon cluster anions were trapped in the FTMS cell and allowed to react with oxygenated compounds such as oxygen, nitrous oxide, water, and carbon dioxide. While carbon cluster cations react readily with O_2 to abstract an oxygen [1], carbon cluster anions are completely unreactive with oxygen, suggesting that the bond energy $D°$ (C_n^- - O) may be quite weak. The lack of charge exchange between C_n^- and O_2 verifies that the electron affinities of carbon clusters [2] are larger than the electron affinity of O_2. No ion-molecule reactions were observed between C_n^- and N_2O. C_3^- is the only cluster ion that will react with CO_2, and forms C_3O^- as the ionic product. This implies that the bond energy $D°$ (C_3^- - O) must exceed the energy required to abstract an oxygen atom from CO_2. Collisional dissociation of C_3O^- results in formation of C_2^-, indicating facile CO loss for this cluster.

Carbon cluster anions do react with water to form C_nOH^- and $C_nH_2O^-$ as the ionic products. The formation of C_nOH^- dominates when n is an odd number while $C_nH_2O^-$ is observed when n is an even number, as shown in Figure 2. Note that hydroxide abstraction and condensation are the only reactions observed (oxide abstraction from water does not occur). In order to verify that these are gas phase reactions, each cluster anion was isolated and allowed to react with water. Collisional dissociation of these product ions revealed facile losses of CO and (C_2OH), as shown below in Table I. For example, C_6^- reacts with H_2O to form $C_6H_2O^-$ as the only ionic product. Collisional activation of this $C_6H_2O^-$ ion results in formation of $C_5H_2^-$ (elimination of CO) and C_4H^-

(loss of C_2OH). Loss of CO as a neutral is readily observed for all these hydrated products. Note that H_2O is <u>not</u> eliminated as a neutral product from any of the $C_nH_2O^-$ species. This implies that the water molecule is not loosely bound to the cluster anion, but may in fact be completely dissociated into oxygen and hydrogen atoms on the cluster ion. The formation of C_4H^- as a fragment from $C_6H_2O^-$ corresponds to a loss of (C_2OH), which may be sequential loss of CO + CH. Note that most of these hydrated carbon cluster anions preferentially fragment to form C_nH^-, where n is an even number. These ions most likely are polyacetylenes and should be quite stable. Further work is in progress to investigate the structures of these hydrogen-containing cluster anions and to examine deuterium-exchange reactions for the determination of isotope effects.

REFERENCES

1. S.W. McElvany, B.I. Dunlap, and A. O'Keefe, J. Chem. Phys. 86, 715 (1987).
2. S. Yang, K.J. Taylor, M.J. Craycraft, J. Conceicao, C.L. Pettiette, O. Cheshnovsky, R.E. Smalley, Chem. Phys. Lett. 144, 431 (1988).

Table I. Collision-induced dissociation of $C_nOH_x^-$

Parent ion	Fragmentation products
$C_4H_2O^-$	$C_3H_2^-$ + CO
C_5OH^-	C_4H^- + CO
$C_6H_2O^-$	$C_5H_2^-$ + CO
	C_4H^- + (C_2OH)
C_7OH^-	C_6H^- + CO
$C_8H_2O^-$	$C_7H_2^-$ + CO
	C_6H^- + (C_2OH)

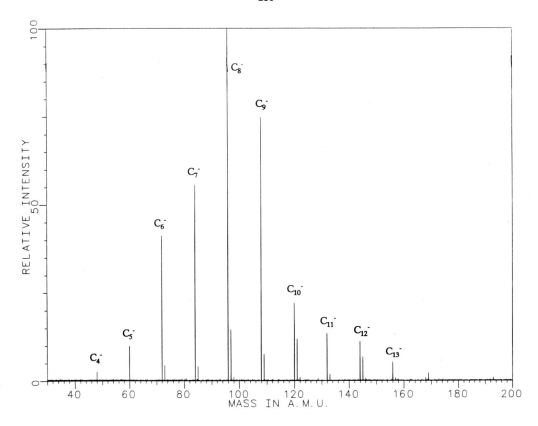

Figure 1. Negative ion laser ablation spectrum of graphite, depicting carbon cluster anions C_4^- - C_{13}^-.

284

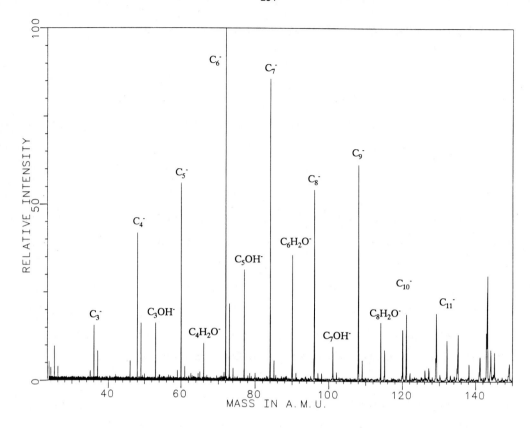

Figure 2. Ion-molecule products generated by reacting C_n^- with H_2O.

DOUBLY CHARGED NEGATIVE IONS OF BUCKY BALL - C_{60}^{2-} *

R.L. Hettich [1], R.N. Compton [1,2], and R.H. Ritchie [1,2]

[1] Oak Ridge National Laboratory, Oak Ridge, TN 37831-6120

[2] University of Tennessee, Knoxville, TN 37996-1600

INTRODUCTION

The discovery of special stability for C_{60}^+ in the laser ablation mass spectrum of graphite [1] has initiated a number of investigations into the properties and structures of large carbon clusters. In particular, the recent relatively facile synthesis of gram quantities of C_{60} has led to a rapid examination of many of the fascinating physical properties of this new form of carbon [2]. The predicted high symmetry (I_h) of this soccerball-shaped molecule [3] has been verified by the observation of the expected four infrared active vibrational modes [2], vibrational Raman spectra [4], and the occurrence of a single NMR line for the $^{13}C_{60}$ molecule [5,6]. Scanning tunnelling microscopy images of the C_{60} molecule reveal spherical species (the carbon atoms are unresolved, presumably due to rotations) [7,8]. Carbon clusters containing seventy atoms also exhibit stability similar to that of C_{60}. The five line NMR spectra reported for $^{13}C_{70}$ support an elongated "rugby ball" shape, as predicted earlier [6].

The isolation of macroscopic quantities of <u>neat</u> multi-atom clusters also allow for the unprecedented study of many cluster phenomena which previously have been complicated by distributions in cluster size. Among many possibilities are the investigation of the properties of surface plasmon excitations, cage effects in photoabsorption, and the electrical properties of microspheres.

Limited studies of the ionic properties of C_{60} have appeared. The ultraviolet photoelectron spectrum of C_{60}^- shows bands at 3, 4.5, and 6 eV with a threshold at 2.8 eV (electron affinity of C_{60}) [9]. The 3 eV peak is attributed to removal of the "extra" electron occupying the threefold degenerate t_u lowest unoccupied molecular orbital.

---

* Sponsored by the Oak Ridge National Laboratory Exploratory Studies Program and the Office of Health and Environmental Research, U.S. Department of Energy under contract DE-AC05-84OR21400 with Martin Marietta Energy Systems, Inc.

In this study, we report the observation of the doubly charged ions, C_{60}^{2-} and C_{70}^{2-}, in the gas phase. Theoretical modelling of potential energy surfaces and speculation on possible formation mechanisms are also presented. For comparison, cyclic voltammetry experiments of C_{60} in methylene chloride indicated three reversible half-wave reduction waves [10], providing evidence for stable C_{60}^-, C_{60}^{2-}, and C_{60}^{3-} in solution. Long-lived doubly charged carbon cluster anions in the gas phase have been reported from the sputtering of a graphite surface by Cs^+ ions, revealing clusters C_n^{2-} where n = 7-28 [11]. In this case, the even-numbered clusters were more intense than their neighboring odd-numbered clusters.

EXPERIMENTAL

Two methods were used to synthesize the C_{60}/C_{70} samples. One method employed laser ablation (xenon chloride excimer laser, 308 nm) of a graphite rod in a quartz vial maintained under 300 Torr of helium. A second method employed an arc (80 amp at 30 volts) between two 1/4-inch graphite rods (Poco graphite) in 100 Torr of helium followed by soxhlet extraction of the particulate into benzene. Commercial samples of C_{60}/C_{70} were also obtained from the MER Corporation (Tucson, Arizona). These samples were deposited onto stainless steel disks and subsequently examined by laser desorption Fourier transform mass spectrometry (FTMS). During the course of the study, at least seven different procedures for sample preparation and collection were examined, and in all cases yielded intensity ratios for C_{60}^{2-} / C_{60}^- and C_{70}^{2-} / C_{70}^- in the range of 0.02 - 0.20.

All mass spectra were acquired with an Extrel FTMS-2000 Fourier transform mass spectrometer equipped with a Quanta Ray DCR-11 Nd:YAG pulsed laser. One laser shot (266 nm, $10^7 - 10^8$ W/cm^2) of the previously prepared samples generated ions (either positive or negative) which were trapped for 3-20 ms at a base pressure of 8×10^{-8} Torr and subsequently detected in the FTMS ion cell. Medium resolution conditions (m/Δm = 2000) and accurate mass measurements were used to identify the ions.

RESULTS

Laser desorption of samples generated by the two methods outlined above revealed carbon clusters of C_n where n = 60-200, with C_{60} and C_{70} observed as the most abundant species. Laser desorption of a pure sample of C_{60}/C_{70}, shown in Figure 1, revealed abundant ions at m/z 720 (C_{60}^-) and 840 (C_{70}^-) as well as minor ions at m/z 360 and 420, which we believe to be C_{60}^{2-} and C_{70}^{2-}, respectively. The insert in Figure 1 is an expansion of the carbon isotope region for C_{60}^- and reveals the abundant ^{13}C

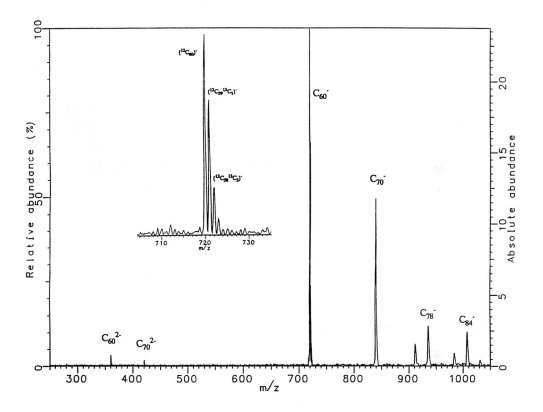

Figure 1. Negative ion 266 nm laser desorption spectrum of purified fullerene sample (primarily C_{60} and C_{70}). The insert is an expansion of the m/z 710-730 region and illustrates the isotopic abundances for C_{60}^-.

isotopic pattern for these clusters. Figure 2 illustrates an expansion of the m/z 350-430 region of Figure 1 and reveals the isotopic pattern for the m/z 360 and 420 ions. Note that the isotopes in both cases appear at one-half integer mass units. In addition, the

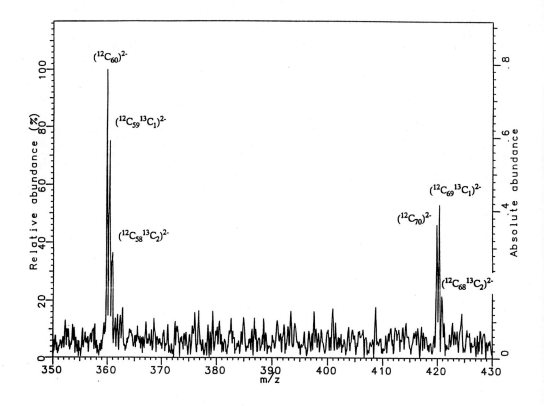

Figure 2. Expansion of the m/z 350-430 region from Figure 1. Note that the isotopes occur at half-integer mass units and have abundances which match those observed for the singly charged ions.

isotopic abundance for m/z 360 is very similar to that observed for m/z 720 (C_{60}^-). These isotopic patterns are precisely the distributions expected for the doubly-charged anions and verify that m/z 360 and 420 are not due to C_{30}^- and C_{35}^-, respectively. Several other experiments (ion ejection, trapping voltage shifts) were performed in order to verify that the ions at m/z 360 and 420 were not due to second harmonic artifacts of the abundant ions at m/z 720 and 840 [12]. The C_{60}^{2-} and C_{70}^{2-} ions observed here have lifetimes which exceed 3 milliseconds, which is the ion detection time for the FTMS experiments. Although a minor ion was also observed at m/z 240, which could correspond to C_{60}^{3-}, experimental investigation revealed that this signal was due to the third harmonic of the abundant m/z 720 ion and <u>not</u> C_{60}^{3-}. Doubly charged positive ions of C_{60} and C_{70} were also detected and identified.

The interaction potential between a charge q and an insulating conducting sphere containing a charge Q is given by the following equation [13];

$$V(\xi) = q/a \left[\frac{1}{\xi^2(1-\xi^2)} + \frac{Q}{q\xi} \right] \qquad (1)$$

where ξ is the distance between q and the center of the sphere (R), divided by the radius of the sphere (a). Classically, such a sphere charged with any number of electrons can bind an additional charge. Figure 3 illustrates the potential of interaction

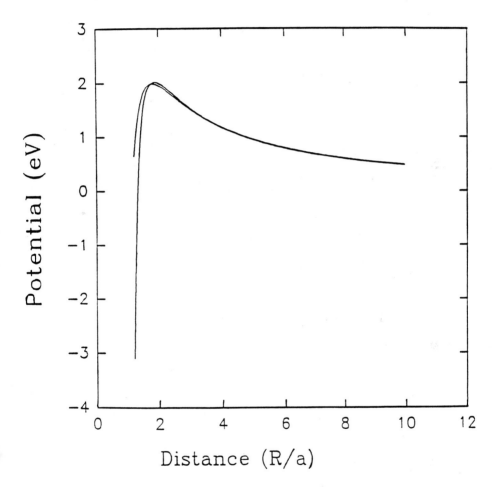

Figure 3. Interaction potential between an electron and a singly charged sphere of radius a derived from Equation (1), shown as the full curve, compared with a potential derived from a pure repulsive plus attractive polarization potential, i.e. $V(R) = e/R - \alpha e/R^4$, shown as the short curve.

between an electron and a sphere containing a single negative charge. Also plotted is the potential derived from a pure repulsive e/R potential plus an attractive polarization potential, $-\alpha e/R^4$. The polarizability, α, of the C_{60} molecule is calculated to be ≈ 80 Å using the empirical method of Miller and Savchik [14]. Quantum mechanically, it is not evident that a charged sphere such as C_{60}^- can accept an additional charge. In fact, restricted Hartree-Fock calculations indicate that the C_{60}^{2-} is unbound [15].

We have taken a semi-empirical approach to the problem of estimating the energy of the second electron on the C_{60}^- ion. We assume that the closed-shell C_{60} molecule is approximately spherically symmetric, and that one may represent the interaction of an added electron with the C_{60} molecule by a pseudopotential. We assume that this pseudopotential may be written in the form

$$V(r) = -\frac{e^2\alpha}{(c+(r-r_c)^2)^2} \qquad (2)$$

where r_c is a radius which is chosen to be ≈ the radius (a) of the C_{60} molecule. This potential is chosen to be attractive at large distances with an asymptote given by $-\frac{e^2\alpha}{r^4}$.

In this view, the electron of the C_{60}^- negative ion is in an image-potential-induced surface state in analogy with the much-studied states at a metallic surface where a band gap exists [16]. However, the pseudopotential we use has an significant component due to the closed shell electronic system and nuclei of the C_{60} molecule so that the characterization of C_{60}^- ion in terms of image states may not be unambiguous, although the polarization potential probably plays an important part in determining the states of the negative ion. We have chosen this pseudopotential such that, upon solving Schrödinger's equation, the binding energy of the added electron is found to agree with the measured value of 2.65 ± .02 eV.

To treat the added electrons of the C_{60}^{2-} ion, we have solved the Hartree-Fock equations for two electrons in a spherically symmetric, singlet ground state, accounting for exchange and for the Coulomb interaction between them, assuming the same pseudopotential for the electron-C_{60} core interaction as was used in determining the binding energy of the one-electron ion. We find that in this approximation, the binding energy per electron in the singlet state may lie in the range from 0.4 eV to -0.3 eV, depending on the values chosen for the parameter r_c. We note, however, that even in the case where this energy is -0.3 eV, the state found is metastable and that the lifetime of the state is expected to be quite long enough to make possible the measurement of the ion in the experiments described here.

Figure 4 shows the pseudopotential, the total effective potential seen by one of the singlet electrons in the ground state, and the square of the wave function of one of the singlet electrons. Here r_c was chosen to be 10 a.u. and the resulting binding energy was found to be $\approx$ 0.4 eV. As one sees, the singlet wave function is well-localized, as it is for all values of r_c that we have tried.

We plan to extend further these theoretical estimates of the properties of the C_{60} ions so as to make the prediction of binding energies more definite. We will explore a self-energy model [17] in which the long-range part of the pseudopotential is characterized in terms of the dynamic response function of an equivalent jellium spherical shell and the short range part is expressed through an exchange-correlation potential.

Based on the results stated above, some interesting speculations can be proposed for the formation of C_{60}^- and C_{60}^{2-}. Unimolecular attachment of slow electrons to C_{60} can lead to formation of very long-lived C_{60}^- ions. Thermal electron attachment to C_{60}^- to produce C_{60}^{2-}, however, would require tunneling through a long-range Coulomb barrier. The probability of tunneling through such a barrier is quite small. We have examined experimentally the attachment of slow electrons to gas phase C_{60} and C_{70} molecules. Slow electrons will attachment to both of these molecules, forming C_{60}^- and C_{70}^-, but not the doubly charged anions. A more likely mechanism for formation of C_{60}^{2-} and C_{70}^{2-} involves negative (Saha) surface ionization, which is induced by rapid laser heating of the surface to a temperature of a few thousand degrees.

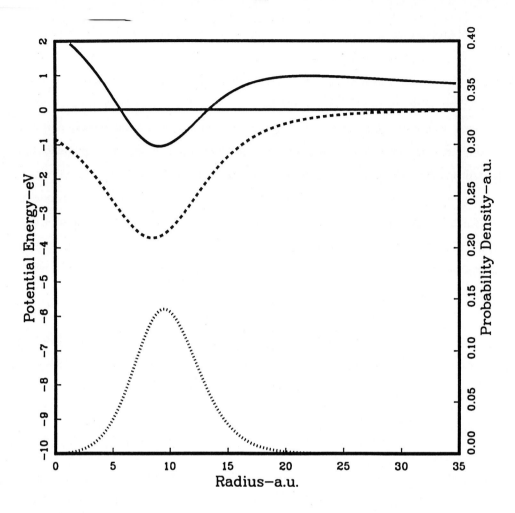

Figure 4. The pseudopotentials and probability function for the singlet state of the C_{60}^{2-} ion. The dashed curve shows the pseudopotential assumed for a single electron in the vicinity of a C_{60} molecule, where r_c in Equation 2 is taken to be 10 a.u. The solid curve is the potential experienced by one of the electrons of the singlet pair in the C_{60}^{2-} ion. The dotted curve shows the wave function squared of one of the electrons in the singlet state.

REFERENCES

1. H.W. Kroto, J.R. Heath, S.C. O'Brien, R.F. Curl, and R.E. Smalley, Nature (London) 318, 162 (1985).

2. W. Kratschmer, L.D. Lamb, K. Fostiropoulos, and D.R. Huffman, Nature 347, 354 (1990).

3. C_{60} is a truncated icosahedron with 12 pentagons and 20 hexagons containing 60 equivalent carbon atom vertices.

4. D.S. Bethune, G. Meijer, W.C. Tang, and H.J. Rosen, Chem. Phys. Lett. 174, 219 (1990).

5. R.D. Johnson, G. Meijer, and D.S. Bethune, J. Am. Chem. Soc. 112, 8983 (1990).

6. R. Taylor, J.P. Hare, A.K. Abdul-Sade, and H.W. Kroto, J. Chem. Soc. Chem. Commun. 20, 1423 (1990).

7. R.J. Wilson, G. Meijer, D.S. Bethune, R.D. Johnson, D.D. Chambliss, M.S. de Vries, H.E. Hunziker, and H.R. Wendt, Nature 348, 621 (1990).

8. J.L. Wragg, J.E. Chamberlain, H.N. White, W. Kratschmer, and D.R. Huffman, Nature 348, 623 (1990).

9. S.H. Yang, C.L. Pettiette, J. Conceicao, O. Cheshnovsky, and R.E. Smalley, Chem. Phys. Lett. 139, 233 (1987).

10. P.-M. Allemand, et.al. J. Amer. Chem. Soc. 113, 1050 (1991).

11. S.N. Schauer, P. Williams, and R.N. Compton, Phys. Rev. Lett. 65, 625 (1990).

12. A.G. Marshall and P.B. Grosshans, Anal. Chem. 63, 215A (1991).

13. J.D. Jackson, "Classical Electrodynamics", second edition, John Wiley and Sons, Inc., (1975), page 59.

14. K. J. Miller and J. A. Savchik, J. Am. Chem. Soc. 101, 7206 (1979).

15. A.H.H. Chang, W.C. Ermler, and R.M. Pitzer, "C_{60} and Its Ions: Electronic Structure, Ionization Potentials, and Excitation Energies," (submitted).

16. P.M. Echenique and J.B. Pendry, Progress in Surface Science 32, #2, 111 (1989).

17. G. MO, C. C. Sung and R. H. Ritchie, Chem. Phys. Lett. 176, 433 (1991).

EVAPORATION AS A DIAGNOSTIC TEST FOR HYDRODYNAMIC COOLING OF LASER-ABLATED CLUSTERS

Cornelius E. Klots
Chemical Physics Section
Oak Ridge National Laboratory
Oak Ridge, TN 37831

I. Introduction

The properties of materials laser-ablated from a surface are of considerable interest. The interrogation of these properties inevitably occurs at a point some distance from the surface. One might then ask what processes have occurred in the intervening path length. Immediately, for example, one wonders whether the material was released as such from the surface or was formed as a result of collisions at a distant point. Similarly, one might ask if an observed "temperature" of the material is characteristic of the ablation process or of subsequent events. We will indicate here how measurements of metastable evaporation rates can provide clues which are pertinent to these questions.

II. The Kinetics of Evaporation

Consider an aggregate of material bound together by van der Waals forces. If it is sufficiently hot, it may undergo evaporation. Any such event will cool the residual material, which will then undergo further evaporation at a lower rate. As a first approximation, the aggregate will have cooled to the point where the rate constant for a subsequent emission is equal to the reciprocal of the time since the original "big bang".

A more careful analysis [1] indicates that such an evaporating population, isolated and normalized to unity at time t_o, will be given at a later time by:

$$P(t) = 1 - (C/\gamma^2) \ln \{t/[t_o + (t - t_o) \exp(-\gamma^2/C)]\} \qquad (1)$$

where C is the heat capacity of the aggregate in units of k_B, and γ is given by

$$\gamma = \gamma_g^2/[1 - (\gamma_g/C)^2/12 \ \ldots\ldots] \qquad (2)$$

where γ_g is the Gspann constant. On a microsecond time scale, and in the absence of discontinuities in the heat of evaporation, a "best value" for γ_g equal to 23.5 (±1.0) has been suggested [2].

Other cooling mechanisms exist. Thermionic emission has now been identified in several materials [3,4], but is at best unlikely to occur more than once in a given aggregate. Infrared emission will also occur but only on time scales much longer than envisaged here.

III. Pertinence to Ablation

Equation (1) prescribes what is in effect an upper limit to how hot an aggregate can be. Thus, if it is even hotter at the moment of ablation, or if it is generated later via massive condensation, then it will have undergone subsequent evaporation. On the other hand, if the ablation is more gentle, or if a subsequently condensing aggregate is cooled by an adiabatic expansion, before the cessation of collisions, then it is likely to be less prone to evaporation than equation (1) would prescribe.

IV. Two Experiments

Radi et al. [5] have ablated carbon clusters from a graphite rod and have studied their evaporation probabilities. They are compatible with equation (1) above [6]. When the same carbon clusters were generated by ablation from a polyimide surface [7] using generally lower laser fluences, the measured decay rates are instead about an order of magnitude lower.

This difference cannot be attributed to a laser fluence effect, under the supposition that a hotter laser produces hotter clusters. Campbell et al. [7] found that as they increased the fluence, their clusters became cooler, i.e. less prone to evaporation. These authors suggested an increased role for hydrodynamic cooling when ablation was more massive.

V. Discussion

Ablation of a polyimide necessarily releases non-carbonaceous material into the vapor phase. If it does not condense, then this material would be cooled by any hydrodynamic flow effects and would in turn be available to cool the condensing clusters. This may explain the differing resultant "temperatures".

A second difference exists between the two experiments. Radi et al. [5] carefully rotated their carbon rod during their experiments so as constantly to expose a new surface; Campbell et al. [7] apparently did not. Deep pitting of a surface might tend to accentuate any hydrodynamic flow effects.

Experiments to distinguish between these two possibilities are not hard to envisage. The immediate point is that both possibilities can be avoided, i.e. that clusters can be made which are as hot as they can be, and that a diagnostic test for this condition exists.

Acknowledgment

Clarifying discussions with E. E. B. Campbell are gratefully acknowledged. This research was sponsored by the Office of Health and Environmental Research, U. S. Department of Energy under Contract No. DE-AC05-84OR21400 with Martin Marietta Energy Systems, Inc.

References

[1] C. E. Klots, J. Phys. Chem. 92, 5864 (1988).
[2] C. E. Klots, Int. J. Mass Spectrom. Ion Phys. 100, 457 (1990).
[3] A. Amrein, R. Simpson, and P. Hackett, J. Chem. Phys. 94, 4663 (1991).
[4] E. E. B. Campbell, G. Ulmer, and I. V. Hertel (Unpublished Manuscript).
[5] P. P. Radi, M. T. Hsu, J. Brodbelt-Lustig, M. Rincon, and M. T. Bowers, J. Chem. Phys. 92, 4817 (1990).
[6] C. E. Klots, Z. Phys. D. (Submitted for Publication).
[7] E. E. B. Campbell, G. Ulmer, and H. G. Busmann, and I. V. Hertel, Chem. Phys. Lett. 175, 505 (1990).

DESORPTION OF LARGE ORGANIC MOLECULES BY LASER-INDUCED PLASMON EXCITATION*

I. Lee[†], E. T. Arakawa, and T. A. Callcott[†]
Biological and Radiation Physics Section
Oak Ridge National Laboratory, Oak Ridge, TN 37831-6123

INTRODUCTION

Ejection of large organic molecules from surfaces by laser-induced electronic-excited desorption has attracted considerable interest in recent years.[1,2] In addition to the importance of this effect for fundamental investigations of the ejection process, this desorption technique has been applied to the study of large, fragile molecules by mass spectrometry. In this paper, we present a new method to induce electronic excitation on the metal surface for the desorption of large organic molecules.

EXPERIMENTAL METHODS

Surface plasmons excited by a Nd:YAG laser in an attenuated-total-reflection (ATR) geometry were used to induce electronic-excited desorption of rhodamine B molecules. The second harmonic of the Nd:YAG laser beam has wavelength 532 nm and pulse width 7 ns. Rhodamine B is a dye molecule with molecular weight 479 amu, ionization potential 6.7 eV[3], and the molecular structure shown in Fig. 1.

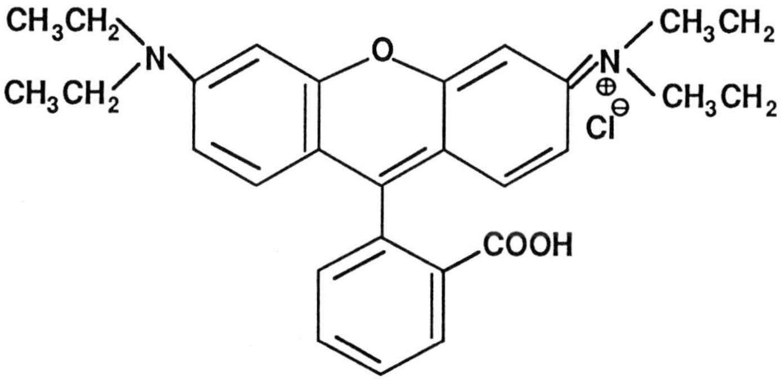

Fig. 1 Molecular structure of rhodamine B.

In our experiment, a 27 nm Al film was deposited on the base of a 45° prism in a vacuum evaporator. A rhodamine B film was next deposited on the Al film by evaporating rhodamine B powder at a very low temperature. The thickness of rhodamine B film was 80 nm. The desorbed neutrals were ionized by a XeCl laser with wavelength 308 nm and pulse width 7 ns. The signals were measured by a time-of-flight mass spectrometer (TOF-MS). The signals from the TOF-MS were recorded by a PC (80286) controlled transient digitizer. The experimental apparatus used in the desorption experiment is shown schematically in Fig. 2. By varying the delay time between the two lasers, we could obtain the delay time distributions of the desorbed neutrals. The kinetic energy distributions could then be transformed by a simple calculation.

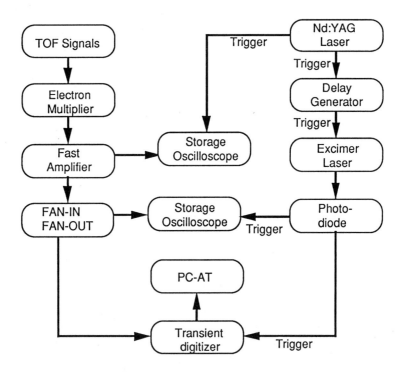

Fig. 2 The experimental apparatus.

RESULTS AND DISCUSSION

We found that at the plasmon resonance angle there were two peaks in the mass spectrum, corresponding to the masses of Al atoms and of rhodamine B molecules. While Al has both a low energy thermal peak at 0.081 eV and a high energy peak at 0.51 eV in the kinetic energy distribution, rhodamine B has only one peak at 0.56 eV kinetic energy. Both the Al and rhodamine B high-energy signals showed peaks at photon incidence angles corresponding to the surface plasmon resonance angle (Fig. 3). The yield of fragment ions of rhodamine B was very close to the background noise and was at least an order of magnitude weaker than the yield of complete molecular ions. The thresholds for both Al and rhodamine desorption were observed at a laser fluence of about $F = 80$ mJ/cm². At laser fluences exceeding about $F = 300$ mJ/cm² the rhodamine B signal disappeared and significant fragmentation signals appeared. The maximum rhodamine B desorption yield at the plasmon resonance angle was obtained at the laser fluence $F = 178 \pm 5$ mJ/cm². When the fluence was increased beyond the maximum, the magnitude of the rhodamine B peak decreased while the magnitude of the Al peak increased. Since Al atoms from the substrate film were always observed together with the rhodamine molecules, it appeared that Al atoms were desorbed over a large area. This was also evidenced by physical damage observable on the metal film.

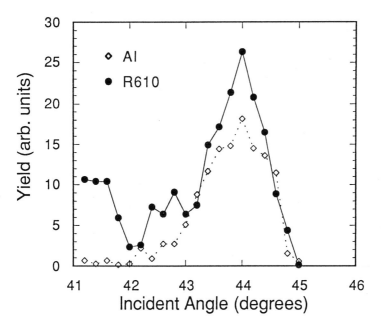

Fig. 3. Desorption yields of rhodamine B molecules (R610) and Al atoms as a function of photon angle of incidence for a laser fluence 178 mJ/cm².

* Work supported by the U.S. Department of Energy under Contract DE-AC05-84OR21400 with Martin Marietta Energy Systems, Inc.

† Also at Department of Physics & Astronomy, University of Tennessee, Knoxville, TN 37996-1200

REFERENCES

[1] T. J. Chuang, Surf. Sci. **178** (1986) 763.
[2] V. G. Movshev and S. V. Chekalin, Sov. J. Quantum Electron. **13** (1983) 925.
[3] M. M. Timoshenko, I. V. Korkoshko, V. I. Kleimenov, N. E. Petrachenko, Y. V. Chizhov, V. V. Rylkov, and M. E. Akopyan, Doklady Physical Chemistry (Consultants Bureau, New York, 1982) p. 800.

SIMULTANEOUS BOMBARDMENT OF WIDE BANDGAP MATERIALS WITH UV EXCIMER IRRADIATION AND keV ELECTRONS*

J. T. Dickinson, S. C. Langford, and L. C. Jensen
Department of Physics
Washington State University
Pullman, WA 99164-2814

In previous work[1] we examined the changes in the surface topography of sodium trisilicate glass ($Na_2O \cdot 3SiO_2$) with exposure to pulsed 248 nm excimer laser light at fluences of 2.6-5 J/cm^2, as well as the character of the products emitted from the glass surface (e.g., +/- ions, electrons, ground state and excited neutral atoms and molecules). At these fluences, ablation readily occurs after a fixed number of preliminary laser pulses (an effect known as incubation). In the current study, we examine the precursors of this high fluence behavior at sub-threshold fluences < 2.6 J/cm^2 and show that the effectiveness of laser bombardment in removing material is strongly dependent on defects produced either by high fluence 248 nm radiation *or* by electron radiation. We show a dramatic synergism in the ablation process by simultaneous bombardment of the glass surface with 0.5-2 keV electrons and laser pulses. A model is discussed involving surface and near-surface defects created by the electron beam that provide single photon absorption centers and free electron--laser heating. We also show that similar results are obtained on single crystal NaCl, LiF, and UV grade fused silica. The potential for performing single photon driven etching/ablation on wide band gap dielectric materials is also discussed.

1. Introduction

The use of lasers and other radiation sources to break bonds as well as induce and promote chemistry via electronic and thermal interactions has greatly facilitated progress in a number of technologies concerning the removal of material from surfaces, e.g., etching, chemical modification, cutting, and drilling. Examples of current areas of high interest involving the use of lasers which obviously benefit from such knowledge include photoetching of surfaces, laser ablation-deposition (e.g., of superconductor thin films), laser microlithography, laser chemical vapor deposition, and various applications to packaging, repair, reflow, and surgical processes. In many instances, lasers have also provided a means of rapidly introducing molecular and ionic fragments of the solid into the gas phase for performing chemical analysis, often associated with mass spectroscopy or gas phase spectroscopy. In this paper, we review our studies of mechanisms[3,4] of the atomic/molecular and ionic emissions from wide bandgap materials, concentrating primarily on a sodium trisilicate glass. It was expected that the excitation of electrons from defects in the sodium trisilicate glass played an important role in the onset of etching and ablation by producing free electrons in the glass conduction band and thus greatly enhancing light-material coupling. We therefore investigated the emission of charged particles over a range of fluences below and near the threshold for etching as well as the influence of defect production by bombardment of the glass surface with energetic electrons.

2. Results and Discussion

Electron Emission. An important result from earlier work the observation of highly energetic ions and excited neutral particles (0-60 eV) accompanying etching of sodium trisilicate glass at above-threshold fluences. To explain these high energies it was necessary to invoke an electrostatic acceleration mechanism, namely, the inverse bremsstrahlung acceleration of electrons in the near-surface region of the vacuum, followed by electrostatic acceleration of ions via a "wake" effect.[5] This process requires substantial electron and ion densities in the plume during the early portion of the laser pulse; subsequent interactions with the laser pulse elevate both the electron and surface temperatures. Thus in the inverse bremsstrahlung model, photoelectron and ion emission intensities play a crucial role in determining the onset of plume formation which coincides with rapid etching. The electron emission intensity vs laser fluence is shown in Fig. 1.

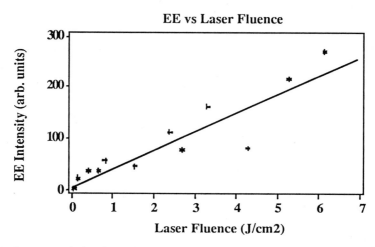

Fig. 1. The electron emission intensity vs laser fluence. Note that it is linear, implying a single photon process.

Over a wide range of fluences, including those sufficient for plume formation, EE is a *linear* function of laser fluence. First order dependence of EE on laser fluence has also been observed during the laser irradiation of fused silica.[6] Given the transparency of sodium trisilicate glass at 5 eV,[7] this suggests that EE is a single photon process involving mid-gap defect states. In our observations of these electrons, we find that the EE intensity is quite stable from pulse to pulse, in contrast to other emissions (e.g., positive ion emission), suggesting that the observed EE at these fluences is not very sensitive to surface conditions. Further, it is important to note that the observed EE was a strong function of laser repetition rate, falling markedly at rates above ~20 Hz. This dependence is due to the depletion of filled traps. It requires times on the order of 50 ms to replenish these traps. The simple first order process observed also suggests that overall charging of the surface was not very extensive; substantial surface charging would be expected to introduce non-linear behavior.[8]

Positive Ion Emission (PIE). Positive ions are also emitted at fluences well below the threshold for plume formation and rapid etching. The low fluence positive ions were mass analyzed with a quadrupole mass spectrometer. Within our sensitivity, all PIE was mass 23, i.e., Na^+. Estimates of E fields along the ion flight paths allow time-of-flight analyses which indicate that the ion energies are substantially less than 1 eV. This is in stark contrast to high fluence PIE which typically displayed energy distributions with peaks in excess of 10 eV. As noted above, we attribute this difference to inverse bremsstrahlung acceleration of electrons coupled with "wake-like" acceleration of positive ions towards the electrons moving away from the surface.

A plot of PIE intensity (averages over 50 laser pulses) vs laser energy is shown in Fig. 2 on a logarithmic intensity scale. The inset shows the PIE intensity at low fluences plotted on a linear intensity scale. Attempts to use simple first order heating and an Arrhenius equation[9] for the PIE emission rate were entirely unsuccessful in fitting the data over the full range of fluences. At lower fluences, the data were consistent with a non-thermal, two-photon process, i.e., ~f^2, where f is the fluence/pulse. [Note that if we assume a "square wave" laser pulse, the photon flux F (photons/cm^2/s) and the fluence/pulse, f, are proportional.] Two photon excitations in this material would yield electronic states above the band gap, producing "conduction" electrons; thus the emission of Na$^+$ under these conditions would most likely involve electronic defects. Ohuchi et. al.[10] have proposed that electron stimulated desorption of Na$^+$ from sodium trisilicate glass involves a trapped hole precursor. Two photon production of related defects has been previously reported.[11,12]

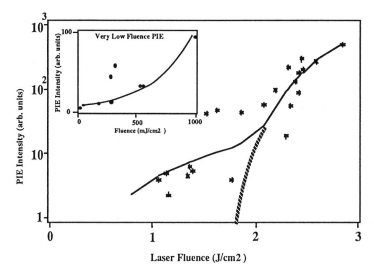

Fig.2 Positive ion emission intensity vs laser fluence/pulse on a logarithmic intensity scale. The inset shows the very low fluence PIE intensities on a linear scale. The solid line in the main graph represents a least squares fit to Eq. 3. The hatched line indicates the consequences of discarding the first term ~f^2.

At higher fluences, the rapidly increasing PIE intensity is consistent with a heating mechanism consisting of *simultaneous* production of free electrons (by single photon excitation from defects) and laser heating of the free electrons by the laser pulse. This is similar to the multiphoton free electron production and heating proposed by Shen et al.[13] for the heating and damage of wide band gap, defect-free materials. An equation describing the temperature rise due to laser–free electron heating has been derived by Epifanov[14,15] and is given by:

$$\frac{\partial T(r,t)}{\partial t} = \frac{C}{\rho c} T^{1/2}(r,t)\, n_c\, F^{3/2}(r,t) \tag{1}$$

where C is a constant involving material parameters, ρ is the mass density of the glass, C is the heat capacity, n_c is the free electron density, and F is the photon flux (photons/cm^2/s). For single photon–free electron generation, n_c would be proportional to $n_d F$ (or $n_d f$), where n_d is the density of defects capable of releasing electrons. Replacing n_d with this dependence and integrating (assuming F to be constant and uniform over the interaction volume *during* the laser pulse), we obtain a temperature rise, ΔT, given by:

$$\Delta T = n_d \left[K_1 \cdot f^{5/2} + \frac{K_1^2}{1200} \cdot f^5 \right] \tag{2}$$

Substituting this non-linear temperature rise into an Arrhenius term added to a term describing the two photon emission process discussed above, we obtain an expression for the ion current, I_{PIE}:

$$I_{PIE} = a_1 f^2 + a_2 \text{EXP}(-b/(n_d \left[K_1 \cdot f^{5/2} + \frac{K_1^2}{1200} f^5 \right])) \tag{3}$$

where a_1, a_2, b, and K_1 are adjustable parameters. Note that at constant f, as n_d increases, I_{PIE} will increase in a non-linear fashion.

The solid line in Fig. 2 represents a non-linear least squares fit of the raw data to Eq. 3. The hatched line shows the effect of ignoring the first term ($\sim f^2$) of this equation. Although there is considerable scatter, the overall fit is quite reasonable and shows the necessity of the f^2 term at low fluence. Additional PIE measurements at fluences as low as 0.05 J/cm^2 using the full gain of the Channeltron electron multiplier are shown in the inset of Fig. 2; we continue to find a strongly non-linear intensity vs fluence dependence, consistent with the two photon emission term of Eq. 3. The the second, laser heating, term becomes important with the very beginning of the onset of plume formation and rapid etching in what appears to be somewhat like an avalanche. In fact, no avalanche (usually attributed to electron multiplication) is required for this PIE behavior; higher order heating (Eq. 2) is sufficient to explain the rapid rise in PIE with fluence. We therefore propose that the dominating factor in these strong laser–surface interactions is the free carrier density, n_c. n_c, in turn, is directly related to the density of defects (n_d) serving as sources for single-photon generated free electrons, consistent with the observed first order electron emission results (see Fig. 1).

For a single photon photothermal process, where all absorbed light goes into heat, then using Eq. 1 we can estimate the temperature rise produced by a single laser pulse. For a short laser pulse of length t (e.g., 20 ns), the characteristic distance heat can flow is $\sim(2k\,t)^{1/2}$; where k is the average heat conductance; this distance is typically $\sim$ a few hundred Å. Thus, for moderately weak absorbers, $1/\alpha$ represents the depth where all of the laser pulse energy is deposited. Thus, if R is the surface reflectivity of the material surface at the incident wavelength, an energy balance (assuming no heat loss by conductivity or radiation) yields a temperature rise, ΔT, of :

$$\Delta T = \frac{F_o(1-R) \cdot \alpha}{\rho C_v} \tag{4}$$

which when substituted into a Maxwellian v-distribution, would produce energy distributions strongly dependent on laser fluence. In contrast to this prediction, until one approaches the threshold fluence, the time interval between the laser pulse and the arrival of the ions at the detector was insensitive to laser fluence. Fig. 3 shows the mass 23 PIE signal vs time (through the QMS) for four different fluences between 0.4 and 1.1 J/cm^2, where most of the ion flight path was in ~zero E field. Little difference in flight times is observed. Such behavior is inconsistent with a thermal emission mechanism. The Na⁺ emission mechanism at low fluences thus is photoelectronic in nature rather than thermal; this is further support for the non-thermal first term included in Eq. 3.

At higher fluences, the Na⁺ energies as well as the associated excited neutral sodium, Na*, do in fact increase with fluence. The latter particle appears above the etching threshold

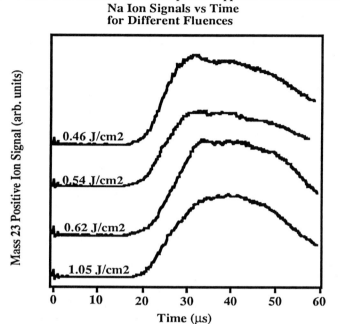

Fig. 3. Mass 23 (Na+) positive ion signals vs time for different fluences (0.46 - 1.05 J/cm^2). The laser pulse struck the surface at t = 0.

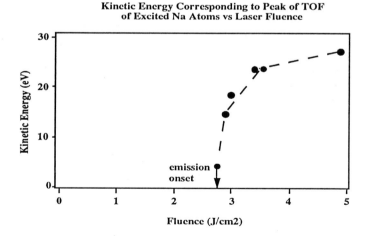

Fig. 4. Excited Na* kinetic energy vs fluence at high fluences.

and is attributed to reneutralization of Na$^+$ in the plume to form a long-lived excited state, i.e., a Rydberg. Figure 4 shows the kinetic energy of these particles as measured by their time-of-flight. This is consistent with acceleration of Na in its ionic form by intense laser-plume interactions, which we claim *start* with the laser heating term of Eq. 3.

Once the surface has been irradiated at high fluences to induce etching and plume formation, a plume could be maintained at lower fluences (e.g., at ~70% $F_{threshold}$). At fluences barely above the threshold for plume formation, the PIE initially decreases (due to depletion of surface Na) but eventually increases dramatically, accompanied by plume formation and the onset of rapid etching. Immediately below the threshold, plume formation was still often observed, but only after thousands of shots with little PIE. At higher fluences (> 2.7 J/cm^2), we observed that a similar but much shorter pre-treatment was required for plume formation. In each case, this induction period is required for the accumulation of damage in the form of near-surface defects. Recently we have demonstrated that intense negative ion emission accompanies ablation at fluences near threshold and higher.[4] This negative ion emission appears to be principally composed of oxygen ions (O$^-$). Oxygen vacancies are precursors to a number of important defects which absorb UV radiation; thus, the observation of oxygen removal from silicate glass by 5 eV photons is of some significance.

Effect of Electron Irradiation on PIE.[3] To determine if electron beam induced states are capable of promoting strong interactions between 248 nm radiation and the glass, we examined the effect of simultaneous and sequential electron and laser irradiation. For this study, we used an arrangement involving a simple Faraday Cup collector positioned 10 cm from the sample surface and biased to collect charge of a desired sign. An electron beam was produced by a Varian Auger gun which provided a spot size of 1.5 mm diameter at energies of 500-3000 eV and currents of 50-300 µA. The laser spot was focused directly in the center of the electron beam spot. The charge striking the cone yielded a current pulse which was capacitively coupled to an amplifier and digitized. Individual charged particle bursts could be detected as a function of arrival time relative to a single laser pulse. Measurements with an unbiased collector show a fast, negatively charged emission peak, and a slower, positively charged emission peak. The fast, negative peak corresponds to low energy photoelectrons. The slower, positive emission peak corresponds to the large, broad peaks detected with both ±bias. Thus the slow negative emission is comoving with a dense cloud of positive ions to which it is electrostatically coupled. The majority of the negative charge consists of electrons with substantial numbers of negative ions (O$^-$, Si$^-$, NaO$^-$, and perhaps NaSi$^-$) are also detected.[4]

In the absence of the e$^-$ beam, no evidence of etching or plume formation is evident at a fluence/pulse of 1 J/cm^2. In Fig. 5(a), we show a sequence of 50 laser pulses (left hand side of the plot) which at this detector sensitivity produced essentially no PIE. At the arrow, a 150 µA (current density of 6 mA/cm^2), 1500 eV electron beam was directed onto the bombarded region. Immediately, PIE was detected. We show the observed peak intensities for 50 additional laser pulses. A Na* plume was visually observed accompanying these PIE bursts. Thus, at a relatively low laser fluence, e$^-$ bombardment can produce a dramatic increase in the extent of the laser interaction with the surface. The e$^-$ beam lowers the threshold for strong coupling to the surface and therefore encourages rapid etching by the excimer laser.

If during a sequence of simultaneous e$^-$ and laser bombardment (with the above parameters), we simply turn off the e$^-$ beam [see Fig. 5(b)], the PIE gradually decreases over a few pulses (1 Hz period), followed by occasional, sporadic bursts; continued laser bombardment for < 100 pulses brings the ion emission down permanently. This suggests that the e$^-$ beam induced damage is stable for at least a few seconds and that multiple pulses can etch the damaged region away, returning the material to its original state. We also conducted experiments in which laser beam struck the surface during a ~100 ms time interval when the e$^-$ beam was temporarily directed away from the sample. The absence of the e$^-$ beam during the 20 ns laser pulse did not significantly affect the resulting PIE. This is consistent with the very small number of incident and secondary electrons produced by the electron gun in the laser-irradiated region of the sample over the duration of the pulse. We estimate ~10^7 electrons in a volume containing roughly 10^{20} atoms of the solid, excluding free charge carriers produced by the laser pulse.

We have made Reflection Electron Energy Loss Spectroscopy (REELS) measurements to detect possible electron beam induced surface states which might participate in the emission process. These REELS measurements show that with increasing electron dose, very distinct energy loss peaks are produced in the -2 to -6 eV energy range (where zero energy is defined by the location of the elastic peak). These results suggest that absorption sites for 5 eV photons are produced by electron-induced surface damage. Similar peaks were induced during Ar ion bombardment. These shallow electronic levels, which may be responsible for light absorption and electron emission at low laser fluences, are associated with surface and/or defect states.

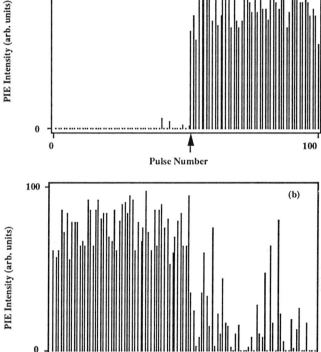

Fig. 5. Peak PIE due to successive laser pulses with and without simultaneous electron irradiation at 1.5 keV. In (a) the electron current density at the interaction region is initially zero and is raised to ~6 mA/cm^2 at the arrow. In (b), the electron current density is initially ~6 mA/cm^2 and is reduced to zero at the arrow.

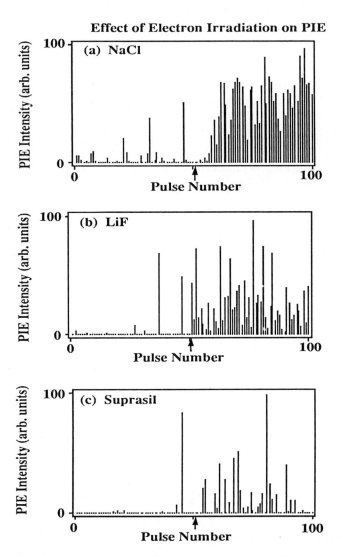

Fig. 6. Peak PIE from (a) NaCl (at 2.8 J/cm^2), (b) LiF (at 6.7 J/cm^2), and (c) fused silica (Suprasil II, at 7.8 J/cm^2) due to successive laser pulses first without, then with simultaneous electron irradiation at 1.5 keV. The arrow marks the first laser pulse after the electron beam was turned on.

REELS studies on amorphous SiO_2 films show loss peaks at 3.5, 5.1, and 7.2 eV which are attributed to a metastable surface phase of nearly monoxide stoichiometry.[16] These energy losses are close to surface levels seen on pure Si (electron affinity: 4.2 eV; band gap: 1.1 eV), suggesting that the responsible levels are indeed within 5 eV of the vacuum level and are capable of participating in photostimulated electron emission, i.e., they are filled states.

The creation of E' centers by excimer laser irradiation at 5 eV has been observed by Stahis and Kastner[17] in high purity amorphous silicon dioxide (Suprasil and Suprasil W) with electron paramagnetic resonance (EPR) techniques. Devine, et al.[12] have reported a 5 eV absorption peak created by irradiation of dry silica at 248 nm and have attributed it to an oxygen vacancy center. Nagasawa, et al.[18] performed calculations on a structure of the form:

$$\mathord{\supset}\text{Si}-\text{Si}\mathord{\subset}$$

which had an absorption at 5 eV, although it involved excitation to a bound state. Because a surface can readily release oxygen without interstitial production, O vacancies have a much higher probability of survival at the surface than in the bulk, and can serve as a shallow electron trap from which electrons are readily promoted to the conduction band. Thus, we suggest that the induction time for the onset of laser ablation in sodium trisilicate glass involves creation of such neutral oxygen vacancy defects containing electrons which can be released by photons of subsequent laser pulses. The laser heating of the resulting "free electrons" leads to lattice heating and the eventual softening or melting of the glass at high fluences, resulting in the observed hydrodynamic sputtering.[1]

Other wide bandgap materials tested in a preliminary fashion are single crystal NaCl, LiF and fused silica (Suprasil II) with 8.6, 13.6, and 8.9 eV bandgaps, respectively. The response to applying an electron beam to these materials at laser fluences below those required for sustained emission were again enhanced ion emission (see Fig. 6) with an e⁻ beam energy of 1.5 keV and current density of 6 mA/cm^2.[3] The relevant defects created by the electrons in the case of the SiO_2 were most likely the same as for the sodium trisilicate. For the alkali halides, F and F" centers are likely candidates.

3. Conclusion

The technological application of low energy electrons to modify the etching and other surface effects of laser radiation may prove extremely useful for a number of materials, including wide band gap dielectrics and strongly absorbing materials such as polymers and perhaps semiconductors. By controlling the energy of the electron beam, the distribution of defects vs distance from the surface can be very easily controlled. The resulting localization of absorption may minimize certain undesired consequences of laser irradiation, such as bulk heating and consequent diffusion. In sodium trisilicate glass, it appears possible that the proper choice of laser and electron beam parameters may yield high etch rates without the thermal stress induced fracture observed at high laser fluences.[1] In the case of photoablative deposition, it may be possible with combined laser/e⁻ beam processing to achieve higher deposition rates without unwanted formation of particulates (ejecta). Chemical modification of material surfaces by laser radiation may also benefit from simultaneous e⁻ irradiation. For instance, Igarashi et al.[19] have recently shown that surface layers of $BaTiO_3$ can be transformed from an insulator to a n-type semiconductor (oxygen deficient) with KrF excimer laser radiation. Combining laser irradiation with electron irradiation may provide useful control of the localization and/or extent of this transformation. Furthermore, well characterized defect chemistry created by controlled electron bombardment may prove useful in elucidating damage mechanisms in optical thin films and other optical materials.

Acknowledgments

The authors wish to thank Larry Pederson and Don Baer, Pacific Northwest Laboratories for providing silicate samples and surface analysis. This work was supported by the Ceramics and Electronics Materials Division of the National Science Foundation under Grant DMR 8912179, and the Washington Technology Center.

References

1. P. A. Eschbach, J. T. Dickinson, S. C. Langford, and L. R. Pederson, J. Vac. Sci. Technol. A7(5), 2943 (1989).
2. P. A. Eschbach, J. T. Dickinson, and L. R. Pederson, MRS Symp. Proc. 129, 385-392 (1989).
3. J. T. Dickinson, S. C. Langford, L. C. Jensen, P. A. Eschbach, L. R. Pederson, and D. R. Baer, J. Appl. Phys. 68, 1831 (1990).
4. S. C. Langford, L. C. Jensen, J. T. Dickinson, and L. R. Pederson, J. Appl. Phys. 68, 4253 (1990).
5. Yu. A. Bykovskii, N. N. Degtyarenko, V. E. Konrashov, and E. E. Lovetskii, Sov. Phys. Tech. Phys. 18, 1597 (1974).
6. G. Tetite, P. Agnostini, G. Boiziau, J. P. Vigouroux, C. Le Gressus, and J. P. Duraud, Optics Comm. 53, 189 (1985).
7. A. Miotello and F. Toigo, Nucl. Instru. Methods Phys. Res. B32, 258 (1988).
8. T. L. Gilton and J. P. Cowin, "Laser Induced Electron Emission: Space Charge and Electron Acceleration," conference presentation, ACS National Meeting 1988, L.A.
9. R. W. Dreyfus, "Are Laser-Ablated Monolayers Accurately Characterized by Their Ion Emission?," in Microbeam Analysis 1989, P. E. Russell, Ed., San Francisco Press, Inc., San Francisco , pp. 261-263 (1989).
10. F. Ohuchi and P. H. Holloway, J. Vac. Sci. Technol. 20, 863 (1982).
11. T. E. Tsai, D. L. Griscom, and E. J. Friebele, Phys. Rev. Lett. 61, 444 (1988).
12. R. A. B. Devine, Phys. Rev. Lett. 62, 340 (1989).
13. X. A. Shen, S. C. Jones, and P. Braunlich, Phys. Rev. Lett. 62, 2711 (1989).
14. A. S. Epifanov, Sov. Phys. JETP 40, 897 (1975).
15. S. C. Jones, P. Braunlich, R. T. Casper, X. A. Shen, and P. Kelly, Optical Engineering (1989).
16. J. E. Rowe, Appl. Phys. Lett. 25, 576 (1974).
17. J. H. Stahis, M.A. Kestner, Phys. Rev. B, 29(12), 7079 (1984).
18. K. Nagasawa, H. Mizuno, Y. Yamasaka, R. Tohmon, Y. Ohki, and Y. Hama, n *The Physics and Techology of Amorphous SiO_2*, R. A. B. Devine, ed., Plenum Press, New York, 1988, pp. 193-198.
19. K. Igarashi, H. Saito, T. Fumioka, S. Fujitsu, K. Koumoto, and H. Yanagida, J. Am. Ceram. Soc. 72, 2367 (1989).

SUPERCONDUCTING TRANSPORT PROPERTIES AND SURFACE MICROSTRUCTURE FOR $YBa_2Cu_3O_{7-\delta}$-BASED SUPERLATTICES GROWN BY PULSED LASER DEPOSITION

David P. Norton, Douglas H. Lowndes, X. Zheng,* R. J. Warmack,*
S. J. Pennycook, and J. D. Budai
Oak Ridge National Laboratory, Solid State Division, *Health and Safety Research Division, Oak Ridge, Tennessee

ABSTRACT

$YBa_2Cu_3O_{7-\delta}/PrBa_2Cu_3O_{7-\delta}$, $YBa_2Cu_3O_{7-\delta}/Pr_{0.7}Y_{0.3}Ba_2Cu_3O_{7-\delta}$, and $YBa_2Cu_3O_{7-\delta}/Pr_{0.5}Ca_{0.5}Ba_2Cu_3O_{7-\delta}$ superlattice structures have been grown by pulsed laser deposition. The superconducting properties of these $YBa_2Cu_3O_{7-\delta}$-based superlattices are shown to depend on the electronic properties of the barrier layers. In particular, the superconducting transition width decreases as the hole carrier density in the barrier layers is increased, while T_c(onset) does not change with barrier layer carrier density. T_c(onset) is apparently determined by the $YBa_2Cu_3O_{7-\delta}$ layer thickness, while the transition width, determined by long-range phase coherence of the superconducting wave function, depends on the electronic properties of the isolating barrier layers. Using scanning tunneling microscopy, we also have investigated the surface microstructure of epitaxial $YBa_2Cu_3O_{7-\delta}$ thin films grown by pulsed laser deposition, and find a highly terraced, granular morphology. This morphology suggests that multilayer thin film structures possess a high density of steps within the individual layers. These steps should be quite important in the transport properties of these superconducting structures.

INTRODUCTION

High-quality epitaxial $YBa_2Cu_3O_{7-\delta}$ thin films have been obtained on a variety of substrate materials using several deposition techniques, including pulsed laser deposition. These films exhibit outstanding superconducting properties including superconducting transition temperatures, T_c, greater than 90 K and critical current densities, J_c, in excess of 10^6 A/cm^2 at 77 K in zero magnetic field and 10^5 A/cm^2 in 8 tesla. Several groups also have reported on the transport properties of high-temperature superconducting/ semiconducting superlattices, including $YBa_2Cu_3O_{7-\delta}/PrBa_2Cu_3O_{7-\delta}$ [1-4], $Nd_{1.83}Ce_{0.17}CuO_\delta/YBa_2Cu_3O_\delta$ [5], and $Bi_2Sr_2Ca_{0.85}Y_{0.15}Cu_2O_8/Bi_2Sr_2Ca_{0.5}Y_{0.5}Cu_2O_8$ [6] structures. For the $YBa_2Cu_3O_{7-\delta}/PrBa_2Cu_3O_{7-\delta}$ (YBCO/PrBCO) superlattices, the superconducting properties are a function of both the superconducting (YBCO) and the barrier (PrBCO) layer thicknesses. T_c

decreases as the YBCO layer thickness is decreased or as the PrBCO layer thickness is increased, but for all YBCO layer thicknesses, including layers one unit cell thick, the superconducting transition temperatures saturate at nonzero values; e.g., for YBCO layer thicknesses of one, two, and three unit cells isolated in a ~20 nm thick PrBCO matrix, the zero resistance transition temperatures, T_{c0}, are ~19 K, ~54 K, and ~70 K, respectively. Other interesting observations can be made regarding the resistive behavior of these structures. First, there appears to be "coupling" between thin YBCO layers that are separated by PrBCO layers only a few unit cells thick. This is evident as, for a given YBCO layer thickness, T_c does not become independent of PrBCO layer thickness until $d_{PrBCO} > 5$ nm. Second, the widths of the superconducting transitions are large, with $\Delta T_c \sim 37$ K for a YBCO layer one unit cell thick isolated in a PrBCO matrix.

Several explanations appear possible for the resistive transitions observed in these superlattice structures, including proximity effect [7], localization effects [8], and hole-filling [9]. However, fundamental questions concerning the role of the barrier layers in these structures must be answered experimentally before specific mechanisms can be considered. In addition, a detailed understanding of the nucleation and epitaxial growth mechanism for these structures has not been developed. An understanding of the growth mechanism will be particularly important as multilayer structures are considered for device applications.

EXPERIMENTAL

The objective of this study is to develop an understanding of the superconducting transport properties of these YBCO-based superlattice structures. One effort is to determine whether the superconducting transitions of YBCO-based superlattice structures depend on the electronic properties of the barrier layers, focusing specifically on the mobile hole concentration (resistivity) of the barrier layers. We also have investigated the surface microstructure of these films, using scanning tunneling microscopy, to observe how close the layered structures are to "ideal". C-axis oriented YBCO epitaxial thin films, as well as superlattice structures, were grown by pulsed laser deposition. The superlattice structures were fabricated using three different barrier layer materials, namely $PrBa_2Cu_3O_{7-\delta}$ (PrBCO), $Pr_{0.7}Y_{0.3}Ba_2Cu_3O_{7-\delta}$ (PrYBCO), and $Pr_{0.5}Ca_{0.5}Ba_2Cu_3O_{7-\delta}$ (PrCaBCO). For the superlattices, stoichiometric polycrystalline targets (YBCO, PrYBCO, PrCaBCO, and PrBCO) were mounted in a multitarget holder which permits target rotation and exchange. A pulsed KrF (248 nm) excimer laser beam was focused to a horizontal line and scanned in the vertical direction across the rotating target. The substrates, positioned ~6.5 cm away from the targets, were bonded to the heater face using silver paint. Substrates used include (001) $SrTiO_3$ and (001) $LaAlO_3$, although (001) $SrTiO_3$ gave the most consistent results. The films normally were grown at a heater (substrate) temperature of 730°C (670°C) in 200 mTorr of oxygen. The laser repetition rate was 1.1 Hz with each pulse depositing ~0.1 nm. After deposition, the films were cooled at 10°C/min in an oxygen pressure of 400

Torr. Total thickness for the YBCO films and YBCO-based superlattice structures was 100–300 nm. This film growth procedure results in YBCO epitaxial thin films with $T_c(R=0) \sim 88\text{--}91$ K, transition widths 1 K or less, and $J_c(77\ \text{K}) \sim 1\text{--}3\ \text{MA/cm}^2$. Both T_c and J_c were determined by standard four-point direct current transport measurements.

RESULTS AND DISCUSSION

Figure 1 shows the resistivity of c-axis perpendicular epitaxial films of the three barrier layer materials. PrBCO is the least conductive with the lowest mobile hole concentration. PrYBCO is more conductive, but still demonstrates a divergent resistivity at low temperatures. The properties of the PrCaBCO thin films, in which divalent Ca is added to introduce holes into an otherwise semiconducting compound, are metallic with very little temperature dependence in the resistivity. A more complete description of the superconducting properties of the PrCaBCO thin film system has been reported elsewhere [10].

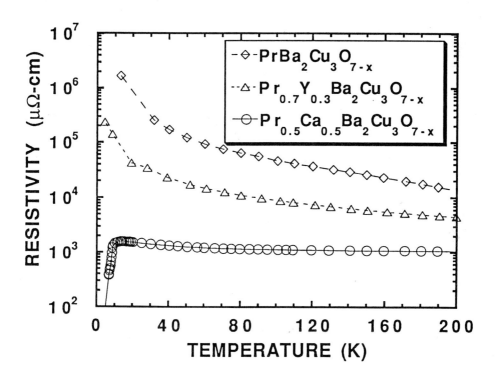

Fig. 1. Resistivity for c-axis perpendicular, epitaxial thin films of $PrBa_2Cu_3O_{7-\delta}$ (◊), $Pr_{0.7}Y_{0.3}Ba_2Cu_3O_{7-\delta}$ (Δ), and tetragonal $Pr_{0.5}Ca_{0.5}Ba_2Cu_3O_{7-\delta}$ (O) grown on (100) $SrTiO_3$.

The crystallinity of the superlattices was studied utilizing four-circle x-ray diffraction. The structures were grown on (100) $SrTiO_3$ and were fully epitaxial with c-axis perpendicular orientation. For the YBCO/PrCaBCO structures, the degree of orthorhombic ordering tends to decrease as the thickness of the PrCaBCO barrier layers increases, in agreement with the near-tetragonal structure for *c*-axis perpendicular PrCaBCO films [10]. For all of the structures, superlattice satellite peaks are present.

Figure 2 shows the R(T) behavior for 1×4 superlattices with either PrBCO, PrYBCO, or PrCaBCO utilized as the barrier layer material. Here, we utilize the nomenclature "$N \times M$" where *N* and *M* are the number of YBCO and barrier layer unit cells per superlattice period, respectively. The resistive transitions of the superlattices clearly depend on the carrier density of the barrier layers. The most interesting effect is a significant increase in T_{c0} with increasing barrier layer carrier density. However, note that T_c (onset) is not significantly influenced by the hole concentration in the barrier layers. This is more clearly seen in Fig. 3 where T_c (onset) and T_{c0} are plotted as functions of barrier layer thickness. Although T_{c0} increases (transition width decreases) significantly as the barrier layer carrier density is increased, T_c (onset) is insensitive to the barrier layer composition, depending only on the thicknesses of the YBCO and barrier layers.

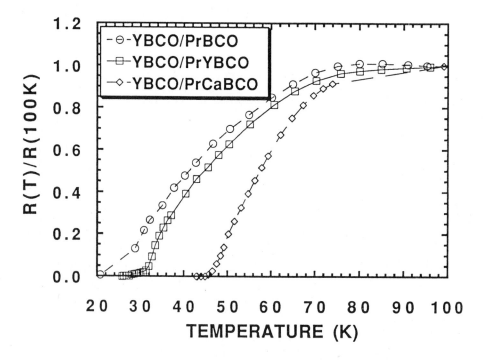

Fig. 2. R(T) for 1×4 YBCO/PrBCO (O), YBCO/PrYBCO (□), and YBCO/PrCaBCO (◊) superlattice structures.

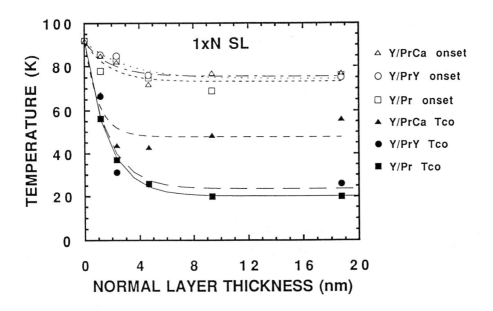

Fig. 3. T_c(onset) (open symbols) and T_{c0} (filled symbols) as a function of normal layer thickness for $1 \times N$ YBCO/PrBCO (□,■), YBCO/PrYBCO (○,●), and YBCO/PrCaBCO (△,▲) superlattice structures.

It recently was proposed that electron transfer from the PrBCO layers into the YBCO layers (resulting in hole-filling in the YBCO layers) can explain the depression of T_c for YBCO/PrBCO superlattices [9]. Hole-filling was previously used to explain the suppression of T_c as Pr is substituted into superconducting "123" oxide materials. In these alloyed systems, the mixed valence of the Pr leads to a reduction of mobile hole density on the CuO_2 planes, with a subsequent reduction in T_c [11,12]. In YBCO/PrBCO superlattice structures, ultrathin (perhaps 1 unit cell thick) YBCO layers with a high mobile hole density are layered alternately with a low carrier density material (PrBCO) of comparable thickness. The possibility of electron transfer from the PrBCO to the YBCO must be considered. If there is a significant reduction of the mobile hole density in the YBCO layers, then a reduction in the transition temperature, T_c (onset), should result. Conversely, by adding holes to the barrier layers (for instance, by doping the PrBCO with divalent Ca), the transfer of holes from the YBCO layers into the barrier layers should be reduced, and T_c (onset) should increase.

Thus, within the hole-filling model, it is difficult to explain why there is no significant change in T_c (onset) as the carrier density in the barrier layers is increased, as seen in Figs. 2 and 3. The hole-filling model predicts that a significant increase in the hole density in the barrier layers should lead to increased hole concentration in the superconducting YBCO layers and an increase in T_c (onset). A second important point is that one does not expect the transition width to be a strong function of the YBCO carrier density. Recent experiments in which the mobile hole concentration was varied directly, by removing oxygen from YBCO thin films and YBCO/PrBCO superlattice structures, show that T_{c0} and T_c (onset) shift together as the mobile hole concentration is decreased, with little or no additional broadening of the transition [13]. Thus, the insensitivity of T_c (onset) to the barrier layer carrier density provides evidence against a simple change in the hole carrier density as the explanation for the depression of T_c in YBCO/PrBCO superlattices.

Although some depression of T_c (onset) is observed in the YBCO-based superlattice structures, significant broadening of the superconducting transition is observed as well. In general, the transition width represents a lack of long-range phase coherence of the superconducting order parameter. Transition broadening has been observed in ultrathin films of low-temperature superconductors, and attributed to the presence of an array of weakly in-plane-coupled Josephson junctions and possibly to 2-D vortex pair unbinding (Kosterlitz-Thouless transition) [14-17]. For very thin films with highly two-dimensional character, it has been suggested that the formation of vortex-antivortex pairs and the thermal breaking of these pairs leads to a broad superconducting transition. In order to observe this Kosterlitz-Thouless (KT) transition, the superconducting thin film should have a sheet resistance greater than 1 kΩ/sq [18]. For the YBCO/PrBCO 1 × 16 superlattice structure, the sheet resistance for an isolated YBCO layer is R_{sq}(100 K) > 3.5 kΩ/sq. In fact, Rasolt et al. recently have shown that the R(T) behavior expected for a KT transition correctly describes the measured R(T) behavior for 1 × 16, 2 × 16, 3 × 16, and 8 × 16 YBCO/PrBCO superlattices [19]. Additional experiments (e.g., I–V measurements) obviously are needed to confirm a Kosterlitz-Thouless transition. If a 2D KT transition is used to describe these superlattice structures, one must resolve how the superconductor/normal interface boundary conditions, specifically adding carriers to the barrier layers, affect the temperature, T_{KT}, at which unbound vortices introduce a finite resistance. Note that evidence for a Kosterlitz-Thouless transition in YBCO thin film and single crystal specimens has recently been reported [20,21].

Similar results would be expected for films with a granular nature, consisting of an array of superconducting islands connected by weak links [15,22]. As the temperature is decreased, the individual islands become superconducting at T_c (onset). However, long-range phase coherence is established by Josephson coupling between islands only at a lower temperature, leading to a broad transition. Although the YBCO-based superlattices are high quality, fully epitaxial structures, some evidence supporting a weak link description has been provided recently by Z-contrast TEM images of these structures [23,24]. In Z-contrast TEM, the intensity of the image is a direct measure of the (atomic number)^2 of the atoms being imaged, providing a distinction between the heavier, Pr-containing layers and the lighter, Y-containing layers [25]. These images indicate that the YBCO layers are not perfectly flat relative to the lattice image; i.e., the YBCO layer shifts up (or down) by one c-axis unit cell increment as one progresses parallel to the *a-b* planes. These one-cell-thick "kinks" in the YBCO layers are due to steps on the growing film surface, which result in a change of composition within the laterally growing cell when the laser ablation target is changed from PrBCO to YBCO. The significance of these "kinks" for current flow in the YBCO layers is that conduction along the c-axis will be necessary at the kinks, in order for a continuous conducting path to be established. They should influence the transport properties of these superlattice structures, contributing to the high resistivity seen in the YBCO layers, and introducing regions of weakened superconductivity. Since the boundaries of these "kinks" are defined by the barrier layers, the properties of the weak links should be influenced by the electronic properties of the barriers.

In addition to transport and Z-contrast TEM characterization of these superlattice structures, we also have investigated the surface microstructure utilizing Scanning Tunneling Microscopy (STM). This technique gives a detailed, microscopic view of the surfaces of these epitaxial films. Figure 4 shows an STM image of a c-axis oriented epitaxial YBCO thin film grown by

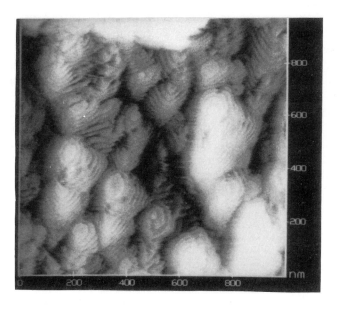

Fig. 4. Scanning tunneling microscopy image of c-axis oriented epitaxial YBCO thin film grown by pulsed laser deposition. Image dimensions are 1000x1000 nm^2.

pulsed laser deposition. This image was obtained at room temperature in air. The dimension of the top image is 1000×1000 nm^2. The epitaxial film is clearly granular, with each grain composed of well-defined terraces. This morphology is indicative of an island growth mode, as opposed to a layer-by-layer growth mode which is more desirable for multilayer fabrication. The presence of these terrace ledges has serious implications when the transport properties of multilayer structures are considered. In particular, each ultrathin YBCO layer within the superlattice structure will possess a high density of steps and kinks due to the microstructure of the growing film surface. Since the superconducting coherence length along the c-axis is significantly smaller than the c-axis lattice parameter, these steps should be important in explaining the transport properties of these superlattice structures, as was pointed out in the discussion of the kinks in the Z-contrast images.

In summary, we have found that the values of T_{c0} (and the transition widths) measured for these superlattice structures are not intrinsic to YBCO layers of a given thickness, but are highly dependent on the boundary conditions and barrier material. T_c (onset), on the other hand, appears to depend intrinsically on the YBCO layer thickness, and only weakly on the barrier layer carrier density. Although additional measurements are needed, the results thus far are consistent with a Kosterlitz-Thouless formalism. However, scanning tunneling microscopy images of the films' surface microstructure suggests that the YBCO layers in the multilayer structures are not microscopically flat and coplanar, but contain numerous steps due to the epitaxial, island-growth mechanism. The presence of "kinks" in the YBCO layers should be important in broadening the superconducting transition, especially for the 1×16 structures, and must be considered.

We would like to thank D. K. Christen, R. F. Wood, and M. Rasolt for helpful comments, and P. H. Fleming for assistance with sample characterization. This research was sponsored by the Division of Materials Sciences, U.S. Department of Energy under contract DE-AC05084OR21400 with Martin Marietta Energy Systems, Inc.

REFERENCES

1. J. -M. Triscone et al., *Phys. Rev. Lett.* **64**, 804 (1990).
2. Q. Li et al., *Phys. Rev. Lett.* **64**, 3086 (1990).
3. D. H. Lowndes, D. P. Norton, and J. D. Budai, *Phys. Rev. Lett.* **65**, 1160 (1990).
4. C. B. Eom et al., *Science* **251**, 780 (1991).
5. A. Gupta et al., *Phys. Rev. Lett.* **64**, 3191 (1990).
6. M. Kanai, T. Kawai, and S. Kawai, *Appl. Phys. Lett.* **57**, 198 (1990).
7. N. R. Werthamer, *Phys. Rev.* **132**, 2440 (1963).
8. S. Maekawa and H. Fukuyama, *J. Phys. Soc. Jpn.* **51**, 1380 (1981).
9. R. F. Wood, *Phys. Rev. Lett.* **66**, 829 (1991).
10. D. P. Norton et al., *Phys. Rev. Lett.* **66**, 1537 (1991).
11. M. E. Lopez-Morales et al., *Phys. Rev. B* **41**, 6655 (1990).
12. J. J. Neumeier et al., *Phys. Rev. Lett.* **63**, 2516 (1989).

13. D. P. Norton and R. Feenstra, unpublished data.
14. S. A. Wolf et al., *Phys. Rev. Lett.* **47**, 1071 (1981).
15. A. E. White, R. C. Dynes, and J. P. Garno, *Phys. Rev. B* **33**, 3549 (1986).
16. A. F. Hebard and A. T. Fiory, *Phys. Rev. Lett.* **44**, 291 (1980).
17. P. A. Bancel and K. E. Gray, *Phys. Rev. Lett.* **46**, 148 (1981).
18. M. R. Beasley, J. E. Mooij, and T. P. Orlando, *Phys. Rev. Lett.* **42**, 1165 (1979).
19. M. Rasolt, T. Edis, and Z. Tesanovic, submitted to *Physical Review Letters*.
20. A. T. Fiory et al., *Phys. Rev. Lett.* **61**, 1419 (1988).
21. N. -C. Yeh and C. C. Tsuei, *Phys. Rev. B* **39**, 9708 (1989).
22. C. J. Lobb, D. W. Abraham, and M. Tinkham, *Phys. Rev. B* **27**, 150 (1983).
23. S. J. Pennycook et al., submitted to *Physical Review Letters*.
24. D. P. Norton et al., submitted to *Physical Review Letters*.
25. S. J. Pennycook and D. E. Jesson, *Phys. Rev. Lett.* **64**, 938 (1990).

"The submitted manuscript has been authored by a contractor of the U.S. Government under contract No. DE-AC05-84OR21400. Accordingly, the U.S. Government retains a nonexclusive, royalty-free license to publish or reproduce the published form of this contribution, or allow others to do so, for U.S. Government purposes."

Monitoring Laser Heating of Materials with Photothermal Deflection Techniques

Mark A. Shannon, Ali A. Rostami[a], and Richard E. Russo
Department of Mechanical Engineering, University of California, and
Applied Science Division, Lawrence Berkeley Laboratory
Berkeley, California 94720

Numerical and experimental work is presented to show the effects on the shape, magnitude, and phase of a photothermal deflection (PTD) signal due to changes in the thermophysical properties of the target and deflecting medium, and the distance of the probe beam from the surface. The PTD signals show qualitative agreement with the temperature gradient normal to the surface, calculated with a numerical model. The results also show that a maximum in time-response profile of the PTD signal occurs in certain instances due to asymmetric changes in the temperature field over time. The effect of a changing temperature field on the PTD signal may be used to detect the onset of phase change in a target.

INTRODUCTION

There has been a great deal of interest to devise experimental techniques to monitor laser-material interactions. These interactions play important roles in applications such as pulsed-laser annealing, laser-induced vapor deposition, various modulated laser-heating techniques, and even laser ablation. Since the development of optical beam deflection techniques in the 1980's [1-3], the deflection of a probe laser beam by thermal and concentration gradients has been demonstrated for a number of applications, including absorption spectroscopy [1-3], thermal diffusivity measurements [4], and imaging [5]. Several of these applications have employed low-energy laser heating, and attempts are now being made to introduce beam deflection techniques to study surface changes during high-power laser irradiation of solids [6,7]. The basis of these techniques is that light will deflect from its initial path when the index of refraction of the medium in which the light propagates changes due to the presence of temperature, particulate, atom, ion, or electron density gradients. Each gradient source, though, has a different effect on the index of refraction of a medium, and thus on the magnitude and direction of the deflection of a beam of light. By examining the transient magnitude and phase of the signal resulting from the deflection of an optical probe-beam one should, in principle, be able to discern the heating, melting, vaporizing, and plasma forming processes which occur at the surface of a solid during intense laser irradiation. However, the signal will be a complex convolution of competing processes making interpretation difficult.

Research supported by the U.S. Department of Energy, Office of Basic Energy Sciences, Division of Chemical Sciences, under contract #DE-AC03-76SF00098. (a) Visiting scholar from Isfahan University of Technology, Isfahan, Iran

In spite of the difficulties, efforts must be made to understand the transient response of the beam deflection in order to study laser-heating mechanisms at the surface of a solid during a single heating cycle. The classical optical beam deflection techniques allow analysis of the steady-state values of magnitude and phase for a periodically heated target. Yet at the onset of melt or vaporization, the deflection response of the probe-beam changes due to variations in thermal properties, reflectance, evolution of vapor, among other factors. Moreover, even without phase change, the shape of the deflection signal and amplitude may change dramatically depending on the boundary conditions, the material properties, the deflecting medium, the modulation frequency of the probe-beam, and the position of the probe-beam with respect to the target surface.

This investigation consists of numerical and experimental work to examine the transient response of photothermally induced optical beam deflection. The effects on the shape, magnitude, and phase of a photothermal deflection (PTD) signal are examined due to changes in 1) the thermophysical properties of the target and deflecting medium, 2) the distance of the probe beam from the surface of the target, and 3) the change of phase of the target material. The effects of thermophysical properties are illustrated by using two different target materials: copper whose thermal diffusivity α_c is much greater than air α_a ($\alpha_c/\alpha_a = 5.286$ @ 293 K), and lead whose diffusivity α_l is approximately the same as air ($\alpha_l/\alpha_a = 1.102$ @ 293K) [8]. The response during linear heating of the target is analyzed and the results are extended to include the numerical results predicting solid/liquid phase change.

NUMERICAL METHOD and EXPERIMENTAL TECHNIQUE

Figure 1 shows a slab of target material, $0 \leq z \leq l$, that is bounded by two semi-infinite media $z < 0$ and $z > l$. The target material is a radiatively absorbing medium while the adjoining semi-infinite media are transparent. The target material is irradiated over a circular area of radius w at $z = 0$ by a laser beam of known spatial distribution and temporal behavior. The intensity of the beam is maximum at the optical axis ($r = 0$), and decreases both radially and axially.

An explicit finite-difference method is used to predict transient conduction in the target and the bounding regions. The spatial and the temporal variation of the beam intensity, the temperature dependence of the optical and the thermophysical properties of the materials, and radiative losses from the surfaces are included in the calculations. The absorbed laser radiation is treated as a variable heat source within the material having any value for the optical penetration depth.

The energy absorbed may be represented as a heat source within the material having a rate of heat generation per unit volume g given by:

$$g = -\frac{dI_a}{dz} = a\, I_s\, (1-R)\, e^{-a\,z} \tag{1}$$

where I_a and I_s are the laser beam intensities inside the target and at the surface respectively and a is the absorption coefficient. The two-dimensional diffusion equation for the temperature T in cylindrical coordinates is

$$\frac{1}{r}\frac{\partial}{\partial r}\left[k_j\, r\, \frac{\partial T_j}{\partial r}\right] + \frac{\partial}{\partial z}\left[k_j\, \frac{\partial T_j}{\partial z}\right] + g_j = \left[\rho c\right]_j \frac{\partial T_j}{\partial t} \tag{2}$$

where k is the thermal conductivity, ρ is the density, c is the heat capacity, and $j = 1, 2, 3$ corresponds to the regions shown in figure 1. Equation (2) was written in an explicit finite difference form and was solved numerically [9].

The apparatus used for our experimental work is described in [10]. The solid target samples are heated in air and a He-Ne probe laser beam adjacent to the surface photothermally deflects due to temperature gradients. A Nd:YAG cw laser with a 1064 nm wavelength modulated by a mechanical chopper is used as the heating source. The probe-beam is aligned so that its centerline orthogonally intersects the centerline of the pump-beam. The deflection of the centroid of the probe beam is measured with a continuous single axis, dual cathode photodetector. A modulation frequency of 500 Hz was used, which correspond to a heating period t_p of 1 ms. The intensity of the laser beam was assumed to decrease exponentially in the radial direction and remain uniform during the heating period. A laser beam with an average power of 0.1 W and a $1/e^2$ beam waist of 430 µ was used. The back surface of the target is assumed to be adiabatic.

RESULTS and DISCUSSION

The results of numerical calculations are shown in figures 2 and 3. Figure 2 shows the time variation of the temperature gradient in air normal to the surface, at $r = 0$ for three different distances from the copper target surface. The copper target in this case is considered to be thermally semi-infinite, because its thickness of 10 mm is much larger than the thermal diffusion length ($\delta_t = 0.63$ mm for $t_p = 1$ ms). Note that the vertical axis actually shows a negative temperature gradient with respect to z. Significant changes in the shape and the amplitude of the temperature gradient are seen as the distance from the surface z_o is increased. For $z_o = 16$ µ, the temperature gradient reaches a maximum, then decreases and asymptotically approaches a steady state value at the end of the heating period. When the power is removed the temperature gradient drops sharply and a change of sign occurs which corresponds to a change in the direction of the heat-flux. The same behavior exists for $z_o = 81$ µ, although the maximum is less distinct and the changes are more gradual. For

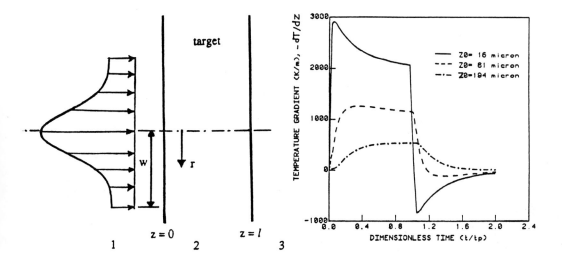

Figure 1. Physical model showing beam irradiation on a target

Figure 2. Temperature gradient in air for 10 mm thick copper target as a function of z_o

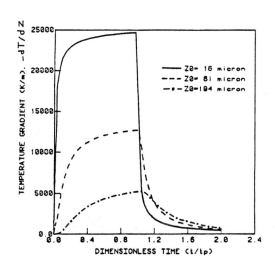

Figure 3. Temperature gradient in air for 10 mm thick lead target as a function of probe-beam offset z_o.

$z_o = 194$ μ, neither a maximum nor a change of sign of the temperature gradient is seen. Also note that the phase lag of the temperature gradient increases as z_o is increased.

Figure 3 shows the temperature gradients for a 10 mm thick lead target. Again as the distance z_o is increased, the peak to peak magnitude decreases and the phase shift of the temperature gradient increases. By comparing the results for copper in figure 2 with those for lead in figure 3, we see the effects of the thermophysical properties of the target material on the gradient. For the lead/air combination, no maxima with respect to time exists for any z_o and the temperature gradients show no change of sign after the removal of laser power. The peak to peak magnitude of the lead/air case is one order of magnitude larger than the copper/air case, due to the small value of the thermal diffusivity of the lead compared to copper.

For the semi-infinite 10 mm thick copper sample, the PTD signals for different z_o distances are shown in figure 4. The photodiode signal is included in the figures as a reference of magnitude and phase for each of the PTD signals. The input power is held constant for all the experiments and the photodiode signal is normalized to one. Note that the temporal profile for the power is not a true square wave, but has a finite rise-time because the modulation is from a mechanical chopper. The rounding-off of the corners is due in part to the finite size of the pump beam and its Gaussian spatial profile. The deviation of the input power waveform from a true square wave causes the PTD signal to have longer rise and fall times and greater rounding at the corners of the signal than that predicted by the numerical model. Of main importance in figure 4 is that the shape of the PTD signal changes as z_o increases from approximately 15 μ to 190 μ. For the semi-infinite 10 mm thick lead sample, the PTD signals for z_o at 15 μ, 80 μ, and 190 μ are shown in figure 5. For the lead sample, no maxima with respect to time are observed and the PTD signal never shows a reversal in direction.

To compare the numerical and the experimental data we use the radial symmetry of the temperature field to relate the deflection with the temperature gradient at $r = 0$. The total angular deflection θ can be expressed as an integrated function of $\partial T/\partial z\,(r=0)$, or $\theta = \int f\,(\partial T/\partial z\,(r=0), z, \cdots)\,dr$. Therefore, the deflection of a centered probe-beam is related to the normal temperature gradient at the center of the pump-beam, and the shape of the PTD signal has a correspondence with the shape of $\partial T/\partial z\,(r=0)$ over time. The effect of the finite size of the probe-beam should change the shape of the PTD signal because $\partial T/\partial z$ changes across the z direction of the beam. The resulting PTD signal will be an integration of θ over z as well as r and should appear qualitatively as a convolution of single-ray deflections corresponding to $\partial T/\partial z\,(r,z)$ at different z_o.

For the case of the copper where the ratio of diffusivities is about 5.3, both the numerical simulation of $\partial T/\partial z$ in figure 2 and the experimental PTD data in figure 4a show a maximum for z_o at approximately 15 μ. The PTD signal will not follow $\partial T/\partial z\,(r=0)$ precisely, but will show rounding and truncation compared to the numerical result. Indeed, in figure 4a the shape appears as a convolution of $\partial T/\partial z\,(r=0)$ for z_o at 16 and 81 µ from figure 2. However, both model and experiment show a maximum and a reversal in the sign of the temperature gradient with time for z_o close to the surface. For z_o at 80 µ, the numerical simulation shows a slight maximum whereas the PTD signal in figure 4b does not, which also can be explained as above. For z_o at 190 µ, neither the numerical nor experimental results show a maximum or reversal. Therefore, in both studies a point is reached at some distance z_m from the surface where a maximum no longer occur for the copper and air domain. For the lead/air case in which the diffusivity ratio is 1.1, neither the numerical nor experimental studies show any maxima or reversal in time. The PTD signal in figure 5a shows that a steady-state value is reached with z_o at 15 µ, whereas the numerical simulation in figure 3 does not. However, the finite-sized probe-beam has regions grazing the surface which should distort the PTD signal as discussed above.

A primary conclusion of these studies is that the shape of the PTD signal can exhibit a maximum in time when there is a small finite heat source, the solid target has different diffusivity than the deflecting medium, and the probe-beam centroid distance z_o is close to the surface. When a maximum occurs at a $z_o\,(r = 0)$, the temperature gradient normal to the surface declines after some point in time. This might suggest that the surface temperature declines or that the heat-flux entering the air declines in time. Yet the numerical solutions for temperature, given in [10], indicate that the temperature increases until the power is removed. Also, energy balance considerations require that the total heat flux not decline with time. For two simple cases it can be shown that maxima in $\partial T/\partial z$ will not occur in time. Expressions for $\partial/\partial t\,(\partial T/\partial z)$ obtained from analytical solutions for temperature from Carslaw and Jaeger [11] for semi-infinite domains with different α heated by an infinite, planar heat source show that a maximum does not occur for any combination of z or α. Also, if the heat source was a point source, no maximum in time would occur if α were the same for each domain. However, when the heating source is finite in size, and the domains have different α, the temperature field changes in time and does not maintain the same spatial symmetry. At small z_o distances for all r > 0 when $\alpha_1 > \alpha_2$, the flux direction shifts with time from the normal to the radial direction. Therefore, the normal components of the flux vectors decrease in time after a maximum is reached [10]. The resulting PTD magnitude will also reach a maximum and decrease. As z_o is increased, the temperature field about a point begins to look more planar, analogous to the curvature of a sphere decreasing as the radius is increased, until at some point z_m the normal flux does not exhibit a maximum in time. While an analytical solution to this problem is not easily tractable, the numerical solutions of figures 2 & 3 show that a relationship exists between z_o, δ_t, α_1/α_2, and a maximum in $\partial T/\partial z$ with time. The effect of a changing temperature field can be used to predict the onset of phase change. Figure 6 shows the temperature gradient in air for an aluminum target which undergoes a solid-liquid phase change . The results were obtained from the numerical calculations using a modified

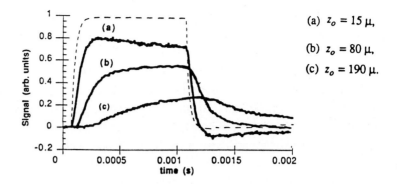

Figure 4. PTD signals for 10 mm thick copper target.

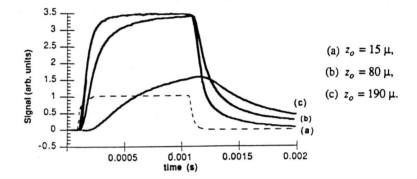

Figure 5. PTD signals for 10 mm thick lead target.

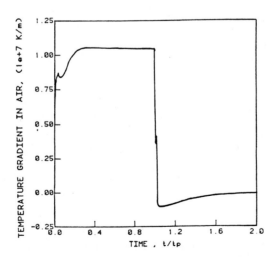

Figure 6. Temperature gradient in air at $z_0=16\mu$ for an aluminum target which undergoes solid-liquid phase change.

enthalpy technique [12]. Unlike the simple heating case where the temperature gradient steadily decreases after the initial peak, it starts to increase after the beginning of melting. This phenomenon may be explained as follows. The thermal diffusivity of the liquid aluminum is almost half that of the solid phase. Therefore the amount of heat penetrating into the target decreases as a result of phase change, which means an increase in the rate of heat transfer into the air. This change of the temperature gradient may be used to detect the initiation of phase change in a target.

REFERENCES

[1] A.C. Boccara, D. Fournier, W. Jackson, and N.M. Amer, Opt. Lett. **5**, 377 (1980).
[2] W.B. Jackson, N.M. Amer, A.C. Boccara, D. Fournier, Appl. Opt. **20**, 1333 (1981).
[3] A.C.Tam, Rev. Modern Phys. **58** n. 2, 381 (1986).
[4] L. Fabbri, Mater. Chem. Physics **23**, 447 (1989).
[5] G.C. Wetsel,Jr. and F.A. McDonald, J. Appl. Phys. **56**, 3081 (1984).
[6] M. Dienstbier, R. Benes, P. Rejfir, and P. Sladky, Appl. Phys. B **51**, 137 (1990).
[7] J.A. Sell, D.M. Heffelfinger, P. Ventzek, and R.M. Gilgenbach, Appl. Phys. Lett. **55**, 2435 (1989).
[8] Y.S.Touloukian and C.Y.Ho, Eds., *Thermophysical Properties of Matter*, (Plenum Press, New York 1972).
[9] A.A. Rostami, R. Greif, and R.E. Russo, ASME Transp. Phen. Material Process. **HTD-146**, 61 (1990).
[10] M.A. Shannon, A.A. Rostami, and R.E. Russo, "Photothermal Deflection Measurements for Monitoring Heat Transfer during Modulated Laser-Heating of Solids", Submitted to J. Applied Physics, Jan. 1991.
[11] H.S.Carslaw and J.C.Jaeger, *Conduction of Heat in Solids 2nd ed* (Oxford University Press, Oxford 1959).
[12] A.A. Rostami, R. Greif, and R.E. Russo, "Modified Enthalpy Method Applied to Rapid Melting and Solidification", Submitted to Int. J. Heat Mass Trans. March 1991.

STUDIES OF LASER ABLATION OF GRAPHITE:
$C_n^{+/-}$ ION KINETIC ENERGY DISTRIBUTIONS

M. J. Shea*
Vanderbilt University
Department of Physics and Astronomy
Nashville, TN 37235

R. N. Compton
Oak Ridge National Laboratory
Oak Ridge, TN 37831

I. Introduction

Laser-ablation of solid carbon is an effective means of generating carbon clusters, both neutral and charged [1][2][3]. In general, molecular structure of the individual clusters determines the equilibrium abundance distributions. However, laser-ablation does not typically yield an equilibrium distribution [4]. We have investigated positively and negatively charged carbon clusters formed by laser ablation by measuring their kinetic energy distributions (KEDs) as a function of laser power. The ionic stabilities of the charged clusters is seen to greatly influence the ion kinetic energy distributions. The data demonstrate both equilibrium and non-equilibrium features.

II. Experimental

The experimental apparatus and techniques employed have been reported previously [5]. Positive and negative ions are ablated from a graphite surface by a 30 ps light pulse from a Nd:YAG laser (λ = 532 mm). The resulting ions are energy analyzed using a spherical sector energy analyzer, and the mass of each cluster ion is determined by its time-of flight through a drift tube at the end of the analyzer. The lower resolution of the TOFMS M/δM~16 was sufficient to separate the $C_n^{+/-}$ clusters for n≤15. The laser spot size was an ellipse of approximately 0.27 cm and 0.19 cm major and minor axes, respectively. The energy per laser pulse was varied over three values: 0.1 mJ, 0.2 mJ, and 0.3 mJ, corresponding to a power density of the order of 10^9 W/cm^2. The target in this experiment was solid polycrystalline graphite of 99.999% purity.

III. Results

Data were taken for clusters C_n^+ for n = 1 to 10. Figure 1 shows kinetic energy distributions KEDs of C_2^+, C_3^+, and C_5^+ generated by a laser pulse with an energy of 0.1 mJ. These KEDs are representative of all clusters measured at this laser pulse energy. Integration of the KEDs for each cluster yields an abundance distribution that agrees with the typical odd-even intensity alternation observed previously (see e.g. Reference 4). The odd-n clusters are more abundant than the even-n clusters for positive ions. Even-n clusters are more abundant for negative ions. The salient feature is that at low laser power all clusters can be represented by the same Maxwell-Boltzmann energy distribution, i.e., the same temperature. Thus, the 0.1 mJ pulse appears to yield a distribution which is the same for all clusters (i.e. an equilibrium distribution), indicating that the abundance ratios reflect the energetics of clustering as opposed to the kinetics of cluster formation (i.e. ion stabilities).

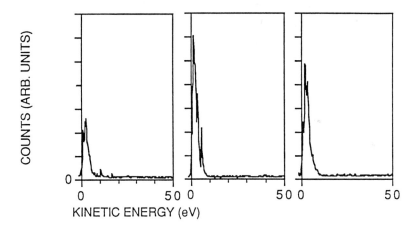

Figure 1. KEDs of C_2^+, C_3^+, and C_5^+ (left to right) for a laser power of 0.1 mJ. The energy distributions are all the same (i.e. same T) indicating that they are in equilibrium.

As the laser pulse energy was increased to 0.2 mJ, the average kinetic energy increased, as shown in Figure 2. The abundance distribution has shifted to lower n, meaning the abundances of C^+, C_2^+, and C_3^+ have grown dramatically compared to that of clusters of higher n. The most noteworthy feature is that the kinetic energy now differs dramatically from species to species, indicating that the apparent equilibrium in energy evident for a 0.1 mJ laser energy was lost. The temperatures derived from the energy distributions for C_n^+ ions are considerably higher for the more stable odd n clusters.

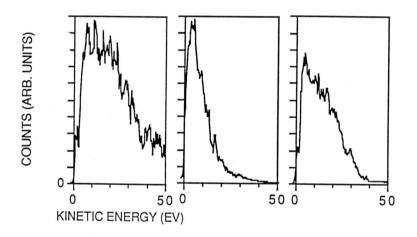

Figure 2. KEDs of C^+, C_2^+, and C_3^+ for a laser power of 0.2 mJ. The ions here all have different temperatures indicating that they are no longer in equilibrium.

For 0.3 mJ per pulse, the mass spectrum is dominated by C^+ which is the result of extensive fragmentation. Again, the more stable ions (odd -n) support a higher average kinetic energy, with the most abundant species, C^+, being the hottest. The average kinetic energy of C^+ at 0.3 mJ per pulse was higher (40 eV) than the average kinetic energy (25 eV) for C^+ at the lower laser pulse energy of 0.2 mJ (C^+ ions take up the increased energy deposited into the system). Surprisingly, now the average kinetic energies of C_2^+, C_3^+, and C_4^+ at 0.3 mJ of laser energy are all lower than their corresponding average kinetic energies at 0.2 mJ. This is shown in Figure 3.

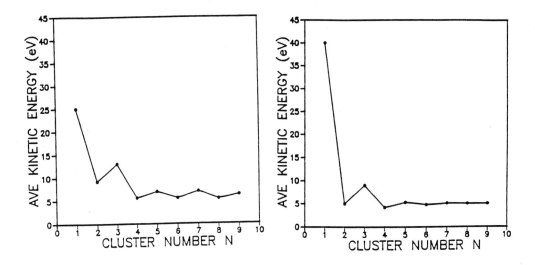

Figure 3. Average kinetic energy vs. cluster number n for carbon cluster ions with a laser energy of (a) 0.2 mJ and (b) 0.3 mJ.for a laser power of 0.2 mJ. The ions here all have different temperatures indicating that they are no longer in equilibrium.

Kinetic energy distributions for negative carbon cluster ions were also determined with similar results. However, in this case the even -n cluster ions are more stable.

IV. Discussion

For the nonequilibrium distributions formed with 0.2 and 0.3 mJ laser pulses, the more abundant (and more stable) clusters supported higher temperatures than did the less abundant (and less stable) species. We propose a link between the kinetic energy of the clusters and their internal energy, which governs bond breaking, in order to rationalize this observation. The translational and internal energies must certainly not be in equilibrium, and the question as to what extent the translational and internal energies are coupled must be considered.

The observation that different clusters have different temperatures can be accounted for by the fact that they have different fragmentation energies (stabilities). We postulate that within a given KED the clusters with a higher kinetic energy also have a higher internal energy. As an example, consider the clusters formed with 0.2 mJ/pulse of laser energy. The 4.7 eV [6] fragmentation energy of C_2^+ is the upper limit on the internal energy for this cluster. At 20 eV, the C_2^+ distribution has diminished to less than 90% of its maximum value. We arbitrarily assign this point on the kinetic energy axis to be that at which the internal energy is equal to the fragmentation energy, i.e., 4.7 eV. Similarly, for C_3^+, with a fragmentation energy of 6.5 eV [6], the distribution is reduced to 90% of its maximum value at a kinetic energy of 4.2 eV. This procedure was applied to clusters n = 2-10 for 0.2, and 0.3 mJ. A linear relationship is found for both cases,

$$E_i = 0.0472 \times K + 3.45 \quad [1]$$
$$E_i = 0.145 \times K + 3.72 \quad [2]$$

where E_i is the internal energy and K represents the translational energy. For 0.3 mJ the coupling becomes stronger, nearly a factor of 4 increase in the slope of the line as compared to the slope for 0.2 mJ.

V. Conclusions

At the lowest laser energy used, 0.1 mJ, the different clusters appear to be in equilibrium, at least in regards to their translational energy. As the laser energy was increased to 0.2 mJ, equilibrium of kinetic energy among the carbon cluster ions was lost. The temperatures of the clusters were related to cluster stability; the more stable clusters exhibited a higher temperature than the less stable clusters. At 0.3 mJ, fragmentation left very few clusters and C^+ is the dominant ion.

Acknowledgments

This research is based on work performed in the Laboratory Graduate Participation Program under Contract No. DE-AC05-76OR00033 between the U.S. Department of Energy and Oak Ridge Associated Universities and the Office of Health and Environmental Research, U. S. Department of Energy under contract No. DE-AC05-84OR21400 with Martin Marietta Energy Systems, Inc.

*Now at GTE, Electrical Products Group, Danvers, MA 01923.

References

[1] McElvany, S. W. et al., "FTMS Studies of Mass-Selected, Large Cluster Ions Produced by Direct Laser Vaporization," Chemical Physics Letters 134 (27 February 1987): 214.
[2] Parent, D. C., and Stephen W. McElvany, "Investigations of Small Ion Structures by Reactions with HCN," Journal of the American Chemical Society 111 (1989).
[3] Hahn, M. Y. et al., "Magic Numbers in C_n^+ and C_n^- Abundance Distributions," Chemical Physics Letters 26 (September 1986): 12.
[4] Heath, James R., "The Physics and Chemistry of Carbon Clusters," Spectroscopy 5 (1990): 36.
[5] Shea, M. J., Compton, R. N., Hettich, R. L., Phy. Rev. A 42, 3579 (1990).
[6] Creasy, William R., "Some Model Calculations of Carbon Cluster Growth Kinetics," Journal of Chemical Physics 92 (15 June 1990): 7223.

INFRARED LASER INDUCED ABLATION AND MELTING IN MODEL POLYMER CRYSTALS

Bobby G. Sumpter, Donald W. Noid,
and Bernhard Wunderlich

Department of Chemistry
The University of Tennessee
Knoxville, TN 37996-1600, USA
and
Chemistry Division
Oak Ridge National Laboratory
Oak Ridge, TN 37831-6182, USA

I. INTRODUCTION

Since their invention the applications of lasers in science and technology has grown at an incredible rate. Presently lasers offer a large range of possibilities in average power, power density, focalization, wavelength capability, efficiency, pulse characteristics; according to the type of lasers (i.e., ion, CO_2, solid-state, dye, excimer lasers, etc.). This wide range of possibilities allow for broad utilizations in important scientific sectors such as medicine, graphic arts, communications, information processing, microelectronics, industrial cutting, and metrology.[1] In addition, lasers have made crucial contributions to spectroscopy, especially to time-resolved spectroscopy.[2] The development of these powerful techniques, which allow real-time probing of energy levels, as well as primary excited state processes has lead to a number of exciting discoveries. For example, the possibility of inducing chemical reactions in a selective way by using lasers to excite a particular vibration of a molecule (called mode selective chemistry) has received a lot of attention from the chemical physics society.[3]

The use of lasers in polymer science has also become an area of considerable interest.[4-7] Many applications have been related to production, that is, the acceleration of the polymerization process and the enhancement of the amount of product which is yielded. Other interests are in using lasers to weld two polymers together and to cleanly cut or burn a hole in a polymer surface (useful in the production of computer circuit boards and for optical information storage). The method of preference has been to employ high-ultraviolet excimer lasers for these processes.[4-7] This is partly due to the fact that the more common intense light sources such as infrared CO_2 or near-ultraviolet Nd:YAG lasers have been observed to cause damage to the material surrounding the laser "hot spot" due to thermal processes such as melting and vaporization. These thermal processes thus do not allow for highly precise etching of the material and are the result of intra- and interpolymer energy redistribution competing effectively with the laser-polymer excitation process. The high-UV excimer laser photons are presumably absorbed rapidly and excite the polymers to higher energy electronic states.[4-7] This excess vibronic energy in the molecules results in rapid local heating of the material in the laser hot spot followed by a little understood[4-7] dynamical process, which in turn leads to the ablation of polymer fragments. It is thought that the UV photons cause considerable bond breakage in the polymers, and repulsive forces between the fragments then dominate the dynamics. Such forces may cause ablation of material.[5] Alternative mechanisms based on thermal processes such as internal conversion or intersystem crossing have also been proposed.[4-7]

In the present report, computer simulation results are presented which provide strong evidence that multiple IR lasers may be used to enhance the energy absorption characteristics of polymer surfaces and hence compete with the high-UV laser process. In the simulations, it was found that a polymer interacting with one laser *always* melted, no matter what laser intensity was used (0.5 - 100 TW/cm^2). In many cases, bond scissions occurred but this was after the crystal had already melted. In contrast, when the model polymer was allowed to interact with *two* laser beams, significantly enhanced energy absorption was observed which led to multiple bond dissociations. Additionally, the "cold" part of the polymer was left essentially unperturbed. The overall process results in cleanly etching many atoms away from the polymer surface.

The results presented in this report suggest that the process(es) may be explained from a nonlinear dynamics perspective. In short, two lasers (or driving forces) in a dynamical system can induce **vibrational chaos**, and enhance energy absorption in the chaotic modes.[9-38] This enhanced absorption occurs on a shorter timescale than the intramolecular energy redistribution (and subsequent melting process in polymers), allowing the breaking of chemical bonds to occur.

In the next section, we describe the methods that have been employed to study the laser-polymer crystal system. The results are discussed in Section III, and the conclusions are given in Section IV.

II. METHODS

Several recent reviews of the molecular dynamics technique and its application[39] exist. The reader is referred to them. Only details relevant to the specific applications discussed here are presented.

In order to study the external interactions of a polymer crystal with high-power infrared lasers, a Hamiltonian is written in terms of: (a) the molecular Hamiltonian, and (b) the external Hamiltonian:

$$H_{system} = H_{molecular} + H_{ext} . \qquad (1)$$

In this equation, the term H_{ext} is for one or two IR laser(s). The molecular Hamiltonian:

$$H_{molecular} = T + \sum_{i=1}^{N-1} V_{2B} + \sum_{i=1}^{N-2} V_{3B} + \sum_{i=1}^{N-3} V_{2B} + \sum V_{NB}^{dyn} + \sum V_{NB}^{sur} , \qquad (2)$$

is composed of kinetic energy terms [first term in Eq. (2), T] and potential energy terms [remaining terms in Eq. (2)]. The potential energy surface is a many-body fit to high level *ab initio* calculations and to experimental data. A notable feature of the many-body potential is that the three- and four-body terms (bending and torsional modes) are multiplied by switching functions which attenuate the bending force constant and the internal rotational barriers to zero as a function of the distances between the atoms describing the internal coordinates. This has been clearly demonstrated as an important addition to the many-body expansion of a potential in order to carry out a realistic simulation of any process which involves large bond extensions or bond breaking [37-42].

The laser field is characterized by the electric field strengths and frequencies of the laser light. Interaction between the molecule and the laser radiation field is included through the dipole moment function. The Hamiltonian for the laser is:

$$H_{ext} = E_1 \cdot \mu(r_i) \cos(\omega_1 t - \theta_1) + E_2 \cdot \mu(r_i) \cos(\omega_2 t - \theta_2) , \qquad (3)$$

where (E_1, E_2) are the electric field strength vectors (polarized in the present case along the z- and x-axis), μ is the dipole moment, t is time, and ω_i, θ_i are the frequency and phase of the laser light. The dipole moment function is:

$$|\mu(r_i)| = Ar_i e^{-\beta r_i^4} \qquad (4)$$

The laser field is focused on a 50-atom segment (6.3 nm^2 area) located at the center of a 200 CH_2 group polymer chain (see Fig. 1). The Hamiltonian given by Eq. (3) describes one or two IR lasers, depending on whether E_i is non-zero. In the present study, we consider the effects of a single IR laser and also of two IR lasers. The parameters E_i and ω_i are optimized in each case to study a range of laser powers and frequencies.

Hamilton's equations are integrated in cartesian coordinates using the vectorized version of a differential equation solver[54] to solve the coupled nonlinear ODEs. The tedious and time-consuming evaluations of the derivatives of the internal coordinates are drastically reduced using <u>our new</u> geometric statement function method.[55]

Ablation was defined as a process where five or more bonds are broken while the molecule remains in the solid state. Bonds were considered to be broken when the bond lengths had exceeded 10 angstroms. The distinction between the solid and melt phase is determined by monitoring the end-to-end distance and radius of gyration which asymptotically approach a constant value as the system melts such that the ratio $EED^2/R_g^2 \sim 6$. Ensembles of 5-20 trajectories were used to compute the lifetimes for the processes.

In order to analyze the abundance of data that is obtained from the molecular dynamics simulations, we have examined the structural, dynamic, and spectral properties. The spectral analysis[56] was carried out by employing novel methods that belong to a class of signal subspace spectral estimators[48,51,57]. Two such methods were used: MUSIC[58] and ESPRIT[59]. Both methods achieve their high-resolution capability by separating the received data into signal and noise subspaces. In short, these methods allow the calculation of a theoretically infinitely resolved spectrum from a very short time series (as short as 1 vibrational cycle). For a recent review of the various applicable spectral techniques to molecular dynamics, see Ref. 57. Presently, the application of MUSIC and ESPRIT is toward determining the dispersion curves for the model polymer. This is used to clarify the frequency shifts in the vibrational modes as the molecule absorbs energy from the interaction with the laser(s). The chacterization of these changes provides the essential information to obtain the mechanistic details of the laser-polymer process(es).

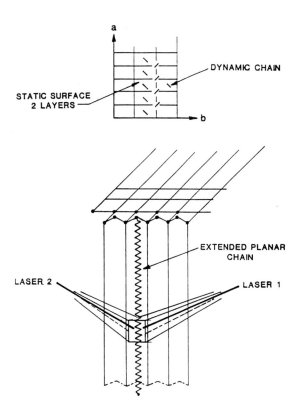

Figure 1. Diagram showing the model of a polymer on a crystal surface. The crystal structure is that or the orthorhombic phase of polyethylene with axes a=0.74 nm, b=0.49 nm, and c=0.255 nm. The two layers of static polymer chains making up the crystal surface consists of 300 CH_2 groups each and the single dynamical chain consists of 200 CH_2 groups. The bottom portion of the figure shows the 50 atom region of the crystal which is irradiated by the laser light of one or two lasers.

III. RESULTS

We have carried out a systematic study of the dependence of the rate of deposition of energy into the modes of a model polymer on a crystal surface.[60] The lifetime for either bond dissociation or ablation is determined as a function of laser intensity and frequency for the case of one and two IR lasers. Figure 2 shows the energy absorbed from a single laser as a function of the laser frequency with a constant laser intensity of 1/2 TW/cm^2.

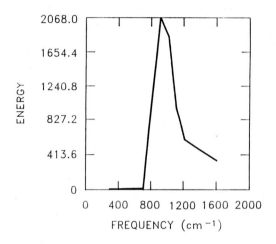

Figure 2. Energy absorbed by the model polymer as a function of the laser frequency for a single laser.

The maximum in the plot(which corresponds to the laser-molecule resonance) is near a frequency of 1000 cm^{-1}. This is slightly red shifted from the actual fundamental vibrational frequency of a C-C bond. There is essentially no energy absorbed from the laser for the lower frequencies (0-700 cm^{-1}). The energy absorption occurs sharply at a frequency very near the laser-molecule resonance (800 cm^{-1}) and then rises to a maximum and slowly decays for frequencies blue shifted from the resonance. For higher intensity lasers it was found that the laser-molecule resonance is shifted even further to the red and there was a slower decay of energy absorption for the higher frequencies. These observations have been noted previously in studies of absorption of energy from lasers in diatomic and small polyatomic molecules[9-38].

Frequencies that are red shifted from the laser-molecule resonance exhibit reversible energy flow between the molecule and laser field that is more or less quasiperiodic and there is no overall accumulation of energy. As the laser frequency is tuned closer to the resonance, the energy absorbed by the molecule increases. After the molecule has absorbed sufficient energy, the quasiperiodic type energy flow between the laser and molecule disappears due to energy relaxation which is faster than the period of the laser-molecule energy flow oscillations. Energy accumulates in the system up to a maximum amount which depends on both the intensity and the frequency of the laser. After the molecule has absorbed a certain amount of energy, the vibrational frequencies detune from the laser and there is an exponential decrease in the kinetic energy that is absorbed. In our simulations, the point at which the energy absorption begins to decrease generally corresponds to the crystal melting and/or to bond dissociations.

From calculation of dispersion curves at various times for the one laser case it was found the frequencies of the optical branch decrease by at least 400 cm^{-1} at the maximum of energy absorption. The frequencies were found to slowly decrease over simulation time up to the point at which the maximum in absorbed energy was obtained. Since the width of the frequency space red shifted from the frequency where the polymer can absorb energy is about 300 cm^{-1} (see Fig. 2), this explains why the molecule does absorb energy for a certain period of time. After the polymer had absorbed enough energy to cause large amplitude vibrations and isomeric transitions, the frequency decreases by 100 cm^{-1} over a 5 ps time. The resulting frequencies are out of the range where energy can be absorbed from the laser. At this energy, the polymer also melts which can further change the frequencies.

The two laser case demonstrates a marked difference in the overall process. While the same type of energy absorption characteristics are present, the rate at which the energy is absorbed is significantly faster (about a factor of two greater than that for the one laser case). The energy that is absorbed does not appear to decrease as it does for the one laser case. The reason for this is that instead of melting the crystal followed by bond dissociations, ablation occurs.

The qualitative differences between the one and two laser cases discussed above is clearly shown in Figs. 3 and 4. In these figures the initial and final configuration of the polymer molecule are shown. Figure 3 is the configurations corresponding to the one laser case. The final configuration (Fig 3b) has undergone significant coiling as well as bond dissociations. In contrast, Fig. 4b shows the corresponding two laser case. In this figure it is clear that the polymer molecule has undergone multiple bond dissociations (10 atoms are removed from the surface) but the polymer molecule has hardly undergone any coiling. The ablation process is more clearly shown in Fig. 5 which is a 3-D plot of the situation similar to that in Fig. 4. Bond dissociations which occurred were in the center of the 200 CH$_2$ group polymer chain, where energy is absorbed from the laser at a competitive rate to internal energy randomization. The ablation products(atoms and polyatomic species) were ejected from the surface with large amounts of translation energy, leaving a cleanly etched region. Within the absorption region, however, the temperature reached values as high as 1490 K, well above the melting point of the model polymer. From examination of numerous cases it appears that there is a minimum threshold to the rate of energy absorption for ablation to occur. An estimate of this threshold rate constant is 8×10^{-2} ps^{-1}.

The use of two lasers can result in the absorption of sufficient energy at a rapid enough rate to cause efficient ablation. In addition, the two laser process can similarly cause more efficient melting and at substantially reduced intensities. Polymer crystal melting can be efficiently induced using either one or two lasers. Compared to the thermal process[49], laser induced crystal melting occurs on a much faster timescale. Laser heating of polymers and the resulting degradation or melting is also more efficient relative to energy input than the associated thermal process. This is due to the timescales for energy absorption and competitive internal energy randomization. While for thermal heating energy is absorbed over a time in the seconds, laser heating occurs on the picosecond or nanosecond timescale.

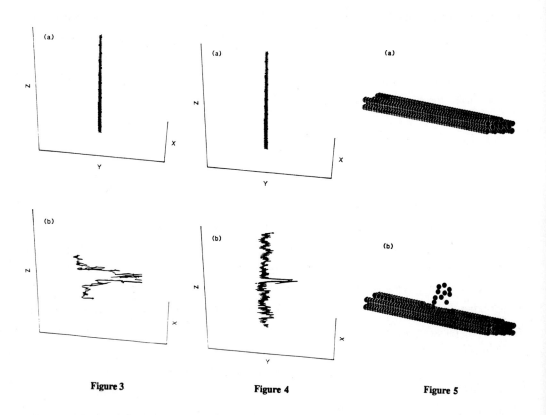

Figure 3   **Figure 4**   **Figure 5**

Figure 3a) Initial Configuration of the 200 CH_2 group model polymer; **3b)** Final configuration of the model polymer after 65 ps irradiation with a single laser.
Figure 4a) Same as Fig. 3a; **4b)** Same as Fig. 3b, except for the two laser process;
Figure 5a) 3-D plot of the initial configurations as in 4a. **5b)** Final configuration as in 4b.

To briefly summarize the various cases studied we have found that the results for the model polymer on a crystal surface can be divided into four different classes: (1) The polymer remains in the solid state; (2) The polymer melts; (3) The polymer undergoes bond dissociations but after there is significant melting. This process may be viewed as an ablation which results in severe charring; or (4) The polymer undergoes ablation, resulting in multiple bond dissociation with no melting.

Finally we note that the ablation process can be significantly enhanced by increasing the initial temperature of the crystal. This in effect places the system into a conformationally disordered state which allows for more efficient energy absorption. A plot of the percent trajectories that ablate as a function of the initial temperature is shown in Fig. 6. As can be seen, there is a definite temperature dependence and that the temperature which corresponds to the maximum ablation is near 400 K.

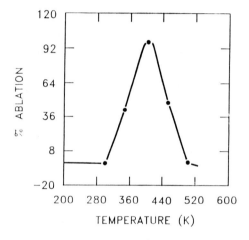

Figure 6. Plot of the temperature dependence of ablation induced by two lasers. The percent trajectories that are ablated are plotted versus the temperature of the system. The lasers have fixed intensities $I_1 = 1/8$, $I_2 = 1/16$ TW/cm^2 and frequencies $\omega_1 = 1000$, $\omega_2 = 800$ cm^{-1}

IV. CONCLUSIONS

We have carried out a detailed molecular dynamics study of the energy absorption and subsequent dynamical changes which occur for the interaction of a model polymer on a crystal surface with infrared lasers.[60] The results have shown that when a single IR laser is used that only melting or ablation with significant melting can occur, regardless of the frequency or intensity of the laser. In contrast, if two IR lasers are used, with one laser red-shifted from the other, efficient and clean ablation occurs. The threshold intensity required to cause ablation of the model polymer was found to be on the order of 3/8 TW/cm^2 at a temperature of 300 K. However, the intensity requirements for the two laser ablation process is also strongly dependent on the initial temperature of the crystal. For higher temperatures, ablation occurs more efficiently. In addition, two lasers were also shown to lead to melting at a much lower threshold intensity than for one laser and at an increased rate.

Damage to the model polymer from the absorption of energy from a single laser appears to occur through a mechanism which involves rapid energy redistribution. The energy absorbed by the polymer leads to vibrational excitation. The rate of energy loss from the absorption site relative to the rate of energy absorbed determines whether the polymer undergoes melting or ablation. If the rate of absorption is high but below the threshold to cause clean ablation, the polymer undergoes bond dissociation but with severe melting. If the rate of energy absorption is low, the polymer will either undergo no transition or it will melt. For sufficiently high absorption rates, those which are faster than the energy relaxation rates of the polymer, energy is rapidly accumulated in a localized region of the polymer and multiple bond dissociation occurs before melting.

The results presented in this report suggest that IR lasers can be used to etch or weld polymers. The key to achieve IR laser induced ablation is to **use two IR lasers** which are appropriately tuned for the particular system.

ACKNOWLEDGMENTS

We would like to thank Dr. Coral Getino for useful discussions throughout the course of this work. Research was sponsored by the Division of Materials Sciences, Office of Basic Energy Sciences, U.S. Department of Energy, under contract DE-AC05-84OR21400 with Martin Marietta Energy Systems, Inc., and the National Science Foundation, Polymers Program, Grant #DMR 8818412. Calculations were carried out on the IBM 3090 at The University of Tennessee and the Cray X(Y)-MP/48 at the National Center for Supercomputing Applications (NCSA).

REFERENCES

[1] *Lasers in Polymer Science and Technology: Applications* Vol. III, Ed. by Jean-Pierre Fouassier and Jan F. Rabek (CRC Press, Boca Raton, Florida, 1990).
[2] P.M. Felker and A.H. Zewail, Adv. Chem. Phys. 70, 265 (1987).
[3] F.F. Crim, Science 249, 1387 (1990).
[4] R. Svrinavasan, Science 234, 559 (1986).
[5] T. Znotins, D. Poulin, and J. Reid, Laser Focues, May 1987.
[6] S. Lazare and V. Granier, Laser Chem. 10, 25 (1989).
[7] R. Srinivasan and B. Braren, Chem. Rev. 89, 1303 (1989).
[8] B.J. Garrison and R. Srinivasan, Appl. Phys. Lett. 44, 849 (1984); J. Appl. Phys. 57, 2909 (1985).
[9] S.K. Gray, J.R. Stine, and D.W. Noid, Laser. Chem. 5, 209 (1985).
[10] D.W. Noid, M.L. Koszykowski, R.A. Marcus, J.D. McDonald, Chem. Phys. Lett. 51, 540 (1977).
[11] D.W. Noid, C. Bottcher, and M.L. Koszykowski, Chem. Phys. Lett. 72, 397 (1980).
[12] D.W. Noid and J.R. Stine, Chem. Phys. Lett. 65, 153 (1979).
[13] J.R. Stine and D.W. Noid, J. Phys. Chem. 86, 3733 (1982).
[14] M.J. Davis and R.E. Wyatt, Chem. Phys. Lett. 86, 235 (1982).
[15] W.H. Miller, J. Chem. Phys. 69, 2188 (1978).
[16] P.W. Milonni, J. Chem. Phys. 72, 787 (1980).
[17] W.H. Miller, J. Chem. Phys. 71, 783 (1980).

[18] A.E. Orel and W.H. Miller, J. Chem. Phys. 70, 4393 (1979).
[19] R.B. Shirts and T.F. Davis, J. Phys. Chem. 88, 4665 (1984).
[20] K.D. Hansel, Chem. Phys. Lett. 57, 619 (1978).
[21] R. Ramaswamy, P. Siders, and R.A. Marcus, J. Chem. Phys. 78, 4418 (1981).
[22] D.L. Martin and R.E. Wyatt, Chem. Phys. 64, 203 (1982).
[23] M.J. Davis, R.E. Wyatt, and C. Leforestier, in *Intramolecular Dynamics*, eds. J. Jortner and B.Pullman, Reidel, 1982 pp 403-427.
[24] P.S. Dardi and S.K. Gray, J. Chem. Phys. 77, 1345 (1982).
[25] R.B. Walker and R.K. Preston, J. Chem. Phys. 67, 2017 (1977).
[26] S.K. Gray, Chem. Phys. 83, 125 (1984).
[27] S.K. Gray, Chem. Phys. 75, 67 (1983).
[28] R.E. Wyatt, G. Hose, and H.S. Taylor, Phys. Rev. A. 28, 815 (1983).
[29] J.R. Stine and D.W. Noid, Chem. Phys. Lett. 77, 287 (1981).
[30] J.R. Stine and D.W. Noid, Opt. Commun. 31, 161 (1979).
[31] K.D. Hansel, in *Laser Induced Processes in Molecules*, Eds. K. L. Kompe and S.D.Smith, Springer, 1979.
[32] R.V. Ambartzumian, Y.A. Brokhov, V.S. Letokhov, G.N. Marakov, A.A. Puretzkii, and N.P. Furzikov, JETP Letters 23, 194 (1976).
[33] D.W. Noid and J.R. Stine, Opt. Commun. 37, 187 (1981).
[34] D.W. Noid and J.R. Stine, J. Chem. Phys. 74, 1724 (1982).
[35] P. Gozel and It. van den Bergh, J. Chem. Phys. 74, 1724 (1981).
[36] K.M. Christoffel and J.M. Bowman, J. Phys. Chem. 85, 2159 (1981).
[37] P.S. Dardi and S.K. Gray, J. Chem. Phys. 80, 4738 (1984).
[38] J.R. Ackerholt, H.W. Galbraith, and P.W. Milonni, Phys. Rev. Lett. 51, 1259 (1983).
[39] M.L. Klein, Ann. Rev. Phys. Chem. 36, 525 (1985).
[40] C. Getino, B.G. Sumpter, J. Santamaria, and G.S. Ezra, J. Phys. Chem. 93, 3877 (1989).
[41] C. Getino, B.G. Sumpter, J. Santamaria, and G.S. Ezra, J. Phys. Chem. 94, 3995 (1990); C. Getino, B.G. Sumpter, and J. Santamaria, Chem. Phys. 145 1 (1990).
[42] R. Duchovic, W.L. Hase, and B.H. Schlegel, J. Chem. Phys. 88, 1339 (1984).
[43] R.J. Wolf, D.S. Bhatia, and W.L. Hase, Chem. Phys. Lett. 132, 493 (1986).
[44] W.J. Lemon and W.L. Hase, J. Phys. Chem. 91, 1596 (1987); B.G. Sumpter and D.L. Thompson, J. Chem. Phys. 87, 5809 (1987).
[45] B.G. Sumpter and D.L. Thompson, Chem. Phys. Lett. 153, 243 (1988).
[46] D.W. Noid, B.G. Sumpter, M. Varma-Nair, and B. Wunderlich, Makromol. Chem., Rapid Commun. 20, 377 (1989).
[47] D.W. Noid, B.G. Sumpter, and B. Wunderlich, Macromolecules 23, 664 (1990).
[48] B.G. Sumpter, D.W. Noid, and B. Wunderlich, Polymer 31, 1254 (1990); D.W. Noid, B.G. Sumpter and B. Wunderlich, Anal. Chem. Acta 235, 143 (1990).
[49] B.G. Sumpter, D.W. Noid, S.Z.D. Cheng, and B. Wunderlich, Macromolecules 23, 4671 (1990).
[50] A. Gelb, B.G. Sumpter, and D.W. Noid, J. Phys. Chem. 94, 809 (1990); A. Gelb, B.G. Sumpter, and D.W. Noid, Chem. Phys. Lett. (1990).
[51] R. Roy, B.G. Sumpter, D.W. Noid, and B. Wunderlich, J. Phys. Chem. 94, 5720 (1990).
[52] D.W. Noid, B.G. Sumpter, and B. Wunderlich, Polymer Commun. 31, 304 (1990).
[53] B.G. Sumpter, D.W. Noid, and B. Wunderlich, J. Chem. Phys. 93, 6875 (1990).
[54] L.F. Shampine and M.K. Gordon, *Computer Solution of Ordinary Differential Equations: The Initial Value Problem*, Freeman: San Francisco, 1975; L.F. Shampine and M.K. Gordon, DEPACK, SAND 79-2374.
[55] D.W. Noid, B.G. Sumpter, B. Wunderlich, and G.A. Pfeffer, J. Comp. Chem. 11, 236 (1990).
[56] D.W. Noid, M.L. Koszykowski, and R.A. Marcus, Ann. Rev. Phys. Chem. 32, 267 (1981).
[57] R. Roy, B.G. Sumpter, G.A. Pfeffer, S.K. Gray, and D.W. Noid, Phys Rep. 000,000 (1991).
[58] S.L. Marple, *Digital Spectral Analysis with Applications*, Prentice-Hall, Englewood Cliffs, 1987.
[59] R. Roy and T. Kailator, IEEE Trans. on ASSP 37, 984 (1989).
[60] B.G. Sumpter, G.A. Voth, D.W. Noid, and B. Wunderlich, J. Chem. Phys. 93, 6081 (1990).

CHEMICAL CHARACTERIZATION OF MICROPARTICLES BY LASER ABLATION IN AN ION TRAP MASS SPECTROMETER

J. M. Dale, W. B. Whitten, and J. M. Ramsey
Analytical Chemistry Division
Oak Ridge National Laboratory
Oak Ridge, TN 37831-6142

We are developing a new technique for the chemical characterization of microparticles based upon the use of electrodynamic traps. The electrodynamic trap has achieved widespread use in the mass spectrometry community in the form of the ion trap mass spectrometer or quadrupole ion trap [1]. Small macroscopic particles (microparticles) can be confined or levitated within the electrode structure of a three-dimensional quadrupole electrodynamic trap in the same way as fundamental charges or molecular ions by using a combination of ac and dc potentials [2]. Our concept is to use the same electrode structure to perform both microparticle levitation and ion trapping/mass analysis. The microparticle will first be trapped and spatially stabilized within the trap for characterization by optical probes, i.e., absorption, fluorescence, or Raman spectroscopy. (We have previously shown that such spectroscopic probes can be extremely sensitive, e.g., a detection limit of one molecule of Rhodamine-6G has been determined in the case of fluorescence spectroscopy [3,4]). After the particle has been optically characterized, it is further characterized using mass spectrometry. Ions are generated from the particle surface using laser ablation or desorption. The characteristics of the applied voltages are changed to trap the ions formed by the laser with the ions subsequently mass analyzed. The work described here focuses on the ability to perform laser desorption experiments on microparticles contained within the ion trap. Laser desorption has previously been demonstrated in ion trap devices by applying the sample to a probe which is inserted so as to place the sample at the surface of the ring electrode [5,6]. Our technique requires the placement of a microparticle in the center of the trap. Our initial experiments have been performed on falling microparticles rather than levitated particles to eliminate voltage switching requirements when changing from particle to ion trapping modes.

Figure 1 shows a schematic diagram of our current apparatus for performing these experiments. The ion trapping device is a modified Finnigan MAT Model 800 Ion Trap Detector (ITD). The trapping electrodes were removed from the ITD vacuum chamber and placed in a 6-inch cube with the rotational symmetry axis of the trap oriented vertically. The cube is attached to a 4-inch oil diffusion pump. The rf voltage from the ITD was reattached to the ring electrode by extending the tranformer tap wire to a high-voltage vacuum feed-through mounted on one of the side ports and retuning the transformer for resonance. The ITD was operated with nominally 1 mtorr of He buffer gas as usual. The ion lens assembly normally used for electron-impact ionization was removed from the top end cap and a

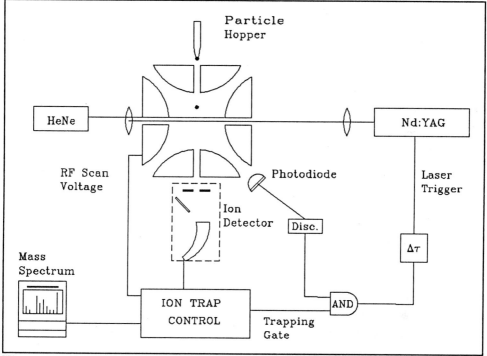

Figure 1

particle dropping device was installed. The particle dropper consisted of a funnel-shaped container with a 450-μm spout. A 300-μm wire attached to a 6-mm rod normally rests in the spout. The rod exits the vacuum chamber through an o-ring sealed fitting. Particles of interest are placed in the container and dispensed by moving the wire attached to the rod. The other three side ports were fitted with 6-inch pyrex windows while the bottom port was covered with a flange mounted with the channeltron electron multiplier. A 5-mW HeNe laser beam is focused into the trap through the windows and opposing 3-mm holes drilled in the ring electrode. The HeNe beam is positioned $\approx$ 1 mm above the center of the trap. The second harmonic laser radiation from a pulsed Nd:YAG laser (Quanta-Ray DCR-2A) propagates in the opposite direction through the trap and is focused by a 1-m lens at the trap center.

Collection of a laser desorption mass spectrum requires synchronization of the Nd:YAG laser with the falling particle and the ITD. The ITD runs continuously through its normal scan sequence of trapping and mass-selective particle ejection. The ITD can have a trapping-mode duty cycle of $\approx$ 40% if appropriate scan settings are used. When particles are dropped they pass through the HeNe probe beam scattering light that is detected by the photodiode. The photodiode signal is converted to TTL, then ANDed with the electron gate signal from the ITD. If a light scattering signal is detected when the ITD is in the trapping mode, the Nd:YAG laser is fired after an adjustable delay time. The laser trigger delay allows for the spatial displacement between the HeNe and Nd:YAG beams. Particles can be reliably illuminated with the 10-ns pulse from the Nd:YAG laser after proper alignment.

Silicon carbide particles (nominal 125-μm diameter) coated with various materials were used in our initial studies. Particles were coated with various quaternary ammonium or phosphonium salts by dissolving them in methanol, combining with a given mass of particles and evaporating the solvent. In all cases 5 mg of salt was evaporated onto 5 gm of particles. This loading would result in roughly 10 monolayers of material on the surface of a particle based upon particle surface area measurements. Compounds investigated include trimethylphenylammonium chloride, triethylphenylammonium iodide, tetrabutylammonium iodide, and tetraphenylphosphonium bromide. All experiments were performed with pulse energies of $\approx$ 1mJ (10^9 W/cm^2) except where noted. Ions were reliably produced from dropped particles with yield correlating with the intensity of the 532-nm light scattered by the particle as detected by the photodiode. The quality of the mass spectra varied primarily due to what appears to be space charge effects. Mass specta of the above compounds all produced intact cations and expected fragment ions. The mass spectra compare favorably with those of Ref. 6 and SIMS data on the same particles.

Figure 2 shows a mass spectrum for a single SiC particle coated with tetraphenylphosphonium bromide irradiated with a single pulse of 532-nm radiation for the Nd:YAG laser. The intact cation peak is observed at m/z 339. Peaks were also observed that correspond to loss of 2 benzene units and loss of 3 phenyl groups at 183 and 105 m/z

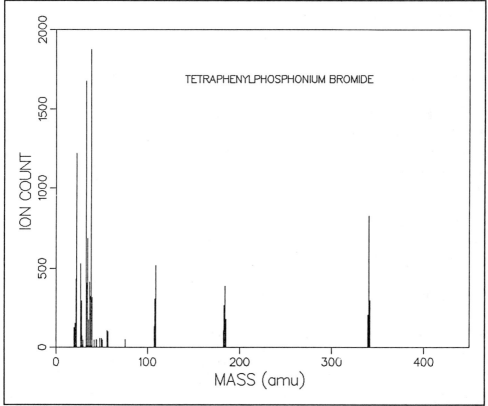

Figure 2

respectively. The loss of a single phenyl group results in an odd electron ion and is occasionally observed at low abundance. This mass spectrum is representative of the better spectra observed. Spectra of this quality are observed with ≈ 50% of the particles. The other %50 of the experiments are either of low ion yield due to poor overlap of the Nd:YAG laser beam with the particle or too high an ion yield resulting in poor mass resolution from space charge effects in the ion trap.

In addition to the above coated SiC particles, uncoated particles of SiC, Fe, and Nb were also investigated. The uncoated SiC particles yielded Na and K ions presumably due to surface contamination. No ions were observed that would be associated with SiC even at intensities of 10^{10} W/cm^2. The iron and niobium particles required slightly higher energies (≈ 3mJ/pulse) to yield ions as would be expected. The iron spectra included the iron isotopes in addition to showing a copper impurity. The Nb spectra also indicated some iron contamination.

Improvements in the current apparatus promise to yield good sensitivity for materials on the surface of microparticles. Submonolayer sensitivities should easily be achieved. Excess ion production leading to space charge effects in the trap is currently more of a problem than lack of signal. Combining particle levitation with ion trapping may allow multiple desorption experiments to be performed on a single particle and thus permit signal averaging.

ACKNOWLEDGEMENTS

The authors would like to express their appreciation to Thomas Rosseel and Werner Christie for obtaining SIMS data on the particles investigated, to Peter Todd for interpretation of some of the mass spectra, and to Gary Glish and Doug Goeringer for suggesting and providing the quaternary salts. Research sponsored by U.S. Department of Energy, Office of Basic Energy Sciences, under contract DE-AC05-84OR21400 with Martin Marietta Energy Systems, Inc.

REFERENCES

1. R. E. March and R. J. Hughes, Quadrupole Storage Mass Spectrometry, Vol. 102, "Chemical Analysis", J. D. Winefordner, Ed., Wiley-Interscience, NY, 1989.

2. R. F. Wuerker, H. M. Goldenberg, and R. V. Langmuir, *J. Appl. Phys.* 30, 441 (1959).

3. W.B. Whitten, J.M. Ramsey, S. Arnold, and B.V. Bronk, Proceedings of the 1989 Scientific Conference on Obscuration and Aerosol Research, CRDEC-SP-026, pg. 45, (1990).

4. W.B. Whitten, J.M. Ramsey, S. Arnold, and B.V. Bronk, *Anal. Chem.* 63, 1027 (1991).

5. D.N. Heller, I. Lys, R.J. Cotter, and O.M. Uy, *Anal. Chem.*, 61, 1083 (1989).

6. G.L. Glish, D.E. Goeringer, K.G. Asano, and S.A. McLuckey, *Int. J. Mass Spectrom. & Ion Proc.*, 94, 15 (1989).

PHOTOPHYSICAL PROCESSES IN UV LASER PHOTODECOMPOSITION OF
$Bi_2Sr_2Ca_1Cu_2O_8$ and $YBa_2Cu_3O_{x+6}$

Lawrence Wiedeman, HyunSook Kim[a] and Henry Helvajian[b]
Laser Chemistry and Spectroscopy Department
Aerophysics Laboratory,
The Aerospace Corporation,
P. O. Box 92957, Los Angeles, California 90009

INTRODUCTION

The laser ablation technique has been used to grow c–axis oriented thin films of both the $YBa_2Cu_3O_{x+6}$ (YBaCuO) [1] and $Bi_2Sr_2Ca_1Cu_2O_8$ (BiSrCaCu:2212)[2] high temperature superconductors (HTc). The technique has also been applied to the growth of layered HTc superlattices [3–4] and in particular tailoring HT_C thin films by site selective substitution [5]. These applications employ high laser fluence ablation for stoichiometric mass transfer. At low laser fluences, we have measured both a wavelength dependence in the removal of specific adsorbates (Fe/Ag) [6], and ion kinetic energy (KE) distributions that suggest a site selective excitation process. These results can be developed to prepare multilayer devices with near contaminant-free interfaces at a fraction of the cost of molecular beam epitaxy (MBE). In addition, by regulating the laser fluence, it is possible to adjust the mass removal between stoichiometric bulk transfer and atomic level species-specific selectivity. The latter process may allow the modification of surfaces with site-specific atomic level control.

In this paper we present the product population and KE distributions for the UV laser photodecomposition of $YBa_2Cu_3O_{x+6}$ and $Bi_2Sr_2Ca_1Cu_2O_8$ compounds as a function of laser fluence and wavelength.

EXPERIMENTAL

Figure 1 shows an idealized experimental chamber incorporating components from two experimental apparatuses. In reality, one chamber has a time-of-flight (TOF) mass spectrometer which measures species with high KE resolution. The other chamber has a quadrupole mass spectrometer (QP), with electron-impact ionization, and could measure both neutral and ion desorbed species. All the experiments are done in high vacuum (< 10^{-9} torr). The two laser sources used are: a) an s–polarized, frequency tripled Nd–YAG laser (pulse width 3 nsec FWHM, 355 nm) and b) an unpolarized excimer laser (pulse width 20 nsec FWHM) operating at 193 nm (ArF), 248 nm (KrF) and 351 nm (XeF) wavelengths. To reduce transient intensity fluctuations in the YAG

---

[a]National Research Council Post Doctoral Fellow.
[b]Person to whom correspondence should be addressed.

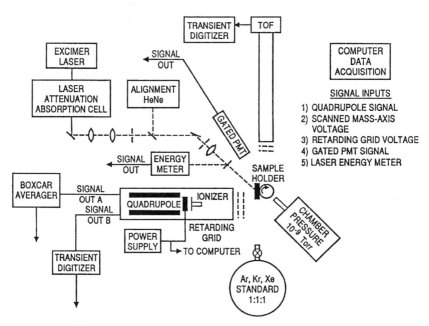

Figure 1. The schematic diagram of the experiment.

laser pulse, the YAG beam is passed through a Raman cell (150 psi of H_2) and only the forward directed beam is used. For both experimental apparatuses the laser beam is collimated, spatially apertured, and directed at an angle of 45 degrees relative to the surface normal. We only use the center portion of the laser beams. The mass spectra are acquired with the laser repetition rate at 2 Hz. A time resolved, multichannel, data acquisition program is used to capture and store the signal from each laser pulse along with the corresponding laser energy, and plate voltages. This method of data acquisition enables subsequent selection and averaging of the data. The measured TOF signal arrival times are converted to nascent photoejected KE by comparison with the calculated energies generated from a model. The model is based on the Simion program (distributed by Idaho Engineering Laboratories) and has been calibrated for our apparatus with a retarding potential experiment. In measuring the KE and population distributions, the laser fluence is maintained near the fluence threshold for ion product formation.

The $Bi_2Sr_2Ca_1Cu_2O_8$:2212 crystal was grown at the Oak Ridge National Laboratory and has a measured transition temperature (Tc) near 80 K. The $YBa_2Cu_3O_{x+6}$ wafer is commercially purchased (99.999% pure) with Tc at 95 K.

RESULTS

A multiphoton absorption process may be described by the following equations:

$$(\text{Solid}) + N \cdot h\nu \rightarrow (\text{Product})$$
$$(\text{Product}) = (\text{Laser Fluence})^n$$
$$\text{Log}(\text{Product}) = n \cdot \text{Log}(\text{Laser Fluence})$$

The slope, n, of the Log-Log plot defines the fluence dependence. Figure 2 shows the photoejected Bi^+ yield as a function of the laser energy from the UV laser excitation of BiSrCaCuO:2212 crystal for three laser wavelengths (351, 248, 193 nm). The data is plotted as the Log(Signal) versus the Log(Laser Fluence). The linear least-squares fit (slope = n) values at the wavelengths 351 nm, 248 nm and 193 nm are 10.9, 6.5 and 5.3 respectively. The results show that the slope decreases with the wavelength. Furthermore, there is some evidence in the 248 nm data of a change in the slope at the higher laser fluences. This trend is not unique. Qualitatively, the slope increases (higher n values) at some laser fluence level above threshold. We also measured the laser fluence dependence for the photoejected Sr^+ ion at 193 nm. The n dependence is lower (4.4) than the fit corresponding to the Bi^+ (n = 5.3). Similar laser fluence dependence experiments have also been done on sintered $YBa_2Cu_3O_{x+6}$. The slope fits to approximately n = 15 for 351 nm laser excitation.

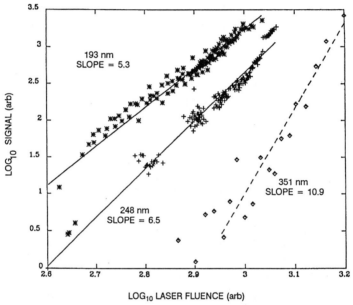

Figure 2. The photoejected Bi^+ yield from BiSrCaCuO (2212) as a function of the laser fluence at three laser wavelengths.

Figures 3 and 4 show high KE resolution TOF mass spectra of the photoejected Cu^+ ions from a sintered $YBa_2Cu_3O_{x+6}$ and a crystalline $Bi_2Sr_2Ca_1Cu_2O_8$ target. The data show the KE distribution is independent of the laser wavelength. Furthermore, the similarity in the Cu^+ KE distribution from the two compounds could be the result of the nearly identical Cu binding site locations. The data in Figs. 3 and 4 show that the most probable Cu^+ kinetic energy (<KE>) is 4.5 eV. The maximum kinetic energy (KEmax) is near 6 eV. We have also measured the KE distributions for the other ejected species. The <KE> and KEmax values are: Sr^+ (<KE> = 5.5 eV, KEmax = 8.5 eV) and Bi^+ 351nm, 248 nm (<KE> = 3.5 eV, KEmax = 5 eV). The small differences between the KEmax values were also observed in the retarding potential experiments. The reported KE distributions were taken at laser energies near threshold for product formation.

This is well below the regime where space charge and plasma acceleration processes become important. Figure 5 shows the change in the single shot TOF FWHM as a function of the laser energy. The data in Fig. 3 and 4 were taken where the FWHM has no dependence on the laser fluence. Not shown is the plasma-induced ion acceleration process, which appears at even higher laser fluences.

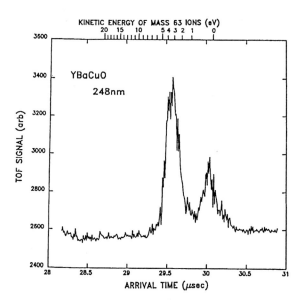

Figure 3. Cu^+ TOF spectrum from a 248 nm irradiated YBaCuO sintered pellet.

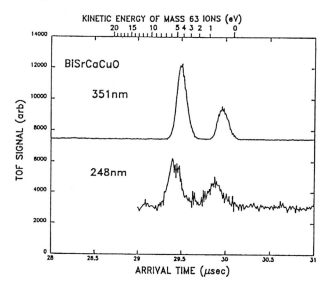

Figure 4. Cu^+ TOF spectra from (a) 351 nm and (b) 248 nm irradiated BiSrCaCuO (2212) single crystal.

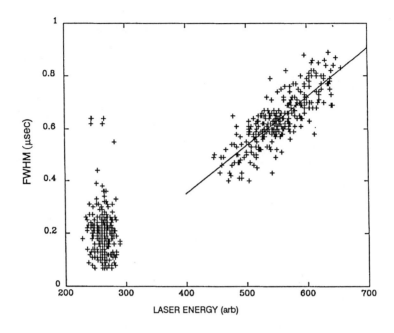

Figure 5. Sr^+ TOF FWHM as a function of laser energy. The target is BiSrCaCuO (2212) and the laser wavelength is 351 nm.

Figure 6 shows four low resolution TOF mass spectra taken at the laser wavelengths 355 nm (YAG), 351 nm and 248 nm. All the spectra show the photoejected products to be predominantly atomic species along with simple diatomic oxides. The top two spectra in the Fig. 6 show virtually identical product distributions though the spectra come from very different BiSrCaCuO targets: polycrystalline (2223) and a (2212) single crystal. The middle two spectra show the insensitivity of the ejected product distribution to the laser pulse duration. In comparison, the bottom two spectra illustrate what appears to be a resonant enhancement in the Bi^+ signal at 248 nm. Though not shown, preliminary results also indicate a significant enhancement in the Sr^+ signal for 193 nm laser excitation. Similar results were also obtained for the sintered $YBa_2Cu_3O_{x+6}$ target though the resonance phenomenon is less dramatic.

At laser fluences above threshold, we measure the appearance of other products. Figure 7a, 7b shows the Y^+/YO^+ ratio as a function of increasing laser fluence for $YBa_2Cu_3O_{x+6}$ ablation at two laser wavelengths. The figure shows the photolysis of the oxide compound. The Fig. 8 shows a QP spectrum of the ejected anion species. The Cl^- is a contaminant added during the manufacturing process. The inset shows the retarding potential results for the O^- anion. The stopping of the anions near zero volts indicates low KE energy and possibly a thermal ejection process.

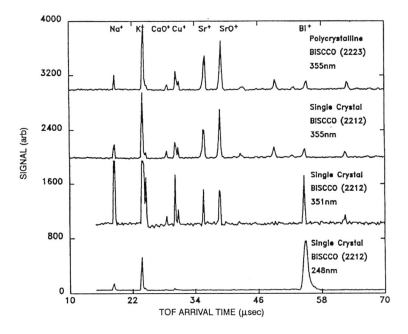

Figure 6. TOF mass spectra for laser photodecomposition of BiSrCaCuO at low KE resolution.

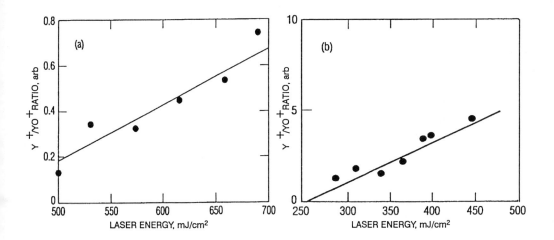

Figure 7. The Y^+/YO^+ yields as a function of laser fluence. YBaCuO target. (a) 351 nm wavelength, (b) 248 nm wavelength.

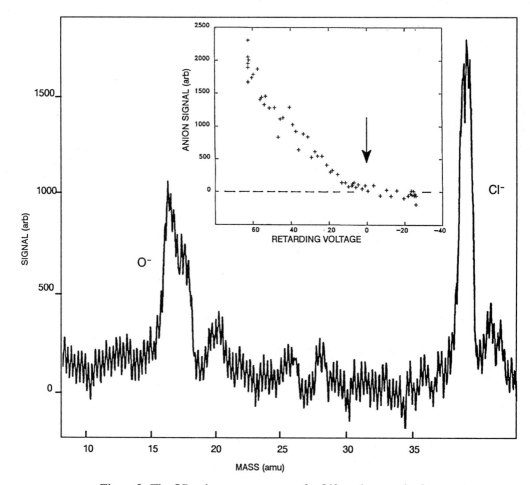

Figure 8. The QP anion mass spectrum for 248 nm laser excitation of YBaCuO. The insert is the O$^-$ anion signal as a function of a retarding potential. The voltage scale represents the difference in voltage between the target (biased negative relative to ground) and the retarding plate.

Also measured at the higher laser fluences were ejected neutral species. Figure 9 is a (QP) mass spectrum of the ejected neutral ejected species for 248 nm laser irradiation of $Bi_2Sr_2Ca_1Cu_2O_8$. The neutral KE distribution which is shown in Fig. 10 is substantially lower than that of the ejected cation KE (see Figs. 3 and 4). In these spectra a retarding potential is used to screen, from the detector, the majority of the nascent Cu^+. Also evident in Fig. 10 is the acceleration of the cation and neutral species with increasing laser fluence.

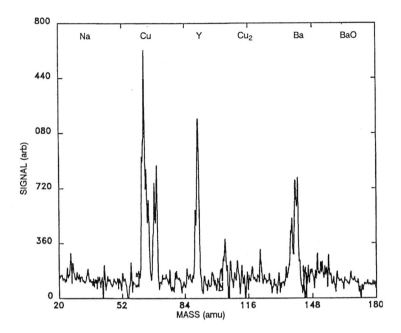

Figure 9. The QP mass spectrum of photoejected neutral species near neutral product detection at 248 nm.

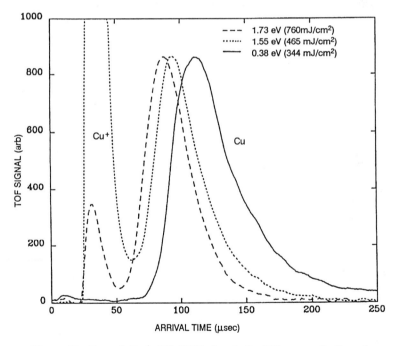

Figure 10. Cu and Cu^+ QP-TOF signals for 248 nm excitation of YBaCuO at three laser fluences.

DISCUSSION

The measured results for the threshold fluence laser photodecomposition of BiSrCaCuO cannot be described by a thermally excited process. The photoejected ion KE distributions do not fit a Boltzmann distribution. A thermal process would not show wavelength dependent behavior in the product population. Although the ion yield from a thermally excited process could be fit over a limited range by a logarithmic dependence the resulting slope would be the same regardless of the radiation wavelength. However, for a multiphoton absorption process, the fluence dependence (n) reflects the number of photons necessary for a single excitation process. The excitation energy, (E_s) is equivalent to the product of the photon dependence n and the photon energy. For our data on Bi^+ the Es values are given in Table 1.

Table 1 shows that for the three laser wavelengths studied, Es is roughly 35 eV. This excitation energy is nearly equivalent to that of the bismuth free atom subshell binding energy: $Bi\ 5d_{3/2}$, $5d_{5/2}$ = 34, 32 eV respectively [7]. A subshell excitation will undergo Auger decay and may create a localized two-hole pair as described by the Knotek–Feibelman theory [8]. The resultant coulombic force could eject either ion with a maximum kinetic energy defined by e^2/R_0 where R_0 is the bond length [9]. Table 2 shows our results for the measured KEmax (BiSrCaCuO:2212), the bond distances as measured by EXAF [10], and the calculated kinetic energy based on the coulomb repulsion model.

The results from Table 1 and 2 show that both the Cu^+ and Bi^+ may be explained by the Knotek–Feibelman desorption process. The Sr^+ does not fit this model. However, using the gas phase SrO internuclear distance, the calculated KEmax agrees more favorably with our measured results. Perhaps the Sr^+ is desorbed from a defect site near an oxygen atom rather than the Sr–O(2) site. This speculation is also supported by our TOF mass spectra (Fig. 6) where strontium is one product which also appears in the form of an oxide.

Table 1. Excitation Energy for Bi^+ Ion Ejection.

Wavelength	Slope	E_s = n∗ (Photon Energy)
351 nm (3.53 eV)	n = 10.9	38.5 eV
248 nm (5.00 eV)	n = 6.5	33.0 eV
193 nm (6.32 eV)	n = 5.3	34.0 eV

Table 2. KE from Coulomb Repulsion.

Bond	Distance (A)	$(e \cdot e)/R_0$	KE max (data)
Cu-O(1)	1.91	7.3 eV	7.0 eV
Bi-O(2)	2.22	6.3 eV	5.5 eV
Si-O(2)	2.74	5.5 eV	8.0 eV
Sr-O(gas)	1.91 [11]	7.3 eV	

A difference is shown in the ejected KE distributions of the desorbed neutral and ion products (Figs. 3 and 10). By comparison, the difference indicates that not only are both the electronic and thermal desorption processes operating, but that partitioning in the ablated matter arrival time occurs (Fig. 10). This phenomenon is fundamental to the laser ablation process and has also been measured in gallium arsenide and in sodium trisilicate systems [12, 13]. This behavior may have implications in defect-free growth of epitaxial superlattices by the laser ablation technique.

In summary, our results show that threshold fluence laser excitation of the $YBa_2Cu_3O_{x+6}$ and $Bi_2Sr_2Ca_1Cu_2O_8$ perovskites leads to electronic excitation via a multiphoton absorption process. We have measured high KE ions ejected which do not fit a thermal desorption model. However, our results can be explained by a Knotek–Feibelman excitation/coulombic explosion process. When the laser fluence is increased above threshold, we measure the interaction of the laser with the plume and the interfering effects of a thermally initiated desorption phenomena. Further increase in the laser energy results in plasma acceleration processes.

REFERENCES

1. M. E. Geusic, W. J. Weber and L. R. Pederson, *Mater. Lett.* 10, 13 (1990).
2. C. R. Guarnieri, R. A. Roy, K. L. Saenger, S.A. Shivashankar and D. S. Yee, J. J. Cuomo, *Appl. Phys. Lett.* 53, 532 (1988).
3. T. Venkatesan, A. Inam, B. Dutta, R. Ramesh, M. S. Hegde, X.D. Wu, L. Nazar, C. C. Chang, J.B. Barner, D.M. Hwang and C.T. Rodgers, *Appl. Phys. Lett.* 56, 391 (1990).
4. X. D. Wu, X.X. Xi, Q. Li, A. Inam, B. Dutta, L. DiDomenico, C. Weiss, J.A. Martinez, B. J. Wilkins, S. A. Schwarz, J. B. Barner, C. C. Chang, L. Nazar and T. Venkatesan, *Appl. Phys. Lett.* 56, 400 (1990).
5. H. Tabata, T. Kawai, M. Kania, O. Murata, and S. Kawai, *Jap. J. Appl. Phys.* 28 L823 (1989).
6. H. Helvajian and R. P. Welle, *J. Chem. Phys.* 91, 2616 (1989).
7. D. Briggs "Handbook of X–ray and UV Photoelectron. Spectroscopy," ed. D. Briggs, Heyden, 153 (1977).
8. M. L. Knotek and P .J. Feibelman, *Phys. Rev. Lett.* 40, 964 (1978).
9. J. I. Gersten and N. Tzoar, *Phys. Rev. B*. 16, 945 (1977).
10. Y. Uehara, N. Kamijo, H. Kageyama, M. Wakata, *Jap. J. Appl. Phys.* 29, 1419 (1990).
11. K. P. Huber and G. Herzberg, "Molecular Spectra and Molecular Structure IV. Constants of Diatomic Molecules," Van Nostrand Reinhold, 630 (1979).
12. A. Namiki, T. Kawai, Y. Yasuda and T. Nakamura, *Jap. J. Appl. Phys.* 24, 270 (1985).
13. P. A. Eschbach, J. T. Dickinson, S. C. Langford, *J. Vac. Sci. Technol.* 7, 2943 (1989).

INFLUENCE OF LIQUEFACTION ON LASER ABLATION: DRILLING DEPTH AND TARGET RECOIL

A.D. Zweig

Wellman Laboratories of Photomedicine
Massachusetts General Hospital
Boston, MA 02114

ABSTRACT

Ablation of materials by highly-absorbed laser pulses (duration 200 μs) is modelled using a steady-state approach that takes into account target liquefaction. Material is removed by a combination of evaporation and ejection of liquid that is caused by radial evaporation-induced pressure gradients at the bottom of the ablation crater. As a result the evaporation rate is reduced and the drilling efficiency enhanced. The model explains experimental drilling-depth and recoil-momentum data.

SUMMARY

Drilling depth and target recoil provide important and easily measurable information on laser ablation. They vary with the distance between the surface of the target and the waist of the laser beam. For laser pulses of several 100 μs duration the observed variation is very similar for biological targets and metals, suggesting that the same physical mechanisms are involved in ablating these materials[1,2].

Targets illuminated with sufficiently intense, highly-absorbed laser radiation are heated at the surface and start to evaporate after a short fraction of the ablating laser pulse. Because of momentum conservation the resulting violent vaporization of material increases the pressure locally. To a first approximation the pressure is proportional to irradiance. The induced pressures vary spatially with the beam profile of the ablating laser. Thus if the target material liquefies prior to evaporation, the pressure-induced shear forces at the surface push material in radial direction towards the crater wall. There the liquid flow changes direction and continues axially, along the wall and out of the hole. This mechanical ejection mechanism becomes more efficient as more liquid moves in the radial direction. Correspondingly, drilling is enhanced by a low viscosity of the molten material, a large thickness of the liquid surface layer, and a small spot size of the drilling beam.

For a given laser pulse the total drilling depth changes with the distance from the surface of the target to the waist of the beam. Under tight focusing conditions the mechanical drilling contribution becomes important, as can be seen from the experimental results presented in Fig. 1a. The solid curve is the result of a model calculation based on a steady-state description of drilling[3]. The calculation took into account the variation of irradiance with distance from the beam waist and used parameters corresponding to the experimental data. Also shown in Fig. 1a is the calculated depth for purely evaporative drilling into a water target (dashed line). As is made clear by the figure,

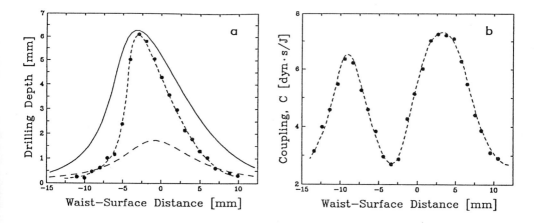

Figure 1: Influence of the initial position of the target surface on the drilling mechanism. Positive (negative) positions correspond to a beam waist located outside (inside) the target. Experimental data for gelatin containing 0.84 g/cm^3 water is indicated by filled circles (•). The short-dashed lines are to guide the eye. All the holes were drilled with single laser pulses ($\lambda = 2.94$ μm, $\tau = 200$ μs, $\omega_0 = 72$ μm, $E_{in} = 40$ mJ). Data from[1]. (a): Drilling depth versus waist-target distance. The solid line is the result of a model calculation that takes into account evaporation and ejection; the dashed line shows the calculated depth for a purely evaporative drilling. (b): Coupling coefficient, C, versus waist-target distance. C is equal to the total recoil momentum normalized by the laser pulse energy. Note that C has a relative minimum when the drilling depth has a maximum.

the experimental values are too large to be caused by evaporation only. The enhanced drilling depth originates from the ejection mechanism.

The recoil momentum, caused by the ejected material during an ablation process on a target, provides another simple means of characterizing the interaction experimentally. The total recoil momentum divided by the pulse energy is called the mechanical coupling coefficient, C, and has units of dyn · s/J. In Fig. 1b its dependence on the beam waist to target distance is shown. The appearance of that curve is similar for a variety of biological targets as well as metals. It can be explained by the generation of liquid material during ablation.

Neglecting the contribution of liquid ejecta to the total target recoil, the coupling coefficient can be estimated roughly by assuming that all the evaporated material leaves the target at the speed of sound. This picture is valid only if the induced gas flow is one-dimensional. For high aspect-ratio holes this can be expected despite the long pulse durations used (200 μs).

Therefore the recoil curves can be interpreted by considering the influence of aspect-ratio and total mass evaporated as follows: For shallow holes the vapor flows mainly parallel to the target surface, with but a small velocity component perpendicular to the surface of the target. As the radial momentum components cancel each other because of symmetry, the coupling coefficient is small. Increasing the aspect-ratio raises the perpendicular velocity component of the vapor flow, and therefore also the coupling coefficient. This raise is seen in Fig. 1b for target positions corresponding to drilling depths smaller than about half a mm. For larger drilling depth the coupling coefficient decreases with increasing depth. This is because the further increase of the hole depth does not enhance the directionality of the vapor flow substantially, but reduces the total mass evaporated. Indeed the energy of the ablating pulse is used to evaporate and liquefy the target. Increasing the contribution of ejection to drilling thus reduces the amount of evaporated material because of energy conservation. As a consequence the coupling coefficient decreases. Therefore, because ejection is more efficient than evaporation with respect to drilling, the maximal drilling depth and the relative minimum of the coupling coefficient in Fig. 1 occur at the same waist-target distance.

CONCLUSION

Liquefaction of target material during ablation changes the laser- target interaction qualitatively. The associated effects are important for ablation by pulses of a duration of 100 μs or longer under tight-focusing conditions. They become insignificant if the liquid cannot transverse the radius of the irradiated spot during the laser pulse, as is typically the case for Q-switched pulses.

Drilling depth and recoil momentum caused by pulsed laser ablation change with distance of the target surface from the beam waist position in a characteristic way, qualitatively identical for all materials that undergo liquefaction prior to evaporation.

The generation of a liquid layer at the phase boundary and its radial displacement during ablation lead to an increase of the drilling efficiency. The associated reduction of the evaporation rate causes a reduction of the temperature at the target surface during ablation.

The recoil momentum induced on the target depends predominantly on the vapor flow pattern. This becomes more and more directional as the aspect-ratio of the drilled holes increases. In that case the coupling coefficient is dominated by the total amount of evaporated mass.

ACKNOWLEDGMENTS

The author is grateful to T.F. Deutsch for the valuable discussions during the preparation of this manuscript. This work has been supported by the Swiss National Science Foundation and by the Air Force Office of Scientific Research.

REFERENCES

[1] M. Frenz, V. Romano, A.D. Zweig, H.P. Weber, N.I. Chapliev and A.V. Silenok, J. Appl. Phys. **66**, 4496 (1989)

[2] M. Bass, M.A. Nassar and R.T. Swimm, J. Appl. Phys. **61**, 1137 (1987)

[3] A.D. Zweig, J. Appl. Phys. (1991), *to be published*